THE POLITICS OF TECHNOLOGICAL CHANGE IN PRUSSIA

THE POLITICS OF TECHNOLOGICAL CHANGE IN PRUSSIA

OUT OF THE SHADOW OF ANTIQUITY, 1809–1848

Eric Dorn Brose

PRINCETON UNIVERSITY PRESS PRINCETON, NEW JERSEY

Published by Princeton University Press, 41 William Street,
Princeton, New Jersey 08540
In the United Kingdom: Princeton University Press, Oxford

Library of Congress Cataloging-in-Publication Data

Brose, Eric Dorn, 1948-
The politics of technological change in Prussia : out of the
shadow of antiquity, 1809-1848 / Eric Dorn Brose.
p. cm.
Includes bibliographical references and index.
ISBN 0-691-05685-4 (acid-free paper)
1. Prussia (Germany)—Politics and government—1806-1848.
2. Industrialization. I. Title.
DD420.B76 1992
943'.07—dc20 92-15232 CIP
Rev.

This book has been composed in Linotron Caledonia

Princeton University Press books are printed on acid-free paper,
and meet the guidelines for permanence and durability of the
Committee on Production Guidelines for Book Longevity of the
Council on Library Resources

Printed in the United States of America

1 3 5 7 9 10 8 6 4 2

For Christian and Dieter

Contents

Illustrations

Acknowledgments

I CONCEIVED this study originally as a short article over the motives behind the modernizing policies of Peter Beuth, Ludwig von Vincke, and Christian [von] Rother. Ten years and over four hundred manuscript pages later, the list of those who have provided indispensable assistance has grown very long indeed. Thanks are due the American Philosophical Society, which funded early research trips to Germany in 1981 and 1983; the National Endowment for the Humanities, which provided a travel grant in 1984; and the American Council of Learned Societies and the German Academic Exchange Service, which supported me with generous funding during the last and most productive stage of research in 1987. Finally, I would like to thank Dean Thomas Canavan of the College of Arts and Sciences at Drexel University, who approved reduced teaching loads throughout much of the 1980s, facilitated matching faculty development funds from Drexel for all of these research tours, and provided college money for one last week of insightful work in Bayreuth in 1991.

Without help and advice from many sides during the past decade, my study would possess far more defects than those which undoubtedly remain. First, I have received invaluable assistance from the staffs of numerous archives and libraries. Special thanks go to Dr. Böhn of the Landeshauptarchiv in Coblenz; to Dr. Herbert Schneider of the Deutsches Freimaurer-Museum; and to Ms. Deidre Harper, Head of the Inter-Library Loan Division at Drexel, whose hard work and persistence over the years procured most of the thick tomes of primary source materials listed in the bibliography. I am also indebted to many friends, family members, and colleagues. My mother, Norma Brose, read and critiqued a manuscript which has more polish as a result. David Barkley and the two external readers at Princeton University Press also performed yeoman service and put me on the path of significant revisions. The empirical historian in me owes thanks, moreover, to two theoreticians, Jane Caplan and Douglas Porpora, for words of encouragement. Finally, I want to thank my wife, Christine Brose, and my colleague, Philip Cannistraro, for emboldening me to take a somewhat different approach to the telling of history.

Note on Proper Names

Throughout the text, [von] for first reference denotes ennoblement during lifetime. Common names will be used for subsequent references except in Conclusion and Index.

THE POLITICS OF TECHNOLOGICAL CHANGE IN PRUSSIA

Introduction

THIS WORK studies some of the political controversies spawned by industrialization in the kingdom of Prussia from the aftermath of the country's defeat by Napoleon to the Revolution of 1848. Central to many of these disputes was the quality of the state's economic and technological leadership—the first major question analyzed in the chapters which follow. I contend that the existence of competing intrastate agencies, cultures, and party-political factions with different ideologies and agendas makes it very difficult to generalize about the state's role in early industrialization. Yet by the 1830s and 1840s it was clear that a multifarious crisis was developing between bureaucrats, soldiers, and businessmen over ownership, control, and promotion of the forces of production—a crisis which was only gradually resolved during the third quarter of the nineteenth century.

By the late 1800s, doubts about the government's economic wisdom had largely disappeared from the historical literature of a prouder and more confident generation of Germans. Beginning with Heinrich von Treitschke's grandiose *History of Germany in the Nineteenth Century* (1879–1889) and Gustav Schmoller's *Acta Borussica* (1892), a multivolume glorification of economic policy under Frederick the Great, German historians depicted Prussia's early technocrats as successful, pioneering, patriotic servants of the modern German nation. Succeeding generations nurtured this tradition of praise for the state's role, witness the third volume of Franz Schnabel's *History of Germany in the Nineteenth Century* (1933–1939); W. O. Henderson's classic study, *The State and the Industrial Revolution in Prussia* (1959); Friedrich Facius's *State and Economy* (1959); and Wilhelm Treue's *Economic and Technological History of Prussia* (1984).[1]

Since Treitschke and Schmoller, many individual bureaucrats and specific institutes have also received a full measure of praise from native-son

[1] Heinrich von Treitschke, *Deutsche Geschichte im 19. Jahrhundert* (Leipzig: Verlag von S. Hirzel, 1879–89), 5 vols.; Gustav Schmoller and Otto Hintze (eds.), *Die Preussische Seidenindustrie im 18. Jahrhundert und ihre Begründung durch Friedrich den Grossen—Acta Borussica-Denkmäler der Preussischen Staatsverwaltung im 18. Jahrhundert* (Berlin, 1892), 3 vols.; Franz Schnabel, *Deutsche Geschichte im Neunzehnten Jahrhundert* (Freiberg: Herder & Co., 1933–37), vol. 3; W. O. Henderson, *The State and the Industrial Revolution in Prussia* (Liverpool: Liverpool University Press, 1958); Friedrich Facius, *Wirtschaft und Staat: Die Entwicklung der staatlichen Wirtschaftsverwaltung in Deutschland vom 17. Jahrhundert bis 1945* (Boppard, 1959); and Wilhelm Treue, *Wirtschafts- und Technik-geschichte Preussens* (Berlin: de Gruyter, 1984).

historians. These studies were often such labors of love and respect for the narrow subject at hand that readers lost sight of the full, complex, and variegated range of Prussian economic and technological policies before 1848. Highlighting—but by no means exhausting—the list of these still valuable works are Gustav Fleck's seminal "Studies in the History of Prussian Railroads" (1896–1899); Paul Schrader, *History of the Royal Seehandlung* (1911); Hermann von Petersdorff, *Friedrich von Motz* (1913); Konrad Wutke's detailed *Out of Silesia's Mining and Metallurgical Past* (1913); Conrad Matschoss's classic *The Great Men of Prussian Industrial Promotion* (1921); Rolf Keller, *Christian von Rother as President of the Seehandlung* (1930); H. J. Straube's *Christian Peter Wilhelm Beuth* (1930) and *Prussian Industrial Promotion in the First Half of the Nineteenth Century* (1933); and Max Schulz-Briesen, *Prussian State Mining* (1933). In recent years, Peter Lundgreen's solid monograph, *Technicians in Prussia* (1975), and Wolfgang Radtke's well-documented *The Prussian Seehandlung* (1981) have kept alive the century-old tradition of the "Acta Borussica."[2]

Images of the Prussian state as a great, modernizing economic leader have proven incredibly persistent. Writing in the 1930s, Antonio Gramsci, an Italian Communist, drew on the empirical findings of German scholars in his theory of "state vanguards" which paved the route for private industry. Composed of semi-independent intellectual elements of the bourgeoisie—or more frequently, of a waning nobility which no longer dominated society—bureaucratic vanguards initiated and supervised industrial growth for many decades before less experienced and determined businessmen were able to assume a leading role themselves. Gramsci referred to this situation as a "passive revolution"—for the pervasive social and economic changes accompanying industrialization had been initiated without violence and by elements other than the bour-

[2] Paul Schrader, *Die Geschichte der Königlichen Seehandlung (Preussische Staatsbank) mit besonderer Berücksichtigung der neuen Zeit* (Berlin, 1911); Hermann von Petersdorff, *Friedrich von Motz* (Berlin: Verlag von Reimar Hobbing, 1913); Konrad Wutke, *Aus der Vergangenheit des Schlesischen Berg- und Hüttenlebens* (Breslau, 1913); Conrad Matschoss, *Preussens Gewerbeförderung und ihre Grossen Männer* (Berlin: Verlag des VDI, 1921); Rolf Keller, *Christian von Rother als Organisator der Finanzen, des Geldwesens und der Wirtschaft in Preussen nach dem Befreiungskriege* (Rostock, 1930); H. J. Straube, *Chr. P. Wilhelm Beuth* (Berlin: Verlag des VDI, 1930); id., *Die Gewerbeförderung Preussens in der ersten Hälfte des 19. Jahrhunderts* (Berlin: Verlag des VDI, 1933); Max Schulz-Briesen, *Der Preussische Staatsbergbau von seinen Anfängen bis zum Ende des 19. Jahrhunderts* (Berlin: Reimar Hobbing, 1933); Peter Lundgreen, *Techniker in Preussen während der frühen Industrialisierung* (Berlin: Colloquium Verlag, 1975); Wolfgang Radtke, *Die Preussische Seehandlung zwischen Staat und Wirtschaft in der Frühphase der Industrialisierung* (Berlin: Colloquium Verlag, 1981); and G. Fleck, "Studien zur Geschichte des preussischen Eisenbahnwesens," *Archiv für Eisenbahnwesen* 19–22, 1896–1899.

geoisie. He believed that vanguard phenomena like these were typical in regions such as Germany, Italy, and Russia which were struggling to overcome the West's industrial lead.[3] Unfortunately, Gramsci did not elaborate further upon the vanguard's motives for giving birth to the bourgeois era.

Marxist scholars of our generation have continued this historiographical tradition. Few conclude, however, that preindustrial bureaucrats were serving the bourgeoisie. Thus Nicos Poulantzas, *Political Power and Social Classes* (1968), argues that absolutist states functioning in the interests of the nobility gave the first impetus to industrialization. Perry Anderson, *Lineages of the Absolutist State* (1974), takes a similar approach but sees state programs as more of a passive aristocratic response to dynamic developments out of its control. Recent East German histories also conceded that government facilitated the coming of industry in Prussia. Begun with the military and public financial interests of the ruling nobility at heart, promotional policies were necessarily limited in nature and paled in significance next to the productive, inventive contributions of the bourgeoisie itself. Like Anderson, historians in the DDR viewed bureaucratic encouragement of industry as part of an inevitable historical process driven forward by the class struggle.[4]

Post–World War Two macroeconomic "growth" analysis has reinforced the Nationalist and Marxist conclusions that Prussian industrialization received tremendous impetus from the chancelleries of Berlin. One of the first economic histories to draw on this new economic subdiscipline was W. W. Rostow's *The Stages of Economic Growth* (1960). Extremely influential, the work used historical examples from eighteenth- and nineteenth-century Europe to articulate the noncommunist credo that capitalist states in the Third World could play a vital role in fostering "industrial take-off." Ulrich Peter Ritter quickly followed Rostow's theoretical lead with *The Role of the State in the Early Stages of Industrialization* (1961), a fairly comprehensive analysis of Prussia's "encouragement" of industry. Wolfram Fischer and Ilja Mieck also reacted positively to the notion that German states stood behind take-off. These empirical contributions received added theoretical underpinning from Alexander Gerschenkron's *Economic Backwardness in Historical Perspective* (1962).

[3] Antonio Gramsci, *Selections from the Prison Notebooks*, edited and translated by Quintin Hoare and Geoffrey N. Smith (New York: International Publishers, 1971), 19, 106–20, 269–70.

[4] Nicos Poulantzas, *Political Power and Social Classes* (London: NLB and Sheed and Ward, 1973), 157–68; Perry Anderson, *Lineages of the Absolutist State* (London: Verso, 1979), 267–78; Andreas Dorpalen, *German History in Marxist Perspective: The East German Approach* (Detroit: Wayne State University Press, 1985), 187–200, and the literature of the DDR discussed there.

The farther behind England continental nations had found themselves, he postulated, the more essential state financing, bank loans, propagandistic justification of industry, and other measures which deviated from the English pattern were to closing the gap.[5]

The strands of a fourth school of analysis, however, are woven into this complex scholarly fabric of state-driven industrialization. For by the 1960s "modernization theory" was attracting widespread support among historians. A distant academic cousin of the Marxist approach to history, modernization theory also posits a linear path along which societies progress from traditional to modern. As originally formulated by pioneering social scientists in the late 1950s, the modernization process begins within the cultural, social, economic, or political subsystem of a premodern country, then gradually permeates all of society and moves it toward a new equilibrium—"modernity."[6]

West German historians of the critical postwar generation have modified this model in two important respects. Still influenced by the Acta Borussica tradition, they have been tempted to place bureaucrats in the forefront of the modernization process. Their works have also overturned the Parsonian, "functionalist" emphasis on the stability and integration of developing social systems (which underlay much of early modernization theory) in favor of left-liberal assumptions about the inevitability of conflict. Reinhart Koselleck's *Prussia between Reform and Revolution* (1966) is a pathbreaking analysis of the reform and restoration eras based on this modified modernization approach. The autonomous, modernizing bureaucracy failed to dislodge the aristocracy in the 1810s, but succeeded in unleashing the dynamic forces of capitalistic industry. By invigorating

[5] R. F. Harrod, "An Essay in Dynamic Theory," *Economic Journal* 49 (1939): 14–33; E. Domar, "Capital Expansion, Rate of Growth and Employment," *Econometrica* 14 (1946): 137–47; Robert M. Solow, "A Contribution to the Theory of Economic Growth," *Quarterly Journal of Economics* 70 (1956): 65–94; W. W. Rostow, *The Stages of Economic Growth: A Non-Communist Manifesto* (Cambridge, Massachusetts: Cambridge University Press, 1960), 6–8, 26–27, 36; Ulrich Peter Ritter, *Die Rolle des Staates in den Frühstadien der Industrialisierung: Die preussische Industrie-Förderung in der ersten Hälfte des 19. Jahrhunderts* (Berlin, 1961), passim; Wolfram Fischer, "Government Activity and Industrialization in Germany (1815–1870)" in W. W. Rostow, *The Take-Off into Self-Sustained Growth* (New York: St. Martin's Press, 1963), 83–94, especially 93–94; Ilja Mieck, *Preussische Gewerbepolitik in Berlin 1806–1844* (Berlin: Walter de Gruyter, 1965), xi, 235; and Alexander Gerschenkron, *Economic Backwardness in Historical Perspective* (Cambridge, Massachusetts: Harvard University Press, 1962), 353–54.

[6] See the discussion of the literature in Samuel P. Huntington, "The Change to Change: Modernization, Development, and Politics," in Cyril E. Black, *Comparative Modernization: A Reader* (New York: The Free Press, 1976), 25–39; Hans-Ulrich Wehler, *Modernisierungstheorie und Geschichte* (Göttingen: Vandenhoeck, 1975), passim; and Hanna Schissler, *Preussische Agrargesellschaft im Wandel* (Göttingen: Vandenhoeck & Ruprecht, 1978), 20–25.

the bourgeoisie, however, the bureaucracy had sown more seeds of self-destruction. Jürgen Kocka projected the same image of the modernizing bureaucracy in a seminal piece on the "Pre-March" (*Vormärz*) era—in other words, the period before revolution erupted in March 1848—and more recently Barbara Vogel's *Comprehensive Freedom of Enterprise* (1983) describes the conflict-ridden course which bureaucratic visionaries steered toward economic modernization. Like so many scholars of the 1960s and 1970s, Wolfram Fischer, Otto Büsch, and Ilja Mieck also felt the tug of modernization theory in their economic histories.[7]

Until the 1960s the concept of state-driven industrialization was a well-entrenched, largely unchallenged orthodoxy underscored, as we have seen, by a rich variety of theoretical and empirical works. But the number of critics was beginning to grow. One of the first western departures from the "Apologia Borussica,"[8] as one historian put it, came with Theodore Hamerow's *Restoration, Revolution, Reaction* (1958). While recognizing the state's role in the German states, Hamerow placed more emphasis on private sector developments. The original version of David Landes's *The Unbound Prometheus* (1965) gave even less play to governmental activity. Subsequent historians began to mount a more direct and serious challenge. Richard Tilly's powerfully analytic *Financial Institutions and Industrialization in the Rhineland 1815–1870* (1966) demonstrated that Prussian monetary policy had been anything but salutary. Similarly, Herbert Kisch assaulted the Acta Borussica tradition with his study of the detrimental effects of Frederician policy on the Crefeld silk industry (1968).

The revolt spread in the 1970s. Friedrich Zunkel questioned the standard view that the Prussian Mining Corps had advanced mining and metallurgical technology, while Frank Tipton published two seminal articles which formulated a general critique of the thesis of state-generated in-

[7] Reinhart Koselleck, *Preussen zwischen Reform und Revolution: Allgemeines Landrecht, Verwaltung und soziale Bewegung von 1794 bis 1848* (Stuttgart: Ernst Klett, 1967). The reader should also consult Jonathan Sperber's excellent reappraisal of Koselleck's book in "State and Civil Society in Prussia: Thoughts on a New Edition of Reinhart Koselleck's *Preussen zwischen Reform und Revolution*," *Journal of Modern History* 57 (1985): 278–96; Jürgen Kocka, "Preussischer Staat und Modernisierung im Vormärz: Marxistisch-Leninistische Interpretationen und ihre Probleme," in Hans Ulrich Wehler, *Sozialgeschichte Heute: Festschrift für Hans Rosenberg zum 70. Geburtstag* (Göttingen: Vandenhoeck & Ruprecht, 1974), 211–27; Barbara Vogel, *Allgemeine Gewerbefreiheit: Die Reformpolitik des preussischen Staatskanzlers Hardenberg 1810–1820* (Göttingen: Vandenhoeck & Ruprecht, 1983); and Mieck, *Preussische Gewerbepolitik*. See also the introduction to Mieck, v–vi, by Wolfram Fischer and Otto Büsch.

[8] Herbert Kisch, *Prussian Mercantilism and the Rise of the Krefeld Silk Industry: Variations upon an Eighteenth-Century Theme* (Philadelphia: American Philosophical Society, 1968), 42.

dustrialization. Rainer Fremdling's *Railroads and German Economic Growth* (1975) made another important contribution to this maturing antithesis. He strengthened the "anti-statist" case by showing both the overriding importance of railroads to German industrial growth and the slight participation of the Prussian government in their promotion. Hanna Schissler's *Prussian Agrarian Economy in Transition* (1978) provided more ammunition by arguing that the Prussian agrarian reforms actually hindered industrialization until about 1850. Hans Ulrich Wehler's *German Social History* (1987) has also departed for the most part from the old pattern.[9] Without waiting to observe what remained of the opposition's arguments, many new-breed economic historians had turned in the meantime to new debates and challenges like regional economic history and proto-industrialization. The "statists" remained vibrant and vigilant, however, as the stream of recent works by Fischer, Radtke, Vogel, and Treue demonstrates.[10]

• • • • •

A major deficiency of many of the studies cited above is their tendency to generalize about the state's contribution without analyzing the full

[9] Theodore S. Hamerow, *Restoration, Revolution, Reaction: Economics and Politics in Germany, 1815–1871* (Princeton, New Jersey: Princeton University Press, 1958); David S. Landes, "Technological Change and Development in Western Europe, 1750–1914," *The Cambridge Economic History of Europe* (Cambridge: Cambridge University Press, 1965), 6:274–601, reprinted in revised form as *The Unbound Prometheus: Technological Change and Industrial Development in Western Europe from 1750 to the Present* (Cambridge: Cambridge University Press, 1969); Richard Tilly, *Financial Institutions and Industrialization in the Rhineland 1815–1870* (Madison: The University of Wisconsin Press, 1966); Kisch, *Prussian Mercantilism*; Friedrich Zunkel, "Beamtenschaft und Unternehmertum beim Aufbau der Ruhrindustrie 1849–1880," *Tradition* 9 (1964); id., "Die Rolle der Bergbaubürokratie beim industriellen Ausbau des Ruhrgebiets 1815–1848," in Wehler, *Sozialgeschichte Heute*, 130–47; Frank B. Tipton, "The National Consensus in German Economic History," *Central European History* 7 (1974): 195–224; id., "Government Policy and Economic Development in Germany and Japan: A Skeptical Reevaluation," *Journal of Economic History* 41 (1981): 139–50; Rainer Fremdling, *Eisenbahnen und deutsches Wirtschaftswachstum 1840–1879: Ein Beitrag zur Entwicklungstheorie und zur Theorie der Infrastruktur* (Dortmund: Ardey-Verlag, 1975); Schissler, *Preussische Agrargesellschaft*, 187. Hans Ulrich Wehler, *Deutsche Gesellschaftsgeschichte* (Munich: C. H. Beck, 1987). Wehler (2:100–102, 132–39) makes some positive remarks about the state's role, but these seem buried in a work which pays little attention to the state. See also the discussion of the literature over regional economic history and protoindustrialization in: P. K. O'Brien, "Do We Have a Typology for the Study of European Industrialization in the XIXth Century?" *Journal of European Economic History* 15 (1986): 297–304; and Schissler, *Preussische Agrargesellschaft*, 40–43. Finally, see Fischer, "The Strategy," 431–42; and the works by Treue, Radtke, and Vogel cited above, respectively, in notes 1, 2, and 7.

[10] See below, chap. 1, n. 4.

range of state programs. Indeed, there were many competing factions and divided authorities within the Prussian state of the early 1800s. Accordingly, there was a wide spectrum of intragovernmental motives for state action as well as great variation in the nature of policies, the types of specific technologies preferred, and the economic effects of the various promotional efforts. It is also important to realize that this complicated mix of motivations, programs, and consequences changed with time. It follows, moreover, that there were many differing public responses to these policies, preferences, and performances—and that public responses also differed over time.

My study attempts to portray and explain this complexity. Chapter 1 introduces Frederick William III, the king of forty-three years (1797–1840) whose eclectic economic views and retiring personality prevented a strictly doctrinaire and ideologically consistent policy toward industry and technology in Prussia. This monarch is in great measure responsible, in other words, for the complexity analyzed throughout the remainder of the book. Frederick William was not an insightful leader or a visionary with the strength to mold the future. But he backed progressive ministers and advisers at many critical moments. From Hardenberg's agrarian and rural-industrial reforms of the 1810s and the rejection of aristocratic efforts to rescind them in the 1820s, to the implementation of the German Customs Union and the acceptance of railroads in the 1830s, the king played a very important role. He refused to limit himself to Hardenberg's recommendations, however, with the result that competing approaches to economic and technological advance were tolerated. Industrialization in Prussia was affected in significant ways by Frederick William III's eclecticism, yet recent monographs make scant mention of him and no full-length biography has yet appeared. While the "Great Man" approach is rarely employed by historians today, Frederick William III seems to be one important personage who warrants more of our attention.

The diversity permitted by the king led naturally enough to conflict—a leitmotif of this work. Chapter 1 focuses on one of the most important struggles which Frederick William was forced to arbitrate—that waged by Chancellor Karl August von Hardenberg and reforming bureaucrats based in the Business Department (*Gewerbedepartment*) against mercantilistic and aristocratic opponents in the 1810s and 1820s. Although there were many bureaucratic divisons in Prussia responsible for the promotion of manufacturing and trade before the Reform Era (1807–1811), the modern Business Department emerged in 1808 during the ministerial reorganization of Baron Heinrich Friedrich Karl vom und zum Stein. The department was responsible for revitalizing business in the broadest sense, not merely factories, and the men who headed it were economic "liberals" who advocated free enterprise, a benign but limited state interference in the economy, and economic advance rooted in the countryside. Hardenberg's reforms were clearly influenced by the men who headed this division.

After departing in chapter 2 from our discussion of economic liberalism—chapter 2 looks at the defeat of political liberalism—chapter 3 returns to the Business Department during the 1820s and 1830s. By this time, the most dynamic personality in the Business Department was Christian Peter Wilhelm Beuth, one of Hardenberg's younger experts and a dynamic advocate of the "appropriate" industries favored by the department since its inception. Born into the family of a Rhenish physician and educated at the University of Halle, Beuth was an economic as well as a political liberal. Like most of the reformers, he believed that constitutionalism and parliamentarism would accompany the social and economic advancement of the middle class. Promotion of industry was thus a political and patriotic act for Beuth, and it was in this promotional area that he made his mark. The Technical Deputation, created in 1811 to promote technological change and transfer, was reorganized and revitalized by Beuth in 1819 to perform this function more effectively. He enhanced his position within the ministries in 1821 by opening the Technical Institute in Berlin. The purpose of the school was to train a new generation of culturally refined, technically competent businessmen. That same year he founded the Association for the Promotion of Technological Activity in Prussia. Modeled after similar organizations in England and France, the Association brought civil servants and industrialists together in an effort to disseminate technical knowledge and propagate the creed of technological progress. By the late 1820s, Beuth had become one of the most influential men in Berlin below the level of the Ministerial Cabinet. During the 1830s, however, he grew bitter over the increasingly heavy-handed nature of Prussian economic leadership.

Contributing to Beuth's disillusionment was the "mercantilistic" empire of the Mining Corps (*Oberberghauptmannschaft*), the subject of chapter 4. First organized in 1769, the "captains" of the Corps managed a vast array of government mines, forges, and foundries by 1800. Corps officials also proudly directed private mines and ironworks with an intricate system of guidelines and regulations. Although the Mining Corps was scheduled for privatization by aides around Hardenberg who opposed mercantilism, it managed to avoid this fate and grow stronger and more influential after 1815. This was partly the result of the integration of Rhenish and Westphalian provincial mine offices into a newly organized central structure. Heading the revitalized mercantile empire was another "plus" for the Corps—Johann Carl Ludewig Gerhard. A second-generation salt mining official and royal favorite, Chief Captain Gerhard (1810–1835) received his technical training in the famous mining academy at Freiberg, Saxony. Like his father, Gerhard lagged behind no one in his devotion to the Corps and its tradition of patriotism, egalitarianism, and the best mining techniques. Chapter 4 culminates with a discussion of the internal feud over deregulation which raged within the Corps during the 1830s and 1840s.

Chapter 5 examines the technical establishment of the Prussian army, another important pillar in the state's economic and technological edifice. Dating back to the time of Frederick the Great, these institutions came into their own during the wars of liberation and subsequent Pre-March era. Staffed largely by bourgeois officers, the complex administrative, educational, and manufacturing structure included the Artillery Department of the War Ministry, the United Artillery and Engineering School, the Artillery Experimental Department and a battery of fourteen powder works, gun and cartridge factories, cannon foundries, and artillery workshops. The economic philosophy of the technical branches was quite different from the other agencies studied here, combining a grudging appreciation of what private companies could produce during wartime emergencies with an anticapitalistic tendency to bully the private sector into meeting army needs. Preference for an autarkical "command economy" strengthened after the war scare of 1830, but so did army ambivalence toward the machines of war and the industrial economy required to produce them. At the apex of the military's industrial edifice stood Prince Wilhelm Heinrich August of Hohenzollern. Somewhat of a maverick among his fellow princes, the king's cousin was an avid progressive who championed everything from Greek independence to gas-lighting. There was no more fitting Chief of the Artillery, a presiding position in the technical wing of the military which he held from 1808 until 1843.

Chapter 6 introduces the Seehandlung (Overseas Trading Corporation). This organization received its charter as a state salt monopoly in 1772. In succeeding decades it added government loans and many other business ventures to its list of investments. An early victim of Hardenberg's programmatic opposition to mercantilistic institutions, the Seehandlung reentered the economic picture in 1820 when the king restored it as an autonomous ministry with considerable freedom of movement. Into the favored position of director stepped Christian [von] Rother, the son of a Silesian peasant. Rother shared the reformers' distinct political liberalism and, like the Young Turks in the Business Department, strove to plant Prussia's manufacturing base in the countryside, not in the cities. But his methods were different. Under Rother's lengthy directorship (1820–1848), the Seehandlung invested in private industrial enterprises and also founded many of its own, emerging as a prodigious neo-mercantilistic force. By the 1830s, Rother was Frederick William's most trusted economic adviser.

Chapters 1, 3, 4, 5, and 6 also include discussion and commentary on the attitudes of private businessmen toward Prussia's variegated economic and technological leadership. With the major exception of low tariffs, the economic programs of Hardenberg and the Business Department were greeted by Prussia's early entrepreneurial elite. As economic liberalism receded before the statism, autarky and antibusiness mentality of the Mining Corps, Army, and Seehandlung during the 1830s and 1840s, industrialists increasingly perceived the state as a threat. Although the Business Department's programs were not as controversial, its continued advocacy of low tariffs added to the friction. Exacerbating the situation and speeding the breakdown of governmental legitimacy was a closely related struggle over the forces of production. Like societies before and after, Prussia became embroiled in the intense politics of technological choice.[11] This controversy raged over the general question of technological modernization in the 1810s and 1820s, then shifted to battles over specific mining, metallurgical, and railroad technologies in the 1830s and 1840s. All too often for the maintenance of harmonious state-societal relations, businessmen perceived themselves to be on the losing end of these arguments.

Chapter 7 traces the interaction between the state's quarreling, intragovernmental citadels and private industrialists over railroad development. While the Mining Corps, Seehandlung, and Business Department waged a very controversial opposition to railroads, Crown Prince Frederick William and the technical branches of the army emerged as major proponents of railways. The resulting legislation in 1838 allowed the railroad lines—and to a great extent, modern industry—to go forward. After his ascension to the throne in 1840, Frederick William IV gave railroads another legislative boost before reversing himself and introducing policies which were very harmful both to railroads and the economy as a whole. My study concludes that the state's role in economic development and the promotion of technology in Prussia was positive on balance in the early 1800s until turning increasingly negative—and therefore more objectionable to businessmen—after the mid-1830s. The alliance between party groupings of political liberals in and out of government who favored parliamentarism was deteriorating well before Frederick William IV's damaging measures escalated differences between elites into a general political crisis.

. . . . .

It was no mere coincidence that so much of Prussia's early technical establishment was politically liberal. For the manner in which social, tech-

[11] For the emphasis on the politics of technological choice in recent histories of technological change, see below, chap. 4, n. 39.

nological, and political phenomena had developed since the eighteenth century initially bound all three closely, logically, and ideologically together. As a technical phenomenon, the industrial transformation of the eighteenth and early nineteenth centuries was largely separate from the scientific revolution of the seventeenth century. There were nevertheless some very important intellectual and attitudinal connections between the two. Baconian empiricism, with its emphasis on categorization, experimentation, and the inevitability of progress, underlay many of the key breakthroughs in crop rotation, steam power, and industrial chemistry. Newtonian physics may also have had important links to early mechanization.[12] The same western mode of rationalist inquiry which challenged older forces of production went much farther, however, by questioning older social and political institutions as well. Armed with the same "methods" as the scientists and entrepreneurs, for instance, eighteenth-century philosophers criticized the irrationality of established religion, serfdom, aristocratic privilege, and the divine right of kings. Thus the new wave of industrial innovation was itself part of a larger intellectual revolution with profound political implications.

And, as the one modern force furthered the other over the 1700s, cause and effect became thoroughly entangled in the minds of contemporaries. Just as the great commercial, industrial, and technological transformations of the modern era "were not possible without intellectual activity," observed Wilhelm von Humboldt, a former minister, in 1823, "so too [these transformations] effect [intellects] in reverse and outlooks become freer and less able to be contained in set forms."[13] David Hansemann, a businessman from Aachen, also believed that "the extraordinary progress made [in the area of] mechanics and ship travel, but especially the application of steam to the art of warfare, are assuring the complete victory of the Enlightenment." Karl August Varnhagen von Ense, a liberal nobleman, took note of the same interconnections in a diary entry of 1825. "America, [revolution in] Haiti, sea travel, commerce and technology," he wrote, "are affairs which arouse souls in an inexorable fashion. The frame of mind which generally arises from these developments is crumbling the old and making way for a new order in the world."[14]

[12] See Margaret C. Jacob, *The Cultural Meaning of the Scientific Revolution* (New York: Alfred Knopf, 1988).

[13] Humboldt to Stein, 4 April 1823, printed in Wilhelm Richter (ed.), *Wilhelm von Humboldt's Politische Briefe* (Berlin: B. Behrs Verlag, 1936), 2:363.

[14] Hansemann to Vogt, 23 December 1825, printed in Alexander Bergengrün, *David Hansemann* (Berlin: J. Guttentag Verlagsbuchhandlung, 1901), 48; Diary of Karl August Varnhagen von Ense, 2 October 1825. The diary was edited by Ludmilla Assing and published for the years 1819–1831 under the title: *Blätter aus der preussischen Geschichte*

The impression given by Humboldt, Hansemann, and Varnhagen of an old order yielding to a new—in short, of an unfolding modernization—is the second major theme addressed in the present work. The notion that the early 1800s represented an unfolding of modern times is strengthened when we consider the elements of continuity between the programs of early industrializers in Prussia and the policies of our own era. These similarities exist for every one of the competing agencies discussed in the following chapters, not merely for Peter Beuth and his dynamic Business Department. Indeed, we should resist the tendency to describe the heavy regulatory hand of mercantilists and army officers as somehow backward. For command economies, despite an ancient tradition, are still an important feature of today's semicapitalist and socialist systems. And mercantilism, born in the Early Modern period, continues to exist in the form of protectionism, subsidies, and government-sponsored monopolies. In fact, it is the laissez-faire approach of the Business Department, most frequently described by historians as modern,[15] which has the most tenuous hold on policy-making in the late twentieth-century world—an observation which is not yet contradicted by recent developments in Eastern Europe and Russia. Seemingly modern, all approaches survive as means to the same end of ongoing economic and technological advancement. In striving to keep abreast of the technical revolution then sweeping the West, Prussia's Business Department, Mining Corps, Seehandlung, and military-technical services were the modernists of their age. As argued in chapter 1, Frederick William III belongs in this category too.

It is certainly a mistake, however, to read too much "modern" thinking into those who fought over policy in the early nineteenth century. The generation under scrutiny here—that between the 1770s and the 1840s—was the first to experience such rapid change that preoccupation with trends of the future became commonplace.[16] Predicting posterity in rapidly changing times was just as uncertain an endeavor in those days, of course, as in ours. Therefore it is not surprising that few Prussians of the early 1800s understood, as we do, where an embryonic process of economic and technological modernization was leading. Their contrasting

(Leipzig: F. A. Brockhaus, 1868–1869) (hereafter cited as Varnhagen Diary I). For this entry, see 3:388.

[15] For a few recent examples, see Wolfram Fischer and Otto Büsch in their introduction to Mieck, *Preussische Gewerbepolitik*, vi; Mieck, *Preussische Gewerbepolitik*, 235–39; and Barbara Vogel, "Die 'Allgemeine Gewerbefreiheit' als bürokratische Modernisierungsstrategie in Preussen," in Dirk Stegmann et al. (eds.), *Industrielle Gesellschaft und politisches System* (Bonn: Verlag Neue Gesellschaft GmbH, 1978), 73, 77–78.

[16] See Reinhard Koselleck, " 'Neuzeit': Zur Semantik moderner Bewegungsbegriffe," in Reinhard Koselleck (ed.), *Studien zum Beginn der modernen Welt* (Stuttgart: Klett-Cotta, 1977), 264–99.

and conflicting visions, unfamiliar to later eyes, were the source of much of the friction in state and society outlined above. Thus the Business Department and many of Hardenberg's aides strove for a gradual, rural—and, increasingly under Beuth—aesthetic industrial process, not the rapid, urban-based and (to them) uglier change which was coming by the 1830s. Rother's designs were also of a rural nature, but he placed much more emphasis on the Freemasonic ideal of social welfare and harmony on earth. The Mining Corps tended to view the mining and metallurgical technologies of the sixteenth and seventeenth centuries as economically and, in a sense, environmentally superior to many of those which emerged in the eighteenth and nineteenth centuries. And until at least the mid-1800s, officers in the Prussian Army, even many in the technical branches, were ambivalent about the military benefits of modern technology.

In striving for objectivity, historians of early nineteenth-century Prussia should begin by avoiding the misleading assumption that our present—or that of the late 1800s—was the future which leaders of this earlier era desired or expected. "It is necessary," as Charles Sabel and Jonathan Zeitlin warn in their study of historical alternatives, "to shift vantage point and imagine a theoretical world . . . that might have turned out differently from the way it did." They predict that "a broader survey of abandoned alternatives to Anglo-American industrialization would likely reveal fossilized combinations"[17] possessing little or no connection with present institutions and mentalities. If we follow this advice, a fascinating panorama unfolds before us. We see, for instance, that many of these early technocrats tended to look into the future—as Marshall McLuhan observed—through the "rearview mirror" of the ancient past. Fascination with ancient Greeks, Romans, and Germans was inextricably linked, in other words, with many Pre-March visions of posterity. Thus classical education, architecture, art, and art-as-such notions were integral to Peter Beuth's goal of an aesthetic industrialization. Although it is more difficult to document, Freemasons like Rother seemed to view mechanization as a return to mysteries and glories which had been lost since antiquity. The Mining Corps' preference for Early Modern techniques represents a variation on this theme of looking forward by looking backward.

Preoccupation with the past was important to the Pre-March in another sense. Because early futuristic exercises were uncertain, they were also laden with anxiety. In fact, many of those who mustered enough

[17] Charles Sabel and Jonathan Zeitlin, "Historical Alternatives to Mass Production: Politics, Markets and Technology in Nineteenth Century Industrialization," *Past & Present* 108 (August 1985): 161, 174.

courage to face the unpredictable industrial age steadied themselves by first looking backward to the heroic examples of the Greeks and Romans. Oddly enough, most of these "modernists" never believed that their own era could measure up in comparison. They remained convinced that their world, although superior to a Medieval Europe idealized by conservatives, was nevertheless overshadowed by the political, military, and artistic accomplishments of the ancients. "Our new world is actually nothing at all," wrote Wilhelm von Humboldt. "It consists of nothing more than a yearning for the past and an always uncertain groping for that which is yet to be created."[18] There was a certain humility to the Pre-March which we do not associate with the confidence—or perhaps overconfidence—of the late nineteenth and twentieth centuries. There were only a few modernists, in fact, who were willing to fully embrace those developments and events which were driving Prussia forward. When that which came was unpleasing or objectionable in relation to preconceived notions of what the future should be, attachments to the Greeks and Romans strengthened. Thoughts of antiquity became for some an escape from reality. Emotional struggles of this sort were particularly evident in a military disturbed by the image of mechanized warfare, but similar reactions were increasingly common among civilian officials whose careers were ending as modern industry began to appear in the 1830s and 1840s. In this sense many of the frustrated modernists held a common crutch with opponents of technological change who remained wedded to the ideal of medieval times. Prussia itself, however, was moving out of the shadow of antiquity.

. . . . .

The third major subject of analysis in my work is the basic "nature" of the state in Pre-March Prussia. Modern scholars who address this question have understandably spent little time justifying the significance of their inquiries. It was obvious to Hans Rosenberg, writing about Early Modern bureaucracy while the ashes of Hitler's terrible creation were still warm, that analysis of the rise of centralized power in Germany's most undemocratic state was indispensable to any understanding of the eventual victory of twentieth-century totalitarianism. Reinhart Koselleck offered his masterful study of Prussia's Pre-March bureaucracy as a prequel to the industrialization, social change, and military violence of the subsequent era, while in the German Democratic Republic, Helmut Bleiber and his colleagues saw the necessary prelude to twentieth-cen-

[18] Humboldt to Wolf, 20 July 1805, printed in Karl August Varnhagen von Ense, *Vermischte Schriften* (Leipzig: F. A. Brockhaus, 1875), 18:242.

tury proletarian unrest in the bowing of a "feudal" state to "bourgeois revolution" between 1789 and 1871. These authors were drawn to the issue of state and society in Prussia before the founding of the Second Empire because they regarded this subject as a necessary part of Germany's eventual tragedy.[19]

The nature of the early Prussian state—which individuals, groups, or social classes controlled, for what purposes and with what results—is thus an important question. Two general theses emerge from the literature. One school stretches backward historiographically from Rosenberg, Koselleck, Jürgen Kocka, and Theda Skocpol to nineteenth-century scholars like Otto Hintze, Gustav Schmoller, and eventually to great Pre-March philosophers and jurists like Friedrich Hegel and his friend, Eduard Gans. This school depicts a state controlled by a *Beamtenstand*, a bureaucratic estate which was socially, legally, and politically separate, distinct, and aloof from outside social classes and estates.[20] Rosenberg and Koselleck argue, however, that a resurgent landowning aristocracy defeated the progressive efforts of the civilian bureaucracy after 1815, eventually infiltrating and inundating the state as well. The thesis of Hans Ulrich Wehler that "pre-industrial elites" survived until 1933 and beyond is a logical extension of these earlier studies of the Pre-March, works buttressed in large part by Wehler's own recent writing on the early nineteenth century.[21]

A second school, of Marxist persuasion, rejects the autonomous state in favor of the class-dominated state. To theorists and historians like Nicos Poulantzas, Perry Anderson, Alf Lüdtke, and many others, Prussia in

[19] Hans Rosenberg, *Bureaucracy, Aristocracy and Autocracy: The Prussian Experience 1660–1815* (Cambridge, Massachusetts: Harvard University Press, 1958); Koselleck, *Preussen*; Bleiber, *Bourgeoisie und bürgerliche Umwälzung*.

[20] See Rosenberg, *Bureaucracy*, 202–28; Koselleck, *Preussen*, 284–559 passim; Kocka, "Preussischer Staat und Modernisierung," in Wehler, *Sozialgeschichte Heute*, 211–27; Theda Skocpol, *States and Social Revolutions: A Comparative Analysis of France, Russia, and China* (New York: Cambridge University Press, 1979), 3–43, 99–111; and Otto Hintze, *Beamtentum und Bürokratie*, edited by Kersten Krüger (Göttingen: Vandenhoeck & Ruprecht, 1981). See also Peter B. Evans, Dietrich Rueschemeyer, and Theda Skocpol, *Bringing the State Back In* (New York: Cambridge University Press, 1985); Hans Branig, "Wesen und Geist der höheren Verwaltungsbeamten in Preussen in der Zeit des Vormärz," in Friedrich Benninghoven and Cècile Lowenthal-Hensel, *Neue Forschungen zur Brandenburg-Preussischen Geschichte* (Cologne: Böhlau Verlag, 1979), 161–71; as well as Gary Bonham's description of the origins of this school, "Beyond Hegel and Marx: An Alternative Approach to the Political Role of the Wilhelmine State," *German Studies Review* 7 (May 1984): 201–2. John R. Gillis, *The Prussian Bureaucracy in Crisis 1840–1860: Origins of an Administrative Ethos* (Stanford, California: Stanford University Press, 1971) accepts the notion of a Beamtenstand for the late 1700s and early 1800s, but argues that this homogeneity broke down in the 1840s as young bureaucrats challenged an older generation.

[21] Wehler, *Deutsche Gesellschaftsgeschichte*, 2:145–61, 297–321.

the Pre-March was—and would remain for many decades after 1848—a *Junkerstaat*. The East German view was similar, although, as noted above, Helmut Bleiber and others believed that the once-dominant feudal behemoth gradually relented to pressure from rising industrialists during the 1830s and 1840s, sought an accommodation with them after 1848, and facilitated the birth of a bourgeois society and the formation of a state which also served these interests during the 1860s. Strands of this thesis are also found in Wehler's new, encyclopedic history of German society in the nineteenth century.[22]

While both schools have greatly advanced the cause of scholarship, it is time to move beyond them. First, the notion that Prussia was the preserve of an autonomous, monolithic bureaucratic estate glosses over the numerous intrabureaucratic fissures along departmental, party-political, and class lines. This model also overlooks the significant ties between bureaucratic factions and extragovernmental allies. Linking historians in both camps, moreover, is the belief that the Prussian aristocracy either dominated or came to dominate government in the Pre-March epoch. Yet this argument stands in stark contrast to the crumbling socioeconomic position of the Prussian nobility, the largely bourgeois composition of the state, and the fact that government frequently served nonaristocratic interests.

The Prussian state possessed a far more complex structure than those depicted in the *Beamtenstand* and *Junkerstaat* models. The state was divided along one plane by intrastate citadels with separate agendas, some of which, outlined above, form the basis of this study. Ties between bureaucratic camps and allies in society were often quite close. Dividing the state on another plane were competing bourgeois professions or cultures like law, cameralism, philology, and science. Cutting across all of these meaningful fault lines, moreover, were numerous political parties grouped loosely into two contending party blocs. Little more than factions or cliques, the parties nevertheless performed the same political function as their modern counterparts. Like the bureaucratic camps and

[22] Poulantzas, *Political Power*, 180–83; Anderson, *Lineages*, 261–78; Alf Lüdtke, *Gemeinwohl, Polizei und "Festungspraxis"* (Göttingen: Vandenhoeck & Ruprecht, 1982); Bleiber, *Bourgeoisie*, passim; id., "Staat und bürgerliche Umwälzung in Deutschland: Zum Charakter besonders des preussischen Staates in der ersten Hälfte des 19. Jahrhunderts," in Gustav Seeber and Karl-Heinz Noack, *Preussen in der deutschen Geschichte nach 1789* (Berlin: Akademie-Verlag, 1983), 82–115; and Wehler, *Gesellschaftsgeschichte*, 2:145–61, 297–321. For a general discussion of the East German view, see Dorpalen, *German History in Marxist Perspective*, 168–203. See also Arno Meyer, *The Persistence of the Old Regime* (New York: Pantheon, 1981); and F. L. Carsten, *A History of the Prussian Junkers* (Hants: Scolar Press, 1989).

the competing professions, the parties also extended from state to society. At the base of this party structure were friendship circles, formal and informal discussion and luncheon clubs, Freemasonic lodges, and other small associations and communities. Presiding over this sometimes chaotic jumble of intriguing departments, professions, and party-political blocs were the monarch and his chief advisers. While evidence supporting this interpretation is woven throughout the book, chapter 1, which introduces Frederick William III, Hardenberg's liberals, and the aristocratic parties; chapter 2, which focuses on friendship circles, liberal party politics in and around the army, and the related issue of public finance; and chapter 3, which includes a discussion of cultural-professional divisions and more on the liberal parties, are especially important to our discussion.

Before presenting the empirical evidence, however, it is necessary to introduce the theories of state and party which, in my opinion, illuminate the weaknesses of previous models and highlight the approach to the nature of the state taken in the chapters below. Gaetano Mosca, an early theorist of elites, is the first. Unaffected by political systems of our era which have less relevance for the Pre-March, he dealt explicitly with the intra- and extra-state forces which swirled around monarchs and their chief advisers. Randall Collins, a contemporary sociologist, offers an innovative and penetrating view of class fragmentation and professional division which shifts focus away from the class cohesion and bourgeois solidarity emphasized of late by many scholars.[23] Moreover, both Mosca and Collins highlight the role of intellectuals who usually staff bureaucracies. Finally, the theory of parties formulated by the pioneering German sociologist, Max Weber, is universal enough to have applicability and explanatory value for nonparliamentary eras like the Pre-March. None of these theories alone is capable of suggesting a thoroughly convincing arrangement of the evidence presented below. Each theory nevertheless identifies significant patterns and relationships which, taken together, amount to a composite image—and a more accurate picture—of the state.

Gaetano Mosca's theory of "the ruling class" attained maturity in 1923

[23] Jürgen Kocka and a team of researchers at the University of Bielefeld emphasize bourgeois cohesiveness more than bourgeois fragmentation. Kocka described the Bielefeld project in a paper given at the German Studies Association conference in Philadelphia, October 7, 1988, entitled "Bürgertum im neunzehnten Jahrhundert: Forschungsergebnisse des ZiF-Projects." A similar emphasis is found in M. Rainer Lepsius, "Zur Soziologie des Bürgertums und der Bürgerlichkeit," in Jürgen Kocka (ed.), *Bürger und Bürgerlichkeit im 19. Jahrhundert* (Göttingen: Vandenhoeck & Ruprecht, 1987), 79–100, although his subsection, "Die Fraktionen des Bürgertums" (86–88), describes the class divisions which find more emphasis in our study.

with publication of his revised *Elements of Political Science*. The Sicilian academician grouped contemporary and historical states along "autocratic-liberal" and "aristocratic-democratic" continuums. The comparative criterion in the "autocratic-liberal" category was the degree of societal participation involved in the creation of sovereign power—hereditary versus broadly elected heads of state. In the more important "aristocratic-democratic" grouping, he divided states according to the social exclusivity of recruitment into what he termed "the highest stratum of the ruling class." But however it was selected, this clique of 20–100 persons existed in every governing system and always represented the arena of real power.[24]

Political and personal relations within this upper stratum were therefore of the greatest analytical importance. Within autocratic systems like that which we will be analyzing below, Mosca concentrated on relationships between monarchs and other elites in the upper echelon of state. He observed that kings usually relied on a major-domo, vizier, prime minister, or chancellor to rule effectively, especially if the sovereign was young, weak, or incompetent. Together they attempted to place loyal persons in all vitally important positions. Opposing them typically were powerful noblemen who vied at court with members of rival great families for preeminence of selection. But when the king and/or chancellor "has talent and strength of will, he sometimes succeeds in breaking the ring of aristocratic cliques that serve him—or, more often, rule him—and he snaps it by elevating to the highest positions persons who are of ordinary birth, who owe him everything and who are therefore loyal and effective instruments of his policies."[25] Combining the "autocratic" state with "democratic" strategies—playing class off against class—was one successful means to dominate competing elites at the top of the governing estate.

Below this decision-making apex came "the lower strata of the ruling class"—the bureaucrats who implemented policy. Civil servants were not, however, a simple tool of leaders in the highest stratum; on the contrary, their initiatives and interpretations of general decisions were as indispensable to the survival of a state as those of junior field commanders to an army. For Mosca assigned them a crucial role in the "juridical defense" of regimes against outside "social forces." If unchecked from above, aggressive movements propelled by (1) martial prowess, (2) religious or scientific zeal ("moral forces"), or (3) landed or monied power

[24] Gaetano Mosca, *The Ruling Class*, edited and revised, with an introduction, by Arthur Livingston (New York: McGraw-Hill, 1939), 394–96, 402.

[25] Ibid., 403.

("material forces") could challenge the ruling class.[26] Moral and material forces were often difficult to distinguish, he observed, because "every moral force tries, as soon as it can, to acquire cohesion by creating an underpinning of interests vested in its favor, and every material force tries to justify itself by leaning upon some concept of an intellectual and moral order."[27] Clever bureaucrats "defended" the state against such threats by cultivating and maintaining respect for reigning customs, mores, laws, and institutions. Wise was the ruling clique, moreover, which coopted and attenuated social forces through recruitment into the bureaucracy and highest offices.[28] And more power to the regent or monarch who manipulated and elevated outside forces in order to block jealous, grasping enemies at court. Such artful, hegemonic games were frequently played in Prussia during our period.

Collins argues that we must study the political economy of culture in order to understand the internal dynamics of modern states and societies.[29] He defines "culture" essentially as the act of shaping images of reality. Informal or "indigenous" culture occurs routinely in the worlds of work, home, and leisure as individuals interact in conversation with others. These conversational experiences "serve to dramatize one's self-image, to set moods, to mentally re-create past realities and to create new ones." In some cases, frequent interactions can lead to "significant ties" based upon "talk containing larger proportions of discussion, ideological debate, entertainment, gossip or personal topics." In such instances, individuals constitute "consciousness communities" where the distinguishing feature is dialogue employing "shared symbols with reality-defining effects for the persons involved." These communities vary in degree of self-consciousness from those with little self-definition to those with "a highly formalized group identity" as manifested in "periodic meetings, regulations, and legal charters."[30]

The more heavily institutionalized and structured communities also produce what Collins terms "formal culture." Churches, schools, professional entertainers, newspapers, occupational and professional groups, and movements of various sorts publicly project their definitions of the nature of the world.

> Formally produced culture is not only more innovative than indigenous culture, but it also allows the creation of much larger and more self-conscious

[26] Ibid., 120–52, 404–5, 438–48.

[27] Ibid., 445.

[28] Ibid., 406, 413, 419–21.

[29] Randall Collins, *The Credential Society: An Historical Sociology of Education and Stratification* (New York: Academic Press, 1979), 49–72.

[30] Ibid., 58–59.

> communities. This type of culture is propagated more widely across different local situations, tends to be made up of more abstract concepts and symbols and it creates more generalized references. Formally produced culture deals less with specific individuals and situations and more with commonalities across many situations. Its images, even if concrete, tend to symbolize every individual or to symbolize the aggregate of individuals as an organized unit. Formal culture is more widely used than indigenous culture; it can relatively quickly negotiate ties among individuals who otherwise have little in common to exchange.[31]

Collins pays particular attention to schools—institutions which not only engage in constant image-shaping activities, but also issue grades and certificates that broadcast "the quantity of cultural goods an individual has acquired."[32] If an academy, institute, university, or even a particular profession or club acquires prestige, ambitious persons will literally "invest" in appropriate educational degrees or memberships. These official papers function as "cultural currency" to purchase a good job or win influence in the right circles ("positional property"). Once past "the gatekeeper," graduates/members with the proper credentials continue to reap handsome returns as entry is facilitated into local consciousness communities. With the right calling cards, in other words, comes an entree to unfamiliar circles where one is "a potential friend because of a common culture."[33]

Inside the gate, indigenous culture reigns supreme as conversations "define both a horizontal bond among equals and a vertical relation among unequals." Usually this is done "by the dominant person . . . picking the topic and influencing what is said about it, affecting the underlying emotional tone and thus controlling the cognitive and moral definitions of reality."[34] In such ways, conversational politics can result in "very strong consensus about common courses of action" in struggles "to control organizations, whether over work pace, gatekeeping criteria, the definition of positional duties and perquisites, assessment of merit, or personal advancement."[35]

Cultural politics is also central to struggles for control of state and society. Class factions, professions, parties, or cliques aspiring for material and political advancement attempt to generate new cultures, penetrate the state, and become "major actors" within the struggle to control that

[31] Ibid., 61.
[32] Ibid., 62.
[33] Ibid., 59.
[34] Ibid.
[35] Ibid.

powerful organization. Traditional elites, on the other hand, try to preserve power by stifling the inventive cultural efforts of subordinate groups or countering defensively with a new definition of reality. In prosperous societies with weak or decentralized states, however, a rapid proliferation of competing cultures and occupational enclaves usually occurs. In such instances, significantly, class conflict becomes "irreparably multisided" and therefore "less . . . revolutionary."[36] In these cases, presumably, weaker states muddle through. As we shall see, Collins's theories are particularly helpful in understanding intrastate conflict and party-political structure in Pre-March Prussia.

In his classic *Economy and Society*, Max Weber included a brief but very insightful passage on class and party. Whereas classes normally fought against other classes in the marketplace over distribution of economic resources, parties usually planned and furthered their goals within "societalized" communities. By this he meant rationally ordered groups or organizations whose hierarchy or "staff" made them capable of affecting changes. "For parties aim precisely at influencing this staff, and if possible, to recruit it from party followers."[37] Whether the goal was realization of "an ideal or material cause" or more "personal" aims like sinecures, political influence, or acquisition of honor, parties operated "in a house of power."[38] By casting his net this widely, Weber allowed for the existence of parties within organizations of all sorts, ranging from social clubs to vastly different kinds of states in greatly differing eras.

It follows from this that party structures will vary accordingly. In a short, undeveloped passage, Weber listed three factors affecting the way parties were shaped and organized. One important determinant was the "structure of domination" within the targeted community, for party leaders "normally deal with the conquest of a community."[39] Mosca's *autocratic-aristocratic* states, for instance, will necessarily spawn a proliferation of small parties scheming and intriguing at court, each usually containing noblemen from outside of the government, while parties in *liberal-democratic* systems will adopt the grass roots apparatus required to win in a parliamentary world of mass politics. In *autocratic-democratic* states like Pre-March Prussia with its upward mobility of recruitment into the ruling class, we would expect to find small, submerged governmental cliques in close contact with party forces in society—in other words, a blend of the two party structures outlined above. Connecting

[36] Ibid., 72.

[37] Max Weber, *Max Weber: Essays in Sociology*, translated by H. H. Gerth and C. Wright Mills (Oxford: Oxford University Press, 1946), 194–95.

[38] Ibid., 194.

[39] Ibid., 195.

party members across the state-society divide were webs of formal and informal culture, as Collins defines it. Parties emanating from different cultures could nevertheless cooperate for a common goal. It is this pattern of cooperation which gives meaning to our model of party blocs.[40]

The second determining factor of party structure according to Weber was the type of "communal action" parties struggled to influence. Presumably, revolutions from below, reforms from within, and military coups from above will require different organizational approaches to the same targeted community. Finally, alternating modes of stratification within the targeted community also influenced party structure. It is significant for our discussion of the Pre-March that Weber felt community stratification by status gave way to stratification by classes during periods of technological and economic transformation.[41] In other words, eras like the eighteenth and nineteenth centuries characterized by accelerating socioeconomic change would transform states from closed, exclusive communities into relatively open structures torn by a variety of social and economic struggles. In the process, parties are drawn into the whirlwind of class *and* status conflict—forces which only resilient party structures can withstand. This whirlwind became particularly evident in Prussia after the Revolution of 1830.

The Conclusion combines the theoretical and empirical strands of analysis into a composite image of the Prussian state. The reader will recall that this final section also makes an assessment of the state's mixed contribution to early industrialization. Finally, the Conclusion attempts to demonstrate how observations about the economic role and basic nature of the state enrich our understanding of Prussia during the historically critical years between the Revolution of 1848 and the founding of the

[40] Ernst Rudolf Huber, *Deutsche Verfassungsgeschichte seit 1789* (Stuttgart: W. Kohlhammer, 1957), 1:121–45, posits two separate parties—a liberal or reform party and a conservative party. Barbara Vogel, "Beamtenkonservatismus: Sozial- und verfassungsgeschichtliche Voraussetzungen der Parteien in Preussen im frühen 19. Jahrhundert," in Dirk Stegmann et al. (eds.), *Deutscher Konservatismus im 19. und 20. Jahrhundert: Festschrift für Fritz Fischer zum 75. Geburtstag und zum 50. Doktorjubiläum* (Bonn: Verlag Neue Gesellschaft, 1982), 1–31, attempts to improve upon Huber by replacing the concept of parties with the notion of liberal and conservative "factions" (*Fraktionen*) within the state. She argues that the state provided the necessary structure for factions in the preparliamentary political system of the Pre-March, but not parties, which require a social base as well. Aided by Weber's theory-fragment and an evident pattern of friendship circles and pluralistic leadership, my study identifies numerous parties grouped into opposing blocs or alliances. The connection between "indigenous culture" (Collins) and party politics also highlights the personal relationships between bureaucrats and nonbureaucrats which account for the extension of party politics from the state into society. In other words, party politics, as defined by Vogel, existed in early nineteenth-century Prussia.

[41] Ibid., 193.

German Empire in 1871. The views expressed there are largely compatible with the theses advanced by Theodore Hamerow, Jeffry Diefendorf, Geoff Eley, and David Blackbourn.[42]

. . . . .

We have placed great emphasis at the outset of our study on Frederick William III. It is appropriate, therefore, that we open chapter 1 by moving "the bourgeois king" to center stage.

[42] Theodore S. Hamerow, *The Social Foundations of German Unification, 1858–1871: Ideas and Institutions* (Princeton, New Jersey: Princeton University Press, 1969), 2 vols.; Jeffry M. Diefendorf, *Businessmen and Politics in the Rhineland 1789–1834* (Princeton, New Jersey: Princeton University Press, 1980); David Blackbourn and Geoff Eley, *The Peculiarities of German History: Bourgeois Society and Politics in Nineteenth-Century Germany* (New York: Oxford University Press, 1984).

I

Rural Industrialization and the Bourgeois King

THE SCHOOLGIRLS had been waiting outside their home village of Weissensee for nearly an hour. Nearby, a deputation of city magistrates and army commanders stood motionless in the cool morning sun, not wanting to dirty their polished boots. At last, a rider galloped to them with the expected announcement, and a moment later, the special travelers appeared. An expressionless Frederick William of Hohenzollern, King of Prussia, rode a few paces in front of his poised queen, Luise of Mecklenburg, sitting more comfortably in her new "lilac" carriage. After the soldiers and officials made their brief remarks, the children handed over a poem of welcome and exultation, and curtsied. Then the regal procession moved out toward the city gates of Berlin where the cannon crews waited for a signal while thousands of loyal subjects waited on the streets, at windows, and on rooftops.

December 23, 1809 was to be a day of pomp and celebration for Berliners. For, after an absence of over three years, their beloved king and queen were finally returning.[1] The festival atmosphere lasted all day and well into the evening. At every corner that night, flickering torches cast an eerie light on great throngs eagerly anticipating another appearance by the royal couple. As the mounted guardsmen cleared a path along Unter den Linden for king and queen, returning from an elegant dinner at Schloss Belevue, jubilant shouts again mixed with the incessant din of church bells and cannon. From the windows of the principal carriage could be seen the tearful face and waving hand of Luise, now publicly moved by the city's welcome. Beside her, the grim, stiff king stared silently ahead. At one halt Luise moved toward her distant husband, a protective, maternal look in her beautiful eyes. Finally she whispered something. But his thoughts were impenetrable.

[1] For detailed descriptions of the reentry to Berlin and subsequent celebrations, *including the emotions of the king and queen*, see the diary of Joseph von Eichendorff, 23 December 1809, printed in Jakob Baxa (ed.), *Adam Müller's Lebenserzeugnisse* (Munich: Verlag Ferdinand Schöningh, 1966), 511; Karl Friedrich Klöden, *Lebens-und Regierungsgeschichte Friedrich Wilhelm des Dritten* (Berlin: Plahn'sche Buchhandlung, 1840), 157–59 (hereafter cited as Klöden, *Friedrich Wilhelm*); Mrs. Charles Richardson, *Memoirs of the Private Life of Louisa, Queen of Prussia* (London: Richard Bentley, 1848), 301–2; and Constance Wright, *Beautiful Enemy: A Biography of Queen Louise of Prussia* (New York: Dodd, Mead & Co., 1969), 218.

The brooding monarch had been visibly ill at ease all day. Never tolerant of unnecessary public appearances, Frederick William had finally agreed to let the events go forward at the repeated urging of city officials. But as the hours wore on, he knew he had been right—the "triumphal" welcome was not only inappropriate, it was in poor taste.[2] Indeed, what possible cause did Prussians have to celebrate after the disasters which had befallen their kingdom? Military collapse before the advancing armies of Napoleon in 1806 had necessitated the flight of the royal family from Berlin to the easternmost towns of Königsberg and Memel. Further defeats were followed in 1807 by the crushing, humiliating Treaty of Tilsit. More than two years had passed since Napoleon had kept him anxiously waiting on the riverbank while the onerous terms of the treaty were negotiated with Tsar Alexander, yet French troops still occupied Prussia's Oder fortresses and no prudent onlooker could see an end to this disgrace—or to Napoleonic hegemony in Europe. No wonder Frederick William was reported grave and preoccupied on the day of his return and throughout the banquets, operas, and church services which highlighted the remainder of the year.[3]

Inside his austere shell, the king of twelve years was still tormented by Prussia's cruel fate. Any head of state would have been deflated and embarrassed by such degrading events, but the nation's tragegy cut much more deeply into a monarch who had never possessed much faith in himself.[4] Early years were spent in the gloomy shadow of his granduncle, Frederick II. The Great One built Sans Souci while the boy's father,

[2] Klöden, *Friedrich Wilhelm*, 157.

[3] Richardson, *Memoirs*, 302.

[4] For character sketches of Frederick William III, see Karl August von Hardenberg's short memoir, n.d. (1808), printed in Georg Winter (ed.), *Die Reorganisation des Preussischen Staates unter Stein und Hardenberg* (Leipzig: S. Hirzel, 1931), 570–75; Klöden, *Friedrich Wilhelm*, passim; Rulemann Friedrich Eylert, *Charakter-Züge und historische Fragmente aus dem Leben des Königs von Preussen Friedrich Wilhelm III* (Magdeburg: Heinrichshofen'sche Buchhandlung, 1842–46), passim; Treitschke, *Deutsche Geschichte*, 1:146–61; *Allgemeine Deutsche Biographie*, 7:700–29 (hereafter cited as ADB); and Eugene Newton Anderson, *Nationalism and the Cultural Crisis in Prussia, 1806–1815* (New York: Farrar & Rinehart, Inc., 1939), 257–97. The short sketch by Hardenberg is reliable; those of Treitschke and Hartmann (in the ADB) are the best we have. The others are useful and informative, but Klöden and Eylert err on the side of praise, while Anderson is critical to the point of cynicism. Carrying on in this latter tradition after the war, Klaus Epstein, *The Genesis of German Conservatism* (Princeton, New Jersey: Princeton University Press, 1966), 387–93, found the king stupid and unimaginative. By this time, in fact, most historians considered Frederick William III such an insignificant figure that their studies barely took note of him. Thus Reinhart Koselleck in his 739-page tome, *Preussen zwischen Reform und Revolution*, made a scant fourteen brief references to Prussia's king of forty-three years. No full biography exists to date (1991), although one by Friedrich Stamm of Hamburg will soon appear with the Siedler-Verlag.

Crown Prince Frederick William, lived in modest, middle class living quarters, saw his education neglected, and was completely excluded from the affairs of state. The young grandnephew grew to adolescence in the same bourgeois accommodations, but, as second-in-line to the succession, existed in even greater political obscurity. It was as if the old man had accepted decline as inevitable or preordained. His last words to the boy in 1786 strengthened this fatalistic impression: "I fear you will have a hard time of it. Stand firm and remember me."[5] Eleven years as Crown Prince in his own powerless right eroded more of the young man's self-confidence. Nor were there any dashing martial exploits to boost spirits and restore inner faith. The prince's only military forray in 1793 saw Prussian troops retake Mainz after a tedious and costly siege, only to have it fall into French hands again a few months later.

Ascending the throne at the age of twenty-seven in 1797, Frederick William was quite unsure of himself, having little more than an ample store of common sense to guide him through the turbulent times. Yet this pragmatism could have served him well. Many far-reaching reforms of state and society—such as reorganization of the bureaucracy, peasant emancipation, a citizens' militia, and economic liberalization—were discussed with loyal bourgeois councillors like Anastasius Mencken and Karl Friedrich Beyme. But the king had neither the self-assurance to trust his instincts and advisers, nor the audacity to challenge conservatives entrenched in the bureaucracy, army, and nobility. It was a largely unchanged Old Regime which went to war in 1806.[6] After Jena, Auerstedt, and Tilsit, Frederick William sank from his normal despondency into a severe depression. He lost faith in God and blamed himself entirely.[7]

However, as 1807 drew drearily on to 1809, a semblance of emotional stability returned. Philip Borowski, a village parson in Neurossgärtner, near Königsberg, restored the king's religious beliefs with supernatural-

[5] Cited in Eylert, *Charakter-Züge*, 1:456.

[6] For reform efforts before 1806, see Treitschke, *Deutsche Geschichte*, 1:148–61; Rudolph Stadelmann, *Preussens Könige in ihrer Thätigkeit für die Landeskultur* (Leipzig: Verlag von S. Hirzel, 1887), 141ff; Max Lehmann, "Ein Regierungsprogramm Friedrich Wilhelm's III," *Historische Zeitschrift* 61 (1889): 441; Otto Hintze, "Preussische Reformbestrebungen vor 1806," *Historische Zeitschrift* 76 (1896): 413–43; G. P. Gooch, *Germany and the French Revolution* (London: Longmans, Green & Co., 1920), 410; William O. Shanahan, *Prussian Military Reforms 1786–1813* (New York: AMS Press, Inc, 1966), 61–87; Hartmut Harnisch, "Die Agrarpolitischen Reformmassnahmen der Preussischen Staatsführung in dem Jahrzehnt vor 1806/07," *Jahrbuch für Wirtschaftsgeschichte* (1977/3), 129–53; and Schissler, *Preussische Agrargesellschaft*, 50–56. The latter three authors are modest—and accurate—in their assessments of the prereform period. Epstein, *Genesis*, 387–93, on the other hand, underestimates both the king and his reforms. Treitschke, again, is still fairly solid.

[7] Kurt von Raumer, *Die Autobiographie des Freiherr vom Stein* (Münster: Verlag Aschendorff, 1955), 33.

istic, deistic sermons which placed the primary responsibility for earthly events on the shoulders of man, not God.[8] From here it was only natural to place less emphasis on "the scourge of God"—and shield a healing psyche—by blaming others. A few months before the court moved back to Berlin, Frederick William penned his now more complex explanation of Prussia's demise.[9] The king freely admitted his own limitations, but argued that he and his trusted advisers had sincerely striven to improve conditions and create a happy, satisfactory existence for people of all classes. That none of this had come to fruition and instead, Prussia had been vanquished, he attributed to society's general unwillingness and inability to improve and progress. The king was particularly angry, however, with the aristocratic elite which had traditionally opposed Hohenzollern power. He lashed out at its depravity, irreligion, demagoguery, belligerence, and above all, its ingratitude. "I singled them out, gave them posts, attached them to my person, gave them honors, decorations, and land," he recalled years later. "When all went well they seemed able to do anything and everything, [but] in misfortune they were exposed and disloyally deserted." Now a few well-meaning, upstanding men strove to revitalize the nation, but the ongoing attempts of these few patriots would founder on the opposition of ungrateful, contentious factions which doubted the wisdom of every plan proposed as a way out of the impasse. To be king of such an arrogant, feuding lot, Frederick William concluded, was a thankless task which he would gladly put aside if circumstances permitted. No Prussian king has had less faith in his country, and in himself, than Frederick William III in 1809. The hurras and vivas of the happy burghers that December could not change this.

· · · · ·

Indeed, the "patriots" faced a difficult challenge. Napoleon was at the height of his power in the winter of 1809/10. Despite the intensifying Spanish uprising, French hosts had overrun Austrian forces at Wagram the previous spring, and in the East, the once-fearsome Russian eagle sat tamely on its perch. The French had also left Prussia in terrible condition. The kingdom which previous Hohenzollerns had forged into a great European power was halved in size, compelled to support a French army

[8] For Borowski, see Eylert, *Charakter-Züge*, 1:216, 220–22; and ADB 3:177–78.

[9] See the king's testament of April, 1809, in BPH, Rep. 49, Nr. 93, GStA Berlin. Anderson, *Nationalism*, 284–85, concludes incorrectly from Max Lehmann's description of this unpublished document that the king was angry with and bitterly opposed to the military and domestic reformers. But the king speaks favorably of their efforts. As Anderson notes correctly elsewhere (p. 278 for quote), the king's anger was directed against the nobility for its ingratitude.

of occupation, and pay a huge indemnity which exceeded the state's pre-war annual revenue. The countryside from Jena to Eylau was ravaged and destroyed. Prussian ports were closed to English commerce under Napoleon's "continental system" and a decade of lucrative agricultural exports to the booming industrial island halted. The subsequent recession in an already devastated countryside quickly spread to the towns where many artisans and manufacturers were forced to close shop. The kingdom's tax base was demolished, making it doubly difficult to pay the French. Prussia's military calamity had been compounded by a severe economic crisis.

As the king's memorandum implied, reform-minded individuals had been hard at work on these problems for many months. In the military, War Minister Gerhard [von] Scharnhorst and a large, intensely loyal entourage of officers and princes of the royal family concentrated on building a new army. They advocated above all an end to the aristocratic privileges promoted and indulged by Frederick the Great. The officer corps was to be opened to all social classes with admission based on competitive examination and promotion on merit. Alongside the regular army would rise a national guard to lend popular fervor to the war of liberation.[10]

In alliance with the military reformers were two circles of talented civilians around Karl August von Hardenberg, chief minister during the gloomy spring of 1807, and Baron Karl vom Stein, Hardenberg's successor until the fall of 1808. Young, idealistic, and politically frustrated until now, these men envisioned a thorough reform of state and society which would charge the nation with energies untapped by the outmoded economic and political institutions of Frederick the Great. In conjunction with a more centralized, ministerial bureaucracy along the lines of Napoleonic France or its German allies, Prussia needed some form of parliamentary representation based on property rights. In addition, all peasants should be freed from serfdom and better-off peasants given land. Finally, class and guild restrictions, respectively, upon the ownership of land and the practice of trades were to be eliminated; equality of taxation between city and countryside established; internal toll barriers removed; and external tariffs lowered. In short, the mercantilistic, "Frederician" system of state subsidies, monopolistic chartered companies, and prohibitive tariff protection should yield to a liberalized economy based on com-

[10] For an introduction to the military reforms, see Shanahan, *Prussian Military Reforms*, 88–149; Gordon A. Craig, *The Politics of the Prussian Army 1640–1945* (New York: Oxford University Press, 1972), 38–53; Peter Paret, *Yorck and the Era of Prussian Reform 1807–1815* (Princeton, New Jersey: Princeton University Press, 1966), 111–53; and more recently, Heinz G. Nitschke, *Die Preussischen Militärreformen 1807–1813: Die Tätigkeit der Militärreorganisationskommission und ihre Auswirkungen auf die preussische Armee* (Berlin: Haude & Spener, 1983), passim.

prehensive freedom of enterprise. The immediate result, they argued, would be more efficient farming, rural manufacturing, and the adoption of technology appropriate to Prussia. Coffers would fill to finance Scharnhorst's dreams of liberation and lay the foundation for a prosperous future.[11]

Little of this had been approved—and even less implemented—when the royal family reentered Berlin. Scharnhorst's military reforms were the farthest along, largely due to the uncharacteristic resoluteness shown by the king in brushing aside the objections of old-line conservatives in the army. Nothing had been executed, however, and one major reform—the national guard—still lacked approval. Military and civilian reform factions were in disagreement over the merits of the proposal; but, almost as important, Frederick William's earlier enthusiasm for the idea had waned. Such overtly aggressive institutions would certainly alarm the French, while the Prussian people, assumed the pessimistic monarch, might not rally to the colors.[12]

Even less had changed in the civilian area. One of Stein's early edicts of October 1807 established the right of all social classes to own land and eliminated the worst personal abuses of serfdom (*Oktoberedikt*). Another of November 1808 granted towns far-reaching rights of self-governance (*Städteordnung*). But none of these ordinances was scheduled to take effect until November 1810.[13] On all other fronts, matters were at an impasse. A national assembly was one reform which Frederick William was extremely reluctant to support. Bureaucratic inertia had increased, moreover, after the civil service cliques which drafted the reforms in Königsberg rejoined the preponderance of conservative officials who had stayed behind in Berlin after Jena. Thus cautious jurists in the new Ministry of Justice expressed objections to undermining the legal rights of guilds and estate owners, while toll and tariff functionaries in the new Ministry of Finance scoffed at the idea of abandoning mercantilism for untested laissez-faire theories.[14] Heads of the state's two economic em-

[11] For an introduction to the domestic reforms, see Vogel, *Allgemeine Gewerbefreiheit*; and Marion W. Gray, *Prussia in Transition: Society and Politics under the Stein Reform Ministry of 1808* (Philadelphia: American Philosophical Society, 1986).

[12] Shanahan, *Prussian Military Reforms*, 100–149; Craig, *Politics of the Prussian Army*, 38–57.

[13] Vogel, *Allgemeine Gewerbefreiheit*, 184.

[14] For the king's antiparliamentarism, see Trietschke, *Deutsche Geschichte*, 1:160–61; for intragovernmental disputes, see G.J.C. Kunth's memorandum, n.d. (September 1808), Rep.151 III, Nr.2176, Bl.5–8v, ZStA Merseburg; Kurt von Rohrscheid, *Vom Zunftzwang zur Gewerbefreiheit* (Berlin: Carl Heymanns Verlag, 1898), 388–94; Walter M. Simon, *The Failure of the Prussian Reform Movement, 1807–1819* (New York: Howard Fertig, 1971), 41–50; and Vogel, *Allgemeine Gewerbefreiheit*, 146–47.

pires, the Mining Corps and the Seehandlung, also came forward to defend their threatened vested interests.[15]

Unfortunately for the reformers, their cause rested at this time with the ineffectual Minister of the Interior, Count Friedrich Alexander von Dohna-Schlobitten. Weak and easily intimidated, the East Prussian nobleman tended to rely on the king for strength and guidance.[16] This was, of course, a dreadful error. Frederick William, unlike many of the zealots around Stein and Hardenberg, was no doctrinaire. His father had taken pains to arrange more tutoring than the uncaring Frederick II had provided his own Crown Prince. But this training, meted out by an army officer, in no way approached the quality and rigor of university instruction. Frederick William's relatively spotty education and consequent need to rely on common sense had molded him by adulthood into an eclectic who considered issues one at a time in a very practical, matter-of-fact fashion. Clever advisers quickly learned to avoid theoretical justifications for their proposals lest the king take offense at such "phrase-making." Institutions which had proven themselves deserved to remain; those which had not, simply did not.[17] And with two major exceptions—guilds, which he believed were largely worthy, and high tariffs, which sometimes proved necessary for survival—he agreed with Dohna's programs.[18] "I had these ideas a long time before,"[19] he said. Rarely forceful himself, however, Frederick William merely grumbled about the criticism and resistance meted out to his patriotic ministers by disobediant subjects and predicted that all present reform efforts, like his own, would come to naught.[20]

And with good reason. There were reactionary forces gathering again by the winter of 1809/10 which added weight to the recalcitrant elements within the civilian bureaucracy. Leading what was to be only the first wave of an aristocratic counterattack were distinguished members of the Brandenburgian nobility like Ludwig von der Marwitz, Otto von Voss-Buch, and Karl von dem Knesebeck. Intellectual force was supplied by two young journalist-philosophers of the growing romantic school, Adam

[15] For the Seehandlung, see Radtke, *Seehandlung*, 5–35, and below, chap. 6; for the Mining Corps, see below, chap. 4.

[16] Rohrscheidt, *Vom Zunftzwang*, 399.

[17] Eylert, *Charakter-Züge*, 1:39–42, 216; and Hardenberg's character sketch (1808), as cited above in n. 4, 572–73. For the king's dislike of fancy talking, see Varnhagen Diary I, 1:205–6.

[18] Rohrscheidt, *Vom Zunftzwang*, 228–38, 316–17. See also Queen Luise to the Duke of Mecklenburg, n.d. (summer 1809), quoted in Richardson, *Memoirs*, 289. Concerning long-overdue reforms in Prussia: "No one perceives this more clearly than the king . . . [who had said] 'This must be changed, we must alter things.' "

[19] Quoted in Treitschke, *Deutsche Geschichte*, 1:150.

[20] See the king's testament of April 1809, in BPH, Rep. 49, Nr.93, GStA Berlin.

Heinrich [von] Müller and Heinrich von Kleist. Voices of dissent were also heard in military ranks from Generals Ludwig Yorck von Wartenburg, Ludwig von Jagow, and August von Kalckreuth.[21] Uniting the aristocratic party was resentment over widespread accusations that the nobility was responsible for Prussia's collapse. This combined with outrage over agricultural and industrial ideas designed to resuscitate the nation at the expense of aristocratic rights of land ownership, control of the labor supply, and ultimately, political prestige, as the bourgeoisie swept into military and political positions formerly reserved for Junkers. The programs were as ill-advised, it was added, as illegal. Peasant emancipation, freedom of land ownership, and the growth of rural industry would destroy the organic harmony and welfare of the countryside, leaving only a cash nexus to govern relations between men. What was worse, materialistic bourgeois values and the lust for profit would emasculate the nation and loosen traditions of loyalty and sacrifice. An ignominious "Jew-State" would replace the heroic Prussia which they knew. The stagnation of the reform movement was thus greeted by adherents of the reactionary party like Yorck, who had always regarded the reformers as "vipers" who would "dissolve in their own poison."[22]

This situation could be easily reversed, however, in an absolute monarchy where final decisions were reserved for the king. And Frederick William, although often reluctant to follow his instincts, was basically sympathetic to many of the reformers' proposals. What was needed in 1810 was a minister with the dynamism to exploit the potential of enlightened despotism in Prussia. The inability of Dohna and his ministerial colleague in Finance, Karl von Altenstein, to stay abreast of French payment demands created the conditions required to revitalize the reform movement. For failing to resolve the state's financial emergency they were dismissed in June 1810. The king elevated the stronger Karl August von Hardenberg to the powerful new office of chancellor.

Hardenberg soon employed the instruments of power bequeathed to him, assuming temporary control of both vacated ministries. This was done as much to paralyze opponents in Finance and Interior as to utilize

[21] For this paragraph, see Wilhelm von Humboldt to Göthe, 10 February 1810, Müller to Marwitz, n.d. (October 1810), and Marwitz to Hardenberg, 11 February 1811, printed in Baxa, *Adam Müllers Lebenserzeugnisse*, 1:525, 569–74, 608–17; Karl August Varnhagen von Ense, *Denkwürdigkeiten des eigenen Lebens* (Leipzig: F. A. Brockhaus, 1843), 2:282; Jakob Baxa, *Adam Müller: Ein Lebensbild aus den Befreiungskriegen und aus der deutschen Restauration* (Jena: Verlag von Gustav Fischer, 1930), 100–111, 136–79; Johann Gustav Droysen, *Das Leben des Feldmarschalls Grafen Yorck von Wartenburg* (Leipzig: Verlag von Veit & Co., 1884), 153–56, 158–63; Guy Stanton Ford, *Stein and the Era of Reform in Prussia, 1807–1815* (Gloucester, Massachusetts: Peter Smith, 1965), 210; Craig, *Politics*, 68.

[22] Quoted in Droysen, *Das Leben*, 162–63.

political forces there, for the chancellor largely ignored the ministries, preferring to use as the engine of change a select party of experts which gathered around his person.[23] Few of the members of Hardenberg's "staff" had held important offices in Prussia before 1806 and thus had no vested bureaucratic attachment to the Old Order. Many, in fact, hailed from areas outside the Prussia of 1810. All were specialists in economic affairs, while most were university-educated, of bourgeois origins, and embued with an economic "liberalism" which combined Physiocratic, Smithian, and Freemasonic ideas. The chancellor alone presented the proposals which emitted from these private counsels to the king. And Hardenberg, who was always polite and respectful to a monarch cognizant of personal failings yet sensitive about royal prerogatives, managed to carry the indecisive Frederick William along with him. In this case the monarchy received sustenance from a party which was positioned strategically near the king and determined to fulfill its own agenda.

The first phase of Prussia's "revolution from above" came in two flurries of legislation during the autumns of 1810 and 1811. Preceding the first decrees and enhancing the atmosphere of rebirth came the dual inaugurations of the University of Berlin and the War School. Here the civil servants and officers of the new Prussia would spring forth. Simultaneously, Stein's emancipation edict terminating the legal subordination of peasants to lords went into full effect. The latter, in fact, necessitated some of Hardenberg's early reforms, for there were loopholes in Stein's decree and the all-important question of property rights remained unresolved.[24] One of the first ordinances therefore suspended local laws sanctioning the servitude of peasant children in noble households (*Gesindeordnung*), while a subsequent edict granted to some 161,000 peasant households—about 10 percent of all farming families[25]—the right to land title in return for payment to lords of a third or a half of peasant land (*Regulierungsedikt*). Capping the first agrarian reforms was an ordinance promoting rural transportation, animal husbandry, and new farming techniques (*Landeskulturedikt*).

These measures reflected the importance of agriculture to Prussia's economy in the early 1800s. Around 60 percent of the population worked on farms, while 73 percent either lived or worked in the countryside. About 70 percent of net investment went into agriculture, moreover, while holders of wealth put a mere 2.2 percent into "industry." But even within the industrial sector farming was important, providing raw mate-

[23] Vogel, *Allgemeine Gewerbefreiheit*, 48–64, 83–97.

[24] See ibid., 165–87.

[25] Schissler, *Preussische Agrarwirtschaft*, 73, 77. Stein estimated the number of families in Prussia around 1806/07 at 2.4 million. With 65 percent of the population engaged in farming, 161,000 equals roughly 10 percent.

rials to linen and woolen operations which absorbed around 71 percent of industrial investment. Capital or producer goods like coal, iron, chemicals, and leather (e.g., for pulleys, machine belts, and harnesses) were almost insignificant at this time, representing a scant 7 percent of net industrial investment or 0.15 percent of the total.[26]

Despite the relative unimportance of industry, Hardenberg and his experts nevertheless had special plans for this sector. "I truly do not consider myself infallible," Hardenberg told one of his party. "I ask for advice [and] gladly listen to every opinion."[27] Accordingly, the persons around the chancellor who brainstormed industrial policy had vastly different visions of what lay ahead. A cautious faction defended the mercantilistic system of urban factories and state-owned enterprises begun by Frederick the Great as the wiser, more modern approach. A more radical group favored privatization and free trade policies which would lead to rural manufacturing.

The principal figures in the latter faction were: Christian Friedrich Scharnweber, Hardenberg's longtime personal secretary and estates manager; Friedrich von Raumer and Peter Beuth, two young counselors from the District Government of Potsdam; and the staff of the Business Department[28] of the Finance Ministry—including Albrecht Thaer, a well-known agronomist; Johann Gottfried Hoffmann, an economist and statistician; and Kaspar Friedrich von Schuckmann, the department head from 1810 to 1812. Another important figure, closer personally to Stein than Hardenberg, was Gottlieb Johann Christian Kunth, Stein's leading technocrat from 1804 to 1806 and the head of the Business Department under Dohna and Altenstein. These men argued that private mines and factories would be much more efficient than state-run complexes. They pointed out, furthermore, that the Frederician system of prohibiting rural enterprises while supporting protected, subsidized, city-based products like cotton, silk, tobacco, leather, and porcelain was economically irrational. Urban plants were subjected to high labor costs and the technologically restrictive influence of the guilds. Raw materials were either imported at great expense (cotton and tobacco), transported long distances from the countryside (leather), or produced with great difficulty (silk), while porcelain could neither compete abroad, nor, as a luxury

[26] Richard Tilly, "Capital Formation in Germany in the Nineteenth Century," in Peter Mathias and M. M. Postan, *The Cambridge Economic History of Europe* (Cambridge: Cambridge University Press, 1978), 7 (1): 419; Friedrich-Wilhelm Henning, *Die Industrialisierung in Deutschland 1800 bis 1914* (Paderborn: Verlag Ferdinand Schöningh, 1973), 20.

[27] Quoted in Friedrich von Raumer, *Lebenserinnerungen und Briefwechsel* (Leipzig: F. A. Brockhaus, 1861), 128. Raumer seems to quote from notes made at the time.

[28] The German word "Gewerbe" was used at the time in the very broad sense of "business," usually including both industry and agriculture.

item, command a large home market. Tariff protection and exorbitant subsidies merely compounded the costs to society of this unnatural system, costs which soared even higher when all of the squandered opportunities were considered. For cottage industry, country textile factories and large-scale (*fabrikmässige*) logging establishments could escape the backward-looking guilds and benefit from cheap labor and proximity to the majority of consumers. What was more, these operations could thrive on inexpensive water power and abundant raw materials such as potatoes for brandy, hops and barley for beer, lumber for furniture and construction, flax and wool for clothing, and dye-crops for finishing textile processes. Legislation should allow these natural, more appropriate "industries" to expand, provide employment for the rural underemployed, and overcome the dire consequences of population growth predicted by the English parson, Thomas Malthus. Discontinuation of subsidies and provision of cheaper foreign commodities through free trade would also represent a tremendous savings to the nation. The older establishments would simply have to compete without state aid, reinvest in more competitive rural ventures, or go under.[29]

Another faction urged Hardenberg to ignore such advice. Heading this group were Johann August Sack, formerly of the Mining Corps, and three moderately conservative public financial experts—Christian von Heydebreck, Heinrich von Begùelin, and Philipp von Ladenberg. Aside from the dubious legality of abolishing the guilds, chartered companies and state monopolies, what sense did it make, they counseled, to allow Prussia's state-run enterprises and older factories to perish? A tremendous investment in mines, buildings, and machinery accumulated over the last sixty years could not reasonably be sacrificed for unpredictable, purely theoretical concepts. In the unlikely event that rural businesses did arise, their progress would have to be measured against the increased costs to the state of collecting taxes outside city walls as well as the catastrophic losses of capital and jobs in cities. Nor could country factories

[29] See Scharnweber's unpublished lecture of 1814, Rep.77 CCCXX 35, Bl.35–37; Dohna to Frederick William, 13 October 1810 (and for "fabrikmässig," the accompanying "Einrichtungsplan" for a Scientific-Technical Institute), Rep.74, H.XV, Nr.2, Bl.10–23; and for ideas within the Business Department, Kunth's memorandum, n.d. (September 1808), Rep.151 III, Nr. 2176, Bl.5–8, ZStA Merseburg. For Kunth, see also Kunth to Stein, n.d. (April 1809), and Kunth to Stein, 11 July 1809, printed in Karl Freiherr vom Stein, *Freiherr vom Stein: Briefe und Amtliche Schriften*, edited by Erich Botzenhart (Stuttgart: W. Kohlhammer Verlag, 1957), 3:108, 157. For Raumer, see his undated article (1811) over the reforms, printed in Rohrscheidt, *Vom Zunftzwang*, 502–7. And for attitudes toward Malthus, see Dohna's memorandum of January 1819, printed in Franz Rühl (ed.), *Aus der Franzosenzeit* (Leipzig: Verlag von Duncker & Humblot, 1904), 325. See also Friedrich and Paul Goldschmidt, *Das Leben des Staatsrat Kunth* (Berlin: Julius Springer, 1881), 60–78; and Vogel, *Allgemeine Gewerbefreiheit*, 137–40.

expect to find foreign markets as unprotected as the Prussian, for the nation's rivals had thus far failed to see the logic of the new laissez-faire theories.[30] "Why should we listen to [Scharnweber, Raumer, et al.]," sneered Heydebreck and Begùelin, "and not to Pitt, Colbert, Struensee and other excellent men?"[31]

The chancellor "gladly listened" to both sides, but his own background and views on political economy predisposed him toward Scharnweber, Raumer, Beuth, and the men of the Business Department. There is a great deal of evidence that Hardenberg adhered to a brand of Physiocratic thinking which relegated "secondary" forms of production like manufacturing to a lesser position behind "primary" forms of production like agriculture and mining.[32] Perhaps the strongest evidence reinforcing this impression of Hardenberg the Physiocrat is found in his famous Riga memorandum of September 1807:

> I am fully convinced that we have promoted the factory system to the detriment of our country, for we have sacrificed the advantages of trade which generally, and especially in Prussia, rest with agriculture. So many export and import bans, so many restrictions through monopolies and other privileges for certain places or persons brought harm instead of gain for the whole. Had we given careful encouragement and support of industry through a policy of commercial and industrial freedom, it can be assumed with reasonable certainty that we would have created a far greater gain than everything brought about artificially at so much expense, produced a stronger, more considerable populace and generated more permanent forms of manufacturing more suitable to the nature of the country. Our balance of payments, moreover, would certainly have improved. I am far from wishing to speak against all factories. The state can and must support this one or that one depending on the circumstances, advancing their facilities and dragging them out of ignorance; but the really useful [factories] arise on their

[30] See Beguèlin and Heydebreck to Hardenberg, 23 August 1810, Rep.92 Schöll 29, Bl.147–51, and Beguèlin to Hardenberg, 22 December 1811, Rep.74 K.VIII.17, Bd.1, Bl.11–13, ZStA Merseburg; and Vogel, *Allgemeine Gewerbefreiheit*, 80–81, 147, 149.

[31] Beguèlin to Hardenberg, 22 December 1811, Rep.74 K.VIII.17, Bd.1, Bl.11–13, ZStA Merseburg.

[32] A. Schwemann, "Friedrich Anton Frh. von Heinitz," *Beiträge zur Geschichte der Technik und Industrie* 12 (1922), 163; ADB, 37:638–39 (Albrecht Thaer); Herbert Pruns, *Staats- und Agrarwirtschaft 1800–1865* (Hamburg: Verlag Paul Perey, 1979), 1:11–12, 224–28; Vogel, *Allgemeine Gewerbefreiheit*, 140. Lending support to this argument are the chancellor's great respect for Frederick the Great's Physiocratic *Oberberghauptmann*, Friedrich Anton von Heynitz; Hardenberg's attention to agricultural improvements while district president of Ansbach-Bayreuth (1792–1806); together with the numerous agrarian enthusiasts he gathered around his person during these early years. See also Wolfhard Weber, *Innovationen im frühindustriellen Bergbau und Hüttenwesen:Friedrich Anton von Heynitz* (Göttingen: Vandenhoeck & Ruprecht, 1976), 181.

own. . . . They are natural, not affected and require no export or import bans to survive.[33]

Hardenberg leaves little doubt here about the primary position of agriculture and the need to establish a different pattern for industry. Unless formerly privileged manufacturing cities could prosper on their own, he noted elsewhere in the document, "it is better that they sink back to villages."[34]

The chancellor received encouragement from the king.[35] For although Frederick William was ambivalent about new technology and harbored doubts about the wisdom of free trade and nonguild enterprise, he had believed since his days as crown prince that the first economic task of the monarch was to promote agriculture. For this reason he looked askance at mining and its Corps, necessary evils which too frequently expropriated land and took it out of production. A second important task was to promote manufacturing rooted in agriculture. Businesses which worked up domestic raw materials were superior to those using foreign materials because "the latter remunerated only the entrepreneurs and their factories, while the former profitted the farmer too, as for example woolen and linen manufacturing."[36] That such operations were usually located in the countryside (linen) or often in small provincial towns (wool) had an additional advantage: king and court, haunted by the precedent of 1789, were convinced that too many workshops in the capital could lead to mob violence or revolution.[37] As early as 1798, in fact, Frederick William had ordered his officials to find ways to shift industrial ventures from Berlin to the provinces.[38] Thus he preferred rural industry for political as well as economic reasons.

Given this hostility toward urban-based manufacturing on the part of king and chancellor, it is not surprising that the agrarian edicts of 1810/11 were accompanied by others designed to implant agricultural side-

[33] Hardenberg to Frederick William, 12 September 1807 (*Rigaer Denkschrift*), printed in Georg Winter (ed.), *Die Reorganization des Preussischen Staates unter Stein und Hardenberg* (Leipzig: Verlag von S. Hirzel, 1931), 332.

[34] Ibid., 331

[35] Frederick William to Hardenberg, 14 November 1811, printed in Eylert, *Charakter-Züge*, 3B:426.

[36] See Frederick William's "Gedanken über die Regierungskunst zu Papiere gebracht im Jahr 1796–97," printed in Max Lehmann, "Ein Regierungsprogramm Friedrich Wilhelm's III," *Historische Zeitschrift* 61 (1889): 454. For his attitude toward mining, see *Protokolle über die Revision des Berggesetzes in Folge der Gutachtlichen Bemerkungen der Provinzialstände Mai 1845 bis Dezember 1846*, 22 May 1845, Rep.84a, 11080, p. 14, GStA Berlin.

[37] Schnabel, *Deutsche Geschichte*, 3:274; see also Kunth's reference to the King's feelings in a report of 25 March 1817, printed in Goldschmidt, *Das Leben*, 289.

[38] Karl Heinrich Kaufhold, *Das Gewerbe in Preussen um 1800* (Göttingen: Otto Schwarz & Co., 1978), 449.

industries in the countryside. The freedom to practice a trade was guaranteed upon payment of a simple licensing fee, thereby eliminating the rights and privileges of the guilds and chartered companies (*Gewerbesteuer Edikt* and *Gewerbepolizei Edikt*). Wage and price supports were also abolished and the legal way opened for investment in an unlimited number of business ventures in towns as well as the countryside, where they had been previously banned. In order for the state to reap the benefit of the anticipated spread of enterprise outside city gates, the excise tax collection system of the towns was extended to the country (*Landkonsumptionssteuern*). It was a clear expression of the reformers' rural orientation, however, that business tax rates descended in four categories from large and middle-sized cities to small towns and the countryside (*Fernerweites Edikt*), giving rural ventures a distinct tax advantage. Finally, the vast commercial empire of the Seehandlung was dismantled and a similar fate prepared for the Mining Corps. Hardenberg wanted to free private mines and finishing establishments from state administration and curtail royal operations to a few model plants which would serve entrepreneurs as a reservoir of mining and metallurgical expertise.[39] By substituting a great deal of economic freedom for the servitude, exclusivity and statism which had heretofore reigned in Prussia, Hardenberg's reformers gambled on quick repayment of the French and a sound, sensible, largely rural economic future.

The cause of economic liberalism had taken great strides forward by January 1812. Further progress with domestic change was halted, however, by the momentous events which began to unfold that year. In June Napoleon grasped for total domination of Europe and invaded Russia with the largest army ever amassed under one command. But, with his Grand Army in disarray by the winter of 1812/13, Napoleonic invincibility was only a memory. Prussia joined Russia and Austria in military campaigns which led from Leipzig (1813) and Paris (1814) to Waterloo (1815), final victory, and considerable territorial expansion. The Rhineland, Westphalia, and a portion of Saxony were added to the six eastern provinces of old Prussia. During the fighting the last of the military reforms, the national guard (*Landwehr*), came into being and proved itself well. The king even promised to establish a parliament. But with Hardenberg and many of his chief aides compelled to leave Berlin for long periods, little additional progress was registered with domestic programs. Not un-

[39] In general, see Vogel, *Allgemeine Gewerbefreiheit*, 165–87. For business taxes, see Vogel (p. 181); and Ritter, *Die Rolle des Staats*, 121. Ritter includes an interesting citation from Adolf Wagner (p. 131): "One can observe that business taxes were actually to constitute a burden only in the economically more significant cities, while in others and in the country they merely functioned more like a moderate surcharge to the class tax."

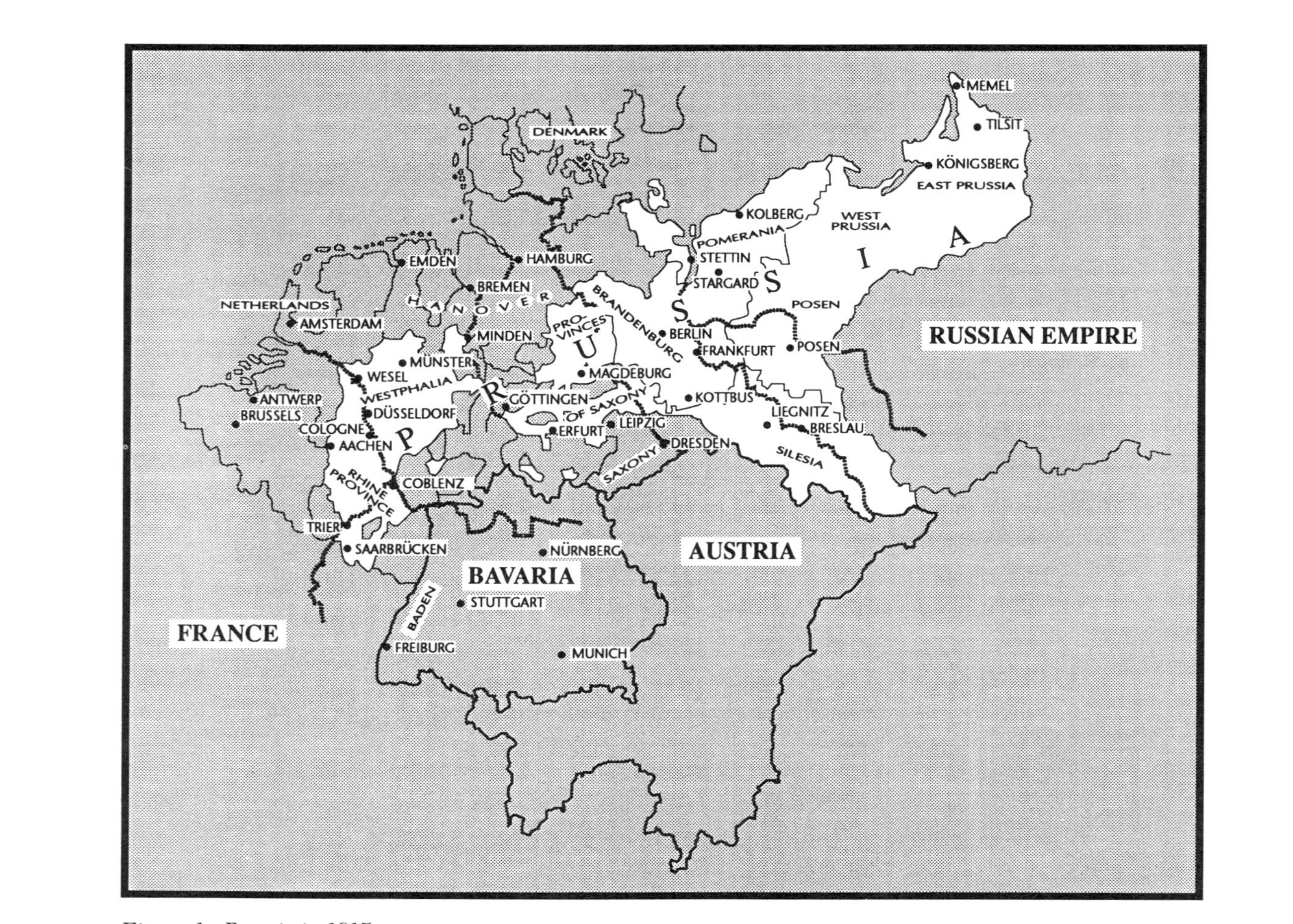

Figure 1. Prussia in 1815.

til the Congress of Vienna drew to a close in 1815 could the chancellor and his staff again pick up the cudgel of reform.

New patent legislation was a sign of renewed momentum in the first months of peacetime.[40] Under discussion since Hardenberg's appointment, an ordinance of October 1815 further liberalized Prussia's economy by introducing the weakest patent protection in Europe. The law subjected inventions and innovations to rigorous testing for newness by state officials. But even when inventors could prove uniqueness, they usually received a patent of no more than a few years' duration. The recipient was also obliged to make public the general nature of his finding, release details to other entrepreneurs if competition was not deemed dangerous by patent officials in the Business Department, or sacrifice patent rights if the patented device was not operational within six months. Deserving innovators received remuneration for their labors under the new system; but the emphasis here was on rapid diffusion of new techniques to every corner of the country, not on protection. The new practices were quite consistent, therefore, with the reformers' campaign to discard monopolies and mercantilism in favor of technological modernization in rural areas.

Political victories did not come as consistently or as easily, however, in postwar Berlin. Hardenberg had grown older, physically more frail, and less adept as a politician. Party enemies in the bureaucracy and old nobility struck back with noteworthy success. Thus in May 1816 aristocratic pressure resulted in a halving of the number of peasant households eligible to obtain noble land. Further complicating the political stage, Frederick William returned from the wars slightly more confident and somewhat more assertive.[41] Although the monarch still harbored thoughts of abdication when party clashes became extremely unpleasant, he became more active and often proved to be an intelligent politician. As a result, the old doubts and psychological inability to act yielded slowly to royal mediation of disputes and occasionally to initiatives which he pursued to successful ends. The king's aversion to all-embracing theories and desire to progress without abandoning programs and institutions which had proven useful became an increasingly important factor in the politics of economic and technological change.[42]

The controversy which raged over tariffs from 1815 to 1817 saw the king begin to resist the consistently antistatist doctrines which motivated Hardenberg's economic liberals. Prior to the outbreak of war in 1806,

[40] Alfred Heggen, *Erfindungsschutz und Industrialisierung in Preussen 1793–1877* (Göttingen: Vandenhoeck & Ruprecht, 1975), 26–41.

[41] Treitschke, *Deutsche Geschichte*, 1:148, 2:184.

[42] For these attributes, see Klöden, *Friedrich Wilhelm*, 5, 66, 67, 388–89; and Eylert, *Charakter-Züge*, 2:125–27.

Prussia maintained a mercantilistic tariff system which banned the import of certain commodities and levied duties of 50–200 percent *ad valorem* on others. The Frederician system went down with the state after Jena, giving way to a series of temporary tariffs which Hardenberg and his staff forced constantly lower. By 1815, the rate for most manufactured items stood at 8 percent.[43] When western textile owners pressed for increases to the 10–25 percent range, Hardenberg's successor as Minister of Finance, Hans von Bülow, told them that it was not tariffs, but rather their own industriousness and attention to technological advance which would eliminate British competition. They should abandon branches of production like cotton manufacture, moreover, "where England and France possess a natural preponderance."[44] Hardenberg exhibited similar indifference to the plight of textile producers by suggesting that after a few years English exporters might grow tired of selling in Germany under cost.[45] Opposed himself to the Frederician system, Frederick William was nevertheless convinced that a middle solution between the old and the new was in order. Thus he informed both ministers in 1816 of the need to cultivate the loyalty of businessmen to Prussia by means of "purposeful measures to protect the factories."[46] These remonstrances led Bülow to propose a tariff package in January 1817 which levied duties of 30 percent on wine and colonial products (e.g., tobacco and sugar), 25 percent on cotton manufactures from England, France, and Austria, 18 percent on iron wares, 12.5 percent on finished cotton goods from other countries, 8–10 percent on other manufactured items, and up to 5 percent on raw materials and intermediate products like pig iron, wool, cotton, and all yarns.[47] Many of these latter items entered duty-free.

This was nevertheless a brazen departure from the economic course set by Frederick the Great. Thus it is understandable that Bülow's proposals prompted renewed criticism from the mercantilists around Heydebreck, Begùelin, and Ladenberg and drew an even icier response from textile producers. Cotton manufacturers in Berlin went so far as to ask the king to allow Heydebreck and his fellow protectionists to investigate

[43] Gustav Schmoller, *Das preussische Handels- und Zollgesetz vom 26. Mai 1818* (Berlin: W. Büxenstein, 1898), 43; Beuth to Vincke, 29 August 1815, Vincke Papers A.III.23, StA Münster. Also see Takeo Ohnishi, *Zolltarifpolitik Preussens bis zur Gründung des deutschen Zollvereins* (Göttingen: Otto Schwarz, 1973), 17–27.

[44] Bülow to Vincke, 20 November 1814, Oberpräsidium B, 2856, StA Münster.

[45] Benzenberg to Gneisenau, 25 December 1816, printed in Julius Heyderhoff (ed.), *Benzenberg: Der Rheinländer und Preusse 1815–1823* (Bonn: Fritz Klopp Verlag, 1928), 58.

[46] Frederick William III to Bülow, 11 June 1816 and 18 October 1816 and to Hardenberg, 31 January 1816, quoted in Goldschmidt, *Das Leben*, 111.

[47] Schmoller, *Das preussische Handels- und Zollgesetz*, 43–44; C.F.W. Dieterici, *Statistische Übersicht der wichtigsten Gegenstände des Verkehrs und Verbrauchs im Preussischen Staate und im deutschen Zollverbande in dem Zeitraum von 1831 bis 1836* (Berlin: E. S. Mittler, 1838), 393; Thomas Banfield, *Industry of the Rhine* (London, 1846–48), 2:11.

the plight of Prussia's urban factories. Worried about a deepening recession in his kingdom, Frederick William granted the petition. The resultant commission, stacked in favor of the mercantilists, recommended against Bülow's proposals and for higher tariffs in April 1817.[48]

At Hardenberg's request, the king permitted two members of the Business Department to participate in the commission's deliberations. The first, Christian Kunth, had been head of the department three times previously, most recently from 1814 to 1816; the second, its current head, was Karl Georg [von] Maassen, the coauthor with Johann Gottfried Hoffmann, of Bülow's tariff. Kunth's minority report, which Maassen seconded, is interesting less for its eloquent Smithian defence of moderately low tariffs than for its Physiocratic advocacy of industrial advance in the countryside. The document lauded the owners of those "truly great factories [which manufacture] wool, linen, metals and minerals in their thousands of branches,"[49] without doles from the state. Kunth was particularly critical, on the other hand, of those manufactured goods like silk and cotton which Prussia could not produce competitively. Such commodities should be imported, he wrote, and the considerable social savings in capital and labor shifted to farming, forestry, construction, cattle-raising, sheep-breeding, and woolen manufacture. Kunth subjected the protariff cotton manufacturers of Berlin to the brunt of his attack. Only a few of them had shown any willingness to mechanize, while most owned marginal, unprofitable operations which presented threats to public safety and political order. Therefore, these cotton factories should be allowed to go under and their proprietors and employees resettled on state domains as farmers. High tariffs would defeat the larger purpose by removing any incentive for these persons to leave Berlin voluntarily and seek employment in small towns and the countryside.

Kunth's report accompanied the findings of Heydebreck's commission to Hardenberg's handpicked Council of State in the summer of 1817. The chancellor had established the body earlier that year as an intrastate parliament to review legislation proposed by the king, the chancellor, or the ministries. Heydebreck, Begùelin, and Ladenberg made no headway here with their arguments for a return to the old prohibitive duties. Outside pressure from the Mining Corps for more tariff protection was similarly unsuccessful.[50] It is significant, however, that thc most avid free traders also failed to alter the compromise nature of Bülow's proposals.

[48] Rohrscheidt, *Vom Zunftzwang*, 514–15, 556–57, 563–64; Goldschmidt, *Das Leben*, 115; Wilhelm Treue, *Wirtschaftszustände und Wirtschaftspolitik in Preussen 1815–1825* (Stuttgart: Verlag von W. Kohlhammer, 1937), 40–41; and Ohnishi, *Zolltarifpolitik*, 28–32.

[49] The report of 25 March 1817 is printed in Goldschmidt, *Das Leben*, 271–306 (quote on p. 274). Maassen signed his name below Kunth's on 6 April 1817.

[50] Treitschke, *Deutsche Geschichte*, 2:213; and for the Mining Corps, see Treue, *Wirtschaftszustände*, 83–84, 229.

While Maassen and Hoffmann adhered to the king's middle line, Kunth, Beuth, and Karl Wilhelm Ferber, their colleague in the Business Department, argued for lower rates to encourage technological innovation and emulation. Tariffs needed to be low, in other words, so that businessmen would be forced to adopt new techniques to survive. They were joined by Theodor von Schön, Friedrich Theodor Merckel, and Ludwig von Vincke, provincial officials in East Prussia, Silesia, and Westphalia. But it was "the principle of free importation with the levying of a moderate duty," as Frederick William's decree put it, which became law in May 1818.[51] The new legislation also swept away all district and provincial toll walls within the kingdom, creating one of the only free trade areas on the continent.

There were other instances when the king's pragmatic eclecticism and sense of fair play manifested itself in the nation's economic policy. Thus he reconstituted the Seehandlung as an autonomous ministry, probably because he trusted the ability of its new director, Christian Rother, to bolster sagging state finances. The corporation received a far-reaching charge which included government lending as well as the right to invest and intervene in the economy (see chap. 6). The reemergence of the Seehandlung, however, was part of a wider mercantilistic trend sanctioned by the king. For, impressed by Ludewig Gerhard and the crucial military role played by the mining "captains" during the Wars of Liberation, Frederick William also moved the Mining Corps out from under the reformers' ax. The proud body of mining and metallurgical experts was elevated alongside the Business Department as a separate division in the Ministry of Finance. After Vienna, the Corps emerged as a consistent opponent of liberal economics, preferring near-monopolistic patent rights, high tariffs, and stricter control of business concessions.[52]

As Peter Beuth put it in 1817, the relationship between the Business Department and the revitalized Mining Corps was characterized by "steady conflict."[53] At the center of the feud was the question of economic growth. To Kunth, Maassen, Beuth—and above them, Hardenberg—mining and metallurgy were indispensable components of rural industrialization. Economic freedom, not state shackles, would fuel this transfor-

[51] For the liberals in the Staatsrat, see Schmoller, *Das preussische Handels- und Zollgesetz*, 37–38; Treitschke, *Deutsche Geschichte*, 2:213–14; and Kunth's report of 25 March 1817, printed in Goldschmidt, *Das Leben*, 296–300. The king's decree of 1 August 1817 is quoted in Goldschmidt, *Das Leben*, 119.

[52] Wutke, *Aus der Vergangenheit*, 473–78; Schulz-Briesen, *Der preussische Staatsbergbau*, 60–62; Varnhagen Diary I, 20 and 25 January 1820, 1:58–59, 63; Radtke, *Seehandlung*, 41, 50.

[53] Beuth to Bülow, 31 December 1817, Rep.120, A.I.1, Nr.2, Bd.1, Bl.64, ZStA Merseburg.

mation in the countryside to the benefit of the treasury as well as the general welfare. Gerhard and his colleagues, on the other hand, were convinced that such an unfettering could only lead to ruinous competition and environmental and public-financial disaster (see chap. 4). In the middle was the king, averse to impractical statism, attracted by the economic and political potential of rural industry, but concerned lest mining attenuate the rights of land owners and inhibit agriculture.[54]

The result was a series of those compromises which would come to typify rule under Frederick William III in the postwar decades. Thus Hardenberg and his supporters wrested control of the licensing of metalworks away from the Corps in the six provinces of pre-1806 Prussia; but Gerhard's comrades were permitted to retain concessioning authority for smelting plants—the first stage of metal production—throughout the new provinces of the Rhineland and Westphalia.[55] The advocates of liberal economic policy were also thoroughly frustrated in their scheme to loosen the Corps' mercantilistic grip over private mining. Balancing this, however, the captains of mining experienced a major setback in their own effort to extend Prussia's patriarchal mining law west of the Rhine where liberal French statutes remained in effect. Under the French system, mines were treated as private property, contradicting Prussian law which viewed them as a temporary privilege to exploit treasures owned by the state. Government regulation was limited to mine safety under French law, whereas Prussian ordinances allowed officials to mandate certain technologies, supervise books, and set prices. But the Corps knew how to persist in a "steady conflict." Thus Gerhard won a major victory against what he perceived as irresponsible overproduction of minerals by blocking all of Hardenberg's proposals to legalize larger, deeper mines. The Mining Corps also foiled the aging chancellor's last attempt to sell Corps mines.[56]

By 1820, therefore, Frederick William was demonstrating uneasiness with the laissez-faire approach of Hardenberg and his staff. The ex-

[54] See *Protokolle über die Revision des Berggesetzes in Folge der Gutachtlichen Bemerkungen der Provinzialstände Mai 1845 bis Dezember 1846*, 4 July 1845, p. 100, GStA Berlin.

[55] Ibid., 24 May 1845, 29; see also the agreement between Interior and Trade of 3 February 1818, Rep.120, A.II.5a, Nr.10, Bd.1, Bl.90–91, ZStA Merseburg.

[56] Hans Arlt, *Ein Jahrhundert preussischer Bergverwaltung in den Rheinlanden* (Berlin: Wilhelm Ernst und Sohn, 1921), 16; Hans Dieter Krampe, *Der Staatseinfluss auf den Ruhrkohlenbergbau in der Zeit von 1800 bis 1865* (Cologne: Johann Heider, 1961), 48–68; M. Reuss, "Mitteilungen aus der Geschichte des Königlichen Oberbergamtes zu Dortmund und des Niederrheinisch-Westfälischen Bergbaues," *Zeitschrift für das Berg-Hütten- und Salinenwesen* 40 (1892): 352–53; Konrad Fuchs, *Vom Dirigismus zum Liberalismus: Die Entwicklung Oberschlesiens als preussisches Berg- und Hüttenrevier* (Wiesbaden: Franz Steiner, 1970), 106–12.

tremely low tariffs of 1815/16 were raised to a moderate level. Older, proven institutions like the Seehandlung and the Mining Corps were revived and provided with opportunities to show their usefulness. Particularly worried about mechanization, the monarch also set limits to bold Business Department schemes designed to further modern productive techniques in country homes and factories. Central to this department's goal of rural industrialization, these bureaucratic efforts warrant more attention at this point.

The program of rural industrialization was in full swing by the time the new tariffs became law in 1818. The Business Department registered some success with the breweries and distilleries which proliferated on the farms of eastern Prussia in the 1810s.[57] It placed a higher priority, however, on spawning regionally dispersed linen and woolen industries based on domestic flax cultivation and sheep farming. Under Kunth, Belgian flax growers were enticed to Silesia to spread their more sophisticated techniques throughout that depressed province.[58] The Technical Deputation of this department, headed by Peter Beuth after 1818, also took direct action to promote the mechanization which low tariffs and weak patent legislation were designed to reinforce. Beuth's division studied the latest foreign techniques for heckling and combing flax and wool, spinning and weaving linen and cloth, and dyeing and shearing each. After closely examining costs and quality, his men distributed sample machines—either procured from abroad or rebuilt from blueprints, models, or simple observation—to select entrepreneurs throughout the kingdom.

The distribution lists for the 1810s included scores of hand- or water-powered devices. From England came "dandy looms" and Lewis & Davis cylinder shearers; from Ireland, Lee's heckling machines; from France, Jacquard looms; and from Baden, Girard's heckling, drawing, and spinning system. The Business Department poured about 60,000 thaler into this effort from 1814 to 1817, a considerable sum at the time. Each manufacturer was required to use the machinery for at least five or six years and demonstrate its usefulness to others in the provinces. Woolen and linen factories like those of Busse in Luckenwald, Alberti in Waldenburg, Oelsner in Breslau, Ruffer in Liegnitz, and Cockerill in Guben, Cottbus, and Berlin were to be the "models," offering skeptical, small-scale producers convincing proof that such investments paid off and that high tariffs or premiums were not needed to compete with England.[59] The

[57] Schnabel, *Deutsche Geschichte*, 3:275–78.

[58] See the ministry's correspondence over this effort (1815–25) in Rep.120, D.V.1, Nr.3, Bd.1, ZStA Merseburg.

[59] Beuth to Bülow, 31 December 1817, 20 May 1821, Rep.120, A.I.1, Nr.2, Bd.1, Bl.64–

Deputation's model and machine collections served a similarly educational and emboldening purpose. The former boasted an array of 1:3 scale models of woolen machinery, while the machine collection itself concentrated on linen spinning devices.[60]

Diffusion of these techniques from the selected factories, however, encountered tremendous obstacles. Undoubtedly the steepest was a shifting pattern of foreign and domestic consumption which reinforced prejudices against new-fangled devices. Since the late 1700s, buyers had been moving away from linen to wool and even more so to cotton, the rage fabric from England. Between 1800 and 1830, linen's rough share of per capita textile consumption in Germany fell from 90 percent to 45 percent, thus greatly reducing any incentives which may have existed to risk scarce capital on new technologies.[61] As the 1810s wore on, opinions in the Business Department were divided over the proper course of corrective action. With a vacant headship after Maassen's promotion to the fiscal branch of the Ministry of Finance in 1818, the debate was a collegial one. Kunth, who wanted to preserve a special place for linen in Prussia's industrial future, prevailed with a scheme that was characteristically rural in its orientation.[62] The old technocrat recommended the resettlement of thousands of marginal linen producers from towns to farms all over the kingdom where they could weave as a secondary occupation. The department wanted 100,000 thaler to buy new "fast-shooting" looms for these prospective cottage producers. This figure would have represented about 4.5 percent of annual net industrial investment in Prussia during these early postwar years.[63] A portion of the funds would also be granted to

70, 73–76, ZStA Merseburg; and 17 July 1822, cited in Straube, "Chr. P. Wilhelm Beuth," 138–43.

[60] See Conrad Matschoss, "Geschichte der Königlich Preussischen Technischen Deputation für Gewerbe," *Beiträge zur Geschichte der Technik und Industrie* 3 (1911): 258–59.

[61] Landes, *Unbound Prometheus*, 171, for 1800; Dieterici, *Statistische Übersicht*, 1838: 394, 399, 410, for 1829/31. According to Dieterici, consumption of linen per pound—as a percentage of the total for linen, wool, and cotton—was 44 percent; consumption per yard was 47 percent.

[62] See the memoranda of 6 November, 16 November, 5 December 1814, 31 December 1815, and 1 October 1817, in: Rep.120, D.V.2c, Nr.3, Bd.1, Bl.74–77, 86–89, 140–42, and Bd.2, Bl.12–13, ZStA Merseburg.

[63] Prussian annual industrial investment figures for 1816–1822 (in 1913 German marks) are found in Tilly, "Capital Formation in Germany," in Mathias and Postan, *Cambridge Economic History*, 7 (1): 427. Using the mark/thaler conversion ratio of 3/1 and inflating for the years 1816–1822, annual investment was 2.29 million thaler. 1913 prices are inflated for the current prices of the earlier period with a price index of investment goods based on Walther Hoffmann's ratio of investment goods prices/iron prices for the 1840s (*Das Wachstum der Deutschen Wirtschaft seit der Mitte des 19. Jahrhundert* [Berlin: Springer Verlag, 1965], 572), extended to the earlier years using a more complete index of iron prices (Alfred Jacobs and Hans Richter, *Die Grosshandelspreise in Deutschland von 1792 bis 1934* [Berlin:

existing smalltown shops which had the potential to modernize and survive. The Ministerial Cabinet, the committee of all ministers which screened legislative proposals before presentation to the king, met to discuss the Business Department's projects in January 1819.

Outside the ministries, however, alternative proposals were coming forth which reached all the way to the privy councils of the king. Baron Hans Ernst von Kottwitz, a Silesian nobleman, authored the most ambitious of these. Although he believed in progress, practicing the new scientific agriculture of Albrecht Thaer on his private estates, technological displacement of helpless artisans by machines went too far for the devoutly religious Kottwitz. In the years after Prussia's military collapse he established a charity workshop in Silesia equipped with simple spinning wheels for the victims of that province's depressed linen and woolen trades. Kottwitz also began to evangelize against the devilish mechanization spreading from England. After 1815 he took his message to Berlin, establishing a home for the poor to save the thousands of workers "of the most depraved sort" who had proliferated in Berlin's "doubly disadvantageous factories"[64] since the time of Frederick the Great. Thus the baron propagated a limited form of anticapitalism: one which eshewed attacks on private ownership of land and progressive horticultural techniques, but condemned urban factories, mechanization, and social misery—all results of the unrestrained industrial growth spawned by capitalism.

During the day Kottwitz spread his social gospel among the upper classes at coffees, luncheons, conservative salons, and by means of a voluminous correspondence with provincial followers of his party—or "pietist sect," as one historian described it.[65] During the evenings Kottwitz invited highly placed courtiers, soldiers, and bureaucrats to his home for the destitute to pray with him. In this way his sect easily penetrated the state and sought to influence it by accusing Hardenberg's liberal bureaucrats of wanting to sacrifice guilds for factories and linen for cotton. Out-

---

Hanseatische Verlagsanstalt Hamburg, 1935, 78]). Using the same method, annual industrial investment for the periods 1822–1831, 1830/31–1840 and 1840–1849, respectively, was 3.6 million, 3.25 million, and 4.71 million.

[64] Kottwitz's remarks to the Hamburg publisher, Friedrich Perthes, are cited in Johannes Kissling, *Der deutsche Protestantismus 1817–1917* (Münster: Aschendorffschen Buchhandlung, 1917), 1:68. For this paragraph, see Schultz to Sack, 2 February 1814, Rep.120, D.V.2c, Nr.3, Bd.1, Bl.60–62, and Kottwitz to Bülow, 4 January 1818, Bd.2, Bl.57; and Kottwitz to Frederick William, 27 February 1819, Rep.74, J.XVI.14, Bd. 2, ZStA Merseburg. See also Jakob von Gerlach (ed.), *Ernst Ludwig von Gerlach: Aufzeichnungen aus seinem Leben und Wirken 1795–1877* (Fr. Bahn, 1903), 1:116–19, 146; Vogel, *Allgemeine Gewerbefreiheit*, 219–23; and Walter Görlitz, *Die Junker* (Limburg: C. A. Starke Verlag, 1964), 184, 209–10.

[65] This is the observation of Johannes Bachmann, *Ernst Wilhelm Hengstenberg: Sein Leben und Wirken nach gedruckten und ungedruckten Quellen* (Güttersloh: C. Bertelsmann, 1876–92), 1:193.

raged by this allegedly foreign orientation, Kottwitz petitioned the king to establish a nationwide system of artisans' colonies in the countryside. The workers, drained from Berlin, were to be housed and retrained as linen workers at state expense.

Frederick William was impressed with the saintly good works and simple, to-the-point approach of Kottwitz. Like the charitable baron, Prussia's head of state also questioned the social value of labor-saving machines which seemed to create unemployment. Frederick William looked askance, moreover, at the antiguild sentiment rampant in the Business Department because of the allegedly healthy role which these traditional institutions played in strengthening the moral fiber and character of apprentices and journeymen. Therefore he granted Kottwitz 15,000 thaler in February 1818 to open a second unmechanized charity workshop in Silesia.[66] His sentiments were also drifting farther away from the liberal economic beliefs of his chancellor of eight years. The king put store in the petitions of "well-meaning men"[67] like Kottwitz who blamed freedom of enterprise for much of the misery among handicraftmen. In a memorandum to Hardenberg of May 1818, Frederick William singled out the antiguild measures of 1810/11 for sacrificing too many practical benefits to the cause of laissez-faire theories which had not yet spread manufacturing to the countryside or thus far produced significant new revenues. "The countryside has not gained what the cities have lost."[68] Accordingly, he ordered Hardenberg to propose "appropriate modifications"[69] of the original antiguild legislation. Then in March 1819 Frederick William alotted the princely sum of 150,000 thaler for the immediate establishment of artisan colonies in rural Brandenburg and Silesia. By year's end the migration was underway with the deportation of hundreds of indigent textile workers and their families from Berlin to surrounding villages.[70]

[66] Stadelmann, *Preussens Könige*, 141–65; Rohrscheidt, *Vom Zunftzwang*, 228–38; Frederick William to Gayl, 12 January 1806, Rep.49, B-P Hausarchiv, Nr.120, GStA Berlin; Varnhagen Diary I, 4 November 1820, 24 September 1825, 1:223, 3:379–80; Frederick William to Braun, 6 September 1812, printed in Hugo Gothsche, *Die Königlichen Gewehrfabriken* (Berlin: Militärverlag der Libelschen Buchhandlung, 1904), 16; Stägemann to Olfers, 26 June 1827, printed in Franz Rühl (ed.), *Briefe und Aktenstücke zur Geschichte Preussens unter Friedrich Wilhelm III* (Leipzig: Duncker & Humblot, 1902), 3:362; Frederick William to Schuckmann, 19 September 1826, Rep.120, D.V.1, Nr.11, Bd.1, Bl.14, ZStA Merseburg; Klöden, *Friedrich Wilhelm*, 91, 162, 256, 267, 285, 312, 352, 368–69; and ADB, 37:639.

[67] Frederick William to Hardenberg, 8 May 1818, cited in Vogel, *Allgemeine Gewerbefreiheit*, 204.

[68] Ibid.

[69] Ibid.

[70] See Frederick William to Bülow, 21 February 1818, Staatsministerium to Hardenberg, 23 January 1819, and Frederick William's decree of 4 March 1819, Rep.120, D.V.2c, Nr.3,

The deurbanization schemes engineered by Kottwitz and his party could not have been completely disappointing to technocrats like Kunth who were also alarmed by the threatening growth of idle labor forces in Prussian cities. But neither of the king's decrees had appropriated monies for new machinery and the coffers were now emptied of the capital anticipated by the Business Department for its own plans. Kunth and his colleagues received the depressing news in May 1820 that they would have to fund these projects out of current budget.[71] Consequently, nothing came of their proposals. The program of rural industrialization had suffered another political setback.

Baron Kottwitz's adroit maneuver against the Business Department was related to broader reactionary developments sweeping Germany and Europe around 1820. Conscious of—and troubled by—the close historical connection between social change, industrial advance, freer attitudes, and political progressivism explained in the Introduction, opponents of change were engaged in a comprehensive effort to smash these links and eliminate all sources of modernist disorder.[72] In Prussia, a resurgent aristocratic movement against modernization of all forms was concentrating efforts on swaying more completely to its side a king who was already alarmed by troubling, frightening events in Europe. These topics are our next subject of analysis.

. . . . .

Europe was engulfed by political violence during the late 1810s and early 1820s. England witnessed the London riots, the Peterloo massacre, and the Cato Street conspiracy, while the Mediterranean was plagued by revolutions, and in France an assassin struck down the second-in-line to the throne. In Germany, activist students twice climbed the Wartburg to protest conservatism in Germany, then a deranged student stabbed August Friedrich Ferdinand von Kotzebue, a conservative poet. His murder shocked the German courts and led to the infamous Carlsbad Decrees. Censorship, tight supervision of university life, and political persecution came to the lands of the German Confederation.

Bd.2, Bl. 59–78, 106,132, 169, ZStA Merseburg; Klöden, *Friedrich Wilhelm*, 267; and Vogel, *Allgemeine Gewerbefreiheit*, 204. For a reference to the resettlements, see Treue, *Wirtschaftszustände*, 216–17.

[71] Vogel, *Allgemeine Gewerbefreiheit*, 78; Hardenberg to Bülow (with Ferber's marginalia), 11 May 1820, Rep. 120, D.V.2c, Nr.3, Bd.2, Bl.169, ZStA Merseburg.

[72] For discussions of later industrialization crises, see Kenneth D. Barkin, *The Controversy over German Industrialization* (Chicago: University of Chicago Press, 1970); and Jeffrey Herf, *Reactionary Modernism: Technology, Culture, and Politics in Weimar and the Third Reich* (New York: Cambridge University Press, 1984).

In Prussia, leading advocates of parliamentary reforms like Wilhelm von Humboldt, Karl Friedrich Beyme, Hermann von Boyen, Karl Wilhelm von Grolman, and Friedrich Theodor Merckel were dismissed or else resigned from their posts. The oft-promised representative assembly was now postponed indefinitely. In 1820 Frederick William promised to convene the estates of the realm (*Reichsstände*) if the state needed to borrow from its people again. As politically significant as this statement was, however, borrowing was a contingency that systematic budget cuts and parsimoniousness could preclude.[73]

Meanwhile, a search for "demagogues" began. As in all scares and witchhunts, measures became extreme. Police opened mail and searched the home of virtually every citizen for evidence of revolutionary activity. They made arrests and uncovered many alleged "Jacobin" and "Carbonari" plots. Blue bloods pushed for more extreme measures, urging the king to purge army, court, and high office of the suspect bourgeoisie. Class relations deteriorated as aristocratic and bourgeois officers hurled insults at one another and fought duels to restore honor.[74]

While political tensions heightened, the conservative parties pushed for an economic purge to match the desired cleansing of political life. We have already seen how Baron Kottwitz assaulted the liberal economic programs of Hardenberg and the Business Department. But he was not alone. Once "legitimate authority" is restored, wrote Adam Heinrich Müller, the ideologue who had helped initiate the aristocratic opposition to Hardenberg in 1810, "we cannot avoid passing over to an *economic* restoration, for the prevailing [tendencies] in state and private administration are leading us inexorably to a decisive crisis." Only economic restructuring would make it "worthwhile to write about national economy." Like the machine-hating anticapitalism of Kottwitz, Müller's dissent concentrated on the emerging factories and mechanical breakthroughs of the industrial revolution. Using criticisms which modern critics in future generations would reemploy, Müller bemoaned the substitution of unnatural, depersonalizing factory jobs for the ennobling, artistic handicraft

[73] For recent, more positive reassessments of the promise of 1820, see Hanna Schissler, "Preussische Finanzpolitik 1806–1820," in Hans Ulrich Wehler (ed.), *Preussische Finanzpolitik 1806–1810: Quellen zur Verwaltung der Ministerien Stein und Altenstein* (Göttingen: Vandenhoeck and Ruprecht, 1984), 55–61; and Alexander von Witzleben, *Staatsfinanznot und sozialer Wandel: Eine finanzsoziologische Analyse der preussischen Reformzeit zu Beginn des 19. Jahrhunderts* (Stuttgart: Franz Steiner, 1985), 230–38. For the first years of the reaction, the reader should consult two recent, well-researched studies: Paul R. Sweet, *Wilhelm von Humboldt: A Biography* (Columbus: Ohio State University Press, 1980), 2:301–46, 383–92; and Hans Branig, *Fürst Wittgenstein: Ein preussischer Staatsmann der Restaurationszeit* (Cologne: Böhlau Verlag, 1981), 124–53.

[74] Varnhagen Diary I, 21 January, 25 January, 11 March, and 4 November 1820, 1:60, 63, 98, 223.

of the small workshop. Man threatened to sink to the level of a machine and the state to that of a "bee-state" comprised of only workers and drones. The remedy was a thorough reorganization of society along estatist lines with nobles, church, and guilds enjoying their medieval corporate privileges and property rights. Private industry, on the other hand, would be drastically reduced "to its proper sphere" and thereby brought "into balance with the other estates." [75]

By the late 1810s Müller and Kottwitz formed the intellectual epicenter of a growing reactionary storm against the industrialization of Prussia.[76] Their influence was most pronounced among the so-called "Pomeranian Pietists." Scions of old aristocratic families like Adolf von Thadden, Ernst von Senfft-Pilsach, and more recently ennobled (1734–1735) families like that of the brothers Leopold and Ludwig von Gerlach regarded their aristocratic heritage as a commission to manage the land, look after souls, and protect the state and working classes from evil influences. Drawn to the practical Christianity of Kottwitz and the romantic medievalism of Müller and other prominent conservatives like Karl Ludwig von Haller, they joined the crusade against a material transformation which they believed was enervating Prussian stock and sweeping away the nobleman's economic and political position in society.

This anti-industrial ideology also extended to Marwitz, Voss-Buch, and

[75] For the "economic restoration" quote, see Müller to Bucholtz, 10 October 1823, quoted in Baxa, *Adam Müllers Lebenserzeugnisse*, 2:630–31. For the "bee-state" and "proper sphere" quotes, respectively, see Müller to Metternich, n.d. (late 1819), printed in ibid., 2:326–30; and Ralph H. Bowen, *German Theories of the Corporative State with Special Reference to the Period 1870–1919* (New York: McGraw-Hill, 1947), 37. For a recent discussion of Müller's thought, see Robert M. Berdahl, *The Politics of the Prussian Nobility: The Development of a Conservative Ideology 1770–1848* (Princeton, New Jersey: Princeton University Press, 1988), 163–81.

[76] This paragraph has been pieced together from the following sources: Wittgenstein to Jung, 16 April 1845, Rep.192 Wittgenstein, BPH, V.8.5., Bl.38–38v, GStA Berlin; Witzleben to Frederick William, 25 January 1818, printed in Wilhelm Dorow, *Job von Witzleben* (Leipzig: Verlag von Bernh. Tauchnitz, 1842), 94–95; diary of Ludwig von Gerlach, 3 May 1817, Otto von Gerlach to Ludwig von Gerlach, 17 November 1821, and Leopold von Gerlach's *Familiengeschichte*, n.d. (1814–1861), printed/quoted in Hans Joachim Schoeps (ed.), *Aus den Jahren preussischer Not und Erneuerung* (Berlin: Haude & Spenersche Buchhandlung, 1963), 227, 300–301, 622; Varnhagen to Oelsner, 11 October 1823, printed in Ludmilla Assing (ed.), *Briefwechsel zwischen Varnhagen von Ense und Oelsner nebst Briefen von Rahel* (Stuttgart: Verlag von A. Kröner, 1865), 143; Caroline von Rochow, *Vom Leben am preussischen Hofe 1815–1852*, edited by Luise von der Marwitz (Berlin: E. S. Mittler, 1908), 116–18, 214; Varnhagen Diary I, 12 November 1821; 29 September 1823; 16 December 1823; 24 February 1824; 13 July 1824; 20 October 1824; 10 June 1826; 13 February 1828; 5 April 1830; 1:366, 2:415; 447, 3:30, 101–2, 153, 4:74–75, 5:31, 280; Treitschke, *Deutsche Geschichte*, 2:229; Schnabel, *Deutsche Geschichte*, 4:385; Bachmann, *Ernst Wilhelm Hengstenberg*, 1:186–94; Görlitz, *Die Junker*, 234–37; and Baxa, *Adam Müller*, 349, 396–97, 403.

General Knesebeck, Müller's Brandenburgian allies during the early fighting against Hardenberg; to Marwitz's brother-in-law, Gustav von Rochow; and Voss-Buch's son, Carl. As Metternich himself observed, "many waverers have been strengthened [by Müller], many strays led back to the true way and also many won for the good cause who, but for the penetrating word of truth, would have adhered to the tirelessly active party of the innovators."[77] Others drawn into this circle of political friendship were Ludolf von Beckedorff, chief of the state's censorship agency, and Heinrich von Kamptz, the leading "persecutor of demagogues." At their favorite gathering place, the Berlin mansion of Gustav and Caroline von Rochow, they discussed "old conservative views" and plotted strategy for influencing people and damming the forces of change. Like Kottwitz's sect, the "party" of the aristocratic "opposition," as Caroline von Rochow described this clique, had easily bridged the narrow divide between aristocratic society and the state.[78]

Other highly placed persons in Prussia were strengthened in their anti-industrial thinking by the politicking disciples of Müller and Kottwitz.[79] Two of the most important were Duke Carl of Mecklenburg, commander of the elite Guard Corps and brother-in-law to the king, and Friedrich Karl zu Sayn-Wittgenstein-Hohenstein, Minister of the Royal House. Another was Crown Prince Frederick William. Trying his untrained hand at politics for the first time during these years, the prince gave every indication of agreeing completely with those around him who wanted to turn back the clock.

Like Kottwitz and his pietists, the aristocratic cliques knew the importance of swaying the monarch. In a lengthy memorandum to the king of late 1817, for instance, Duke Carl accused Hardenberg's reformers of harboring treasonous, revolutionary ideas and included the freedom of enterprise legislation of 1810/11 in his list of indictments. Because Frederick William usually refused to be drawn into political discussions with family members like Carl, however, the Duke worked through intermediaries like fellow Mecklenburger Baron Friedrich von Schilden, the Court Chamberlain, whom the king allowed to give political advice on a regular basis. Possessing the same rare privilege, Wittgenstein represented a far greater asset to this clique because of his great persuasive powers and a web of contacts which extended all the way to Metternich in Vienna.[80]

[77] Metternich's remarks of 1826 are cited in Bowen, *German Theories*, 33–34. For this paragraph, see the sources cited above in n. 76.

[78] For the quotes in this paragraph, see Rochow, *Vom Leben*, 124, 175, 214.

[79] For this paragraph, see the sources cited above in n. 76.

[80] Witzleben to Frederick William, 25 January 1818, printed in Dorow, *Job von Witzleben*, 94–95; Treitschke, *Deutsche Geschichte*, 2:229; Rochow, *Vom Leben*, 125; Varnhagen

By 1821 the various aristocratic parties had fashioned a political tool which they hoped would serve them well. Known popularly as the "Crown Prince's Commission" after its chairman, the conservative body included Wittgenstein, Voss-Buch, and Friedrich Ancillon, the prince's conservative former tutor. The king charged it with formulating proposals for provincial diets, but the final report went much farther, calling for entailment laws to protect aristocratic property from falling into bourgeois hands—and new legislation *to restore guild privileges.* Frederick William III had instructed Hardenberg to modify the guild laws in 1818, but an unsympathetic bureaucracy was successfully dragging its feet despite royal reminders. Among other things, the Crown Prince's Commission was designed to expedite far-reaching guild "reforms." For once they were reinvigorated, the guilds would certainly attempt to inhibit modern technology and reduce industry to its "proper sphere," as Adam Müller put it.[81] The guilds could serve indirectly, in other words, to shore up the nobility's crumbling economic and political position.

The initial postwar offensive of the Müllerites, pietists, and "ultras" was at once infuriating, frustrating, and embarrassing for Prussia's advocates of rural industrialization. This was because the followers of Stein and Hardenberg had their own deep-seated prejudices against urban industries and were themselves increasingly upset about the social consequences of an industrialization which they could observe in a more advanced stage in the newly acquired western provinces. Thus Ludwig von Vincke, Governor of Westphalia and a proponent of mechanization, bemoaned the progress of cotton manufacturing in his province because this was occurring at the expense of the greatly preferred linen industry.[82] Kunth, an outspoken opponent of guilds and the "medieval" faction around Müller, harbored his own concerns, observing with disgust the misery and dehumanization of cotton textile workers who appeared to be sinking to the level of machines. It did not serve higher purposes, he wrote in November 1816, for employers to think of nothing but profits while the nation slowly turned into "one big cotton cloth workshop."[83] Bartold Georg Niebuhr, a Berlin historian and former associate of Hardenberg who taught ancient history at the University of Berlin, noted

Diary I, 5 and 19 March 1822, 10 February 1823, 2:54–55, 68–69, 295–96; and Duke Carl to Wittgenstein, 26 January 1831, Rep.192, B-P Hausarchiv V.1.17, GStA Berlin.

[81] Bertold to Voss-Buch, 27 November 1822, and Voss-Buch to Bertold, n.d. (December 1822), printed in Rohrscheidt, *Vom Zunftzwang*, 577–78; Varnhagen Diary I, 12 and 29 December 1821, 21 November 1823, 1:377–78, 389–90, 2:438; and Herbert Obenaus, *Anfänge des Parlamentarismus in Preussen bis 1848* (Düsseldorf: Droste Verlag, 1984), 173–80.

[82] See Vincke's marginalia to the reports of the Department of Trade and Industry in Münster, 3 March 1840 and 23 March 1841, Regierung Münster 305, StA Münster.

[83] Kunth to Bülow, 12 October 1816, Rep.120, A.V.2, Nr.4, Bl.71–72, ZStA Merseburg.

similarly that western industrialists did not lack markets, but did require "a spiritual renewal . . . especially in the cities."[84]

Johann Gottfried Hoffmann, an early member of the Business Department who became one of Hardenberg's aides after 1815, was also greatly concerned about the rise of big factory towns. In a memorandum drafted for the chancellor and sent to the western governors in 1817, he lamented that textile operatives, conditioned from childhood to perform one mindless task, gradually lost the desire and ability to work at other jobs. Consequently, while entire textile towns suffered from unemployment, there could be shortages of household servants and day-laborers for farming and construction. Nor would these miserable workers have the physical and spiritual strength to defend the Fatherland. It was not his intention to place barriers in the path of technological change. But some "middle road"[85] had to be found between the mechanical backwardness of the guild era and the deplorable conditions of modern textile factories. This new avenue lay in better education for employers and employees[86] and freer economic legislation permitting a healthier, more natural, and economically more appropriate industrialization outside of the cities.

The immediate threat to the nation's technical establishment, however, could not go unanswered. Indeed with Rother hated by the clique at court, Gerhard called a "pig" by one of the Mecklenburgers, the Crown Prince joking about Beuth and his "entourage" of technicians, and the engineers relegated to latrine duty on maneuvers, some positive image-building for technology was in order.[87] The Business Department cleverly opted against attacking the guilds so popular in the king's entourage, choosing a more positive approach instead. So Kunth suppressed his own worries about machine-work and dislike of the cotton industry in a pamphlet emphasizing only the benefits of mechanization in England. Whereas 50,000 operatives produced cotton cloth valued at five million pounds sterling before Arkwright's spinning innovations, he wrote in 1820, 500,000 workers were now gainfully employed producing eighteen times as much. "Machine spinning has affected the population much more advantageously than disadvantageously, in other words exactly the

[84] Niebuhr to Dore Hensler, 31 December 1814, printed in Dietrich Gerhard and William Norwin (eds.), *Die Briefe Bartold Georg Niebuhrs* (Berlin: Walter de Gruyter & Co., 1926–29), 2:543–44.

[85] Hardenberg (Hoffmann) to Solms-Laubach, 5 September 1817, Vincke Papers A.V.92, StA Münster.

[86] Ibid.; see also Scharnweber's unpublished lecture, n.d. (1814), Rep.77 CCCXX 35, Bl. 35–37, and Kunth to Bülow, 12 October 1816 (see n. 83), ZStA Merseburg.

[87] Varnhagen Diary I, 25 January 1820, 21 February 1820, 1:63, 83; Crown Prince Frederick William to Wittgenstein, 12 January 1822, BPH, Rep.192 Wittgenstein, V.1.5, GStA Berlin; Treitschke, *Deutsche Geschichte*, 3:468. See also below, chap. 4.

opposite of what is feared."[88] England possessed no blast furnaces in the 1500s, he continued, produced 17,350 tons of pig iron in 1730, but now turned out more than 500,000 tons. "This is the effect of steam engines."[89] Peter Beuth joined the chorus in the proceedings of the Association for the Promotion of Technical Activity. "While I hear only protest after protest [against machinery] in my Fatherland, sixty-seven new spinning and power-loom establishments were built . . . in Lancashire."[90] Could Prussia afford to lag behind?

By the early 1820s, therefore, a concerted reactionary movement was underway in Prussia to restore political and economic life to allegedly more legitimate forms. Led by the Business Department, proponents of economic liberalism were fighting back. This struggle entered a new and decisive phase after the death of Hardenberg in November 1822. For weeks it appeared that Frederick William would elevate Voss-Buch, one of the leading advocates of reactionary politics and economics, to the chancellery. His sudden death in January 1823, demoralized conservatives, buoyed the spirits of Hardenberg's followers, and opened a renewed intrigue for the ear of the king.

Bureaucratic opponents of the ultras' restorative programs rallied around Job von Witzleben, the privy councillor for military affairs and perhaps the king's only real friend. Although no advocate of unlimited freedom of enterprise or extensive parliamentary experiments like those in southern Germany, Witzleben nevertheless exerted a constant tempering influence on the king, pointing out the exaggerations inherent in reactionary arguments and illuminating the undesirable consequences of feudal excesses. The moderate soldier appealed to the progressive, unpretentious, bourgeois side of Frederick William, repeatedly returning the monarch to views which inside he sensed were fair, just, and meritorious.[91]

With encouragement from progressive friends, Witzleben urged the king in March 1823 to appoint Wilhelm von Humboldt chancellor. Showing considerable political instinct, Frederick William rejected this particular piece of advice, opting to leave the office vacant and bolster royal prerogatives. The reclusive monarch had no intention of becoming his own chancellor, however, with the result after 1823 that he tended to

[88] *Verhandlungen des Vereins zur Förderung des Gewerbefleisses in Preussen* 3 (1824): 55ff, GB/TU Berlin (hereafter cited as *Verhandlungen*, GB/TU Berlin). Kunth's article had appeared in 1820 in pamphlet form.

[89] Ibid.

[90] Ibid., 183ff.

[91] For Witzleben's influence, see Varnhagen Diary I, 1 May 1820; 4 December 1821; 21 February 1823, 1:129, 371, 2:305; Rochow, *Vom Leben*, 97, 102; Dorow, *Job von Witzleben*, 37, 63, 67, 93–96, 269; and ADB, 43:676.

function as an occasional referee preventing any one of many conflicting intrastate forces from upsetting a political balance which both protected Hohenzollern power and permitted him emotionally soothing periods of inactivity.[92] Consequently, the importance of his privy councils increased and the influence of men close to the top—like Duke Carl, Prince Wittgenstein *and especially, Witzleben*—grew.

As Frederick William and Witzleben drew closer together, the likelihood decreased that the reactionary parties would achieve their political demands. Once the courts failed to reveal any real threats to the state, for instance, the "persecution of the demagogues" was called off. By 1824 censorship was looser, opera and theater freer, and leading inquisitors like Heinrich von Kamptz making amends publicly with political victims such as Merckel and Georg Andreas Reimer, owner of an influential publishing firm in Berlin. Aristocratic demands for class purity in the state were also ignored.[93] The king's own relatively humble childhood and adolescence, his disgust with the effete court life of his father and severe disillusionment with the nobility after Jena combined to produce a socially unpretentious monarch who was at ease with members of the lower and middle classes. "Only through the honest folk, the upright burgher and simple peasant," he said, "can things improve."[94] Frederick William "approved of their naturalness and lack of ceremony, their direct speech, their silence, their religious faith," writes one historian.[95] Thus it was not unusual for the king to speak with his burghers on the street and occasionally accept their social invitations. Bourgeois, in turn, were invited to court balls, appointed to bureaucratic posts, and ennobled at a pace that amused him because it infuriated the rank-conscious old nobility. In short, a purge of the bourgeoisie was against the king's nature—instincts which were reinforced, it bears repeating, by Witzleben.

Such a class purge was also against the crown's basic interests. For over a century, in fact, Hohenzollern kings had carefully controlled the organs

[92] Varnhagen Diary I, 15 June 1821; 26 February 1822; 15 December 1822; 12 January 1825; 1 March 1825; 14 May 1829, 1:326, 2:45, 265, 3:212–13, 244–45, 5:205.

[93] For the top half of this paragraph, see Karl Demeter, *The German Officer Corps in Society and State 1650–1945*, translated from the German by Angus Malcolm (London: Weidenfeld and Nicolson, 1965), 127–28; and Varnhagen Diary I, 2:159, 229, 388, 3:204–5, 289, 310, 380 for 18 July 1822; 20 October 1822; 14 August 1823; 3 January 1825; 14 May 1825; 15 June 1825; and 24 September 1825. For the bottom half of this paragraph, see Elise von Bernsdorff, *Aus ihren Aufzeichnungen: Ein Bild aus der Zeit von 1789 bis 1835* (Berlin: E. S. Mittler, 1896), 1:283; Varnhagen Diary I, 21 January 1820; 23 April 1821; 2 February 1824, 1:60, 290, 3:20, and numerous examples from 1826 (4:1–170 passim); Rochow, *Vom Leben*, 135; and Nikolaus von Preradovich, *Die Führungsschichten in Österreich und Preussen (1804–1918)* (Wiesbaden: Franz Steiner, 1955), 104–5.

[94] Quoted in Eylert, *Charakter-Züge*, 3 (1): 117.

[95] Anderson, *Nationalism*, 278.

of government by cleverly playing one class off against another.[96] Under Frederick William I (1713–1740), the old Junker nobility which had dominated earlier kings was largely excluded from office. The gruff monarch staffed the state instead with pliable new nobles and commoners loyal to him—75 percent of his ministers, 84 percent of his privy councillors, and 83 percent of his provincial officials in 1740. Frederick the Great (1740–1786) reversed the process, finding it prudent to gain independence from overreaching civil servants and to rely on the "natural" talents of an *attenuated* nobility.[97] All of his War and Domains Board presidencies and 95 percent of his ministerial appointments went to Junkers. Frederick William III turned tables again. *Nouveaux arrivèes* already held 36 percent of top government positions on the eve of Jena, and after 1815 this percentage rose higher as commoners entered the officer corps (46% in 1818), foreign service (50% in 1830), and top provincial administration (76% in 1820). Bourgeois in the ministries outnumbered the old Prussian nobility by three-to-one. Even higher ratios were reached in technological divisions like the Business Department (100% in 1824), leading Seehandlung staff (88% in 1828), captaincy of the Mining Corps (64% in 1817), and artillery officer corps (58% in 1830). Until the new arrivals became a greater threat, it would remain folly to purge them and alter the "autocratic-democratic" (Mosca) nature of the state. Such a move would only reward the parties which longed to turn back the clock to a day when kings were weak and noblemen strong.

The aristocrats' goal of economic restoration also slipped farther from reach. For Frederick William refused to let "phrase-makers" like Adam Müller sway him into a doctrinaire rejection of modern technology. Rather, the king's undogmatic nature led to a certain ambivalence on this issue. He was alarmed by the prospect of imported English looms throwing his weavers out of business, but the same monarch, on the other

[96] For the statistics cited in this paragraph, see Rosenberg, *Bureaucracy*, 68; Hubert C. Johnson, *Frederick the Great and his Officials* (New Haven: Yale University Press, 1975), 252–53, 289–91; Preradovich, *Führungsschichten*, 79, 105; Demeter, *The German Officer-Corps*, 15; Koselleck, *Preussen*, 435, 689; *Handbuch über den Königlichen Preussischen Hof und Staat*, 1817, 1824, 1828; Kurt Wolfgang von Schöning, *Historisch-biographische Nachrichten zur Geschichte der Brandenburgisch-Preussischen Artillerie* (Berlin: E. S. Mittler, 1845), 514. The statistic for bourgeois in the ministries in the 1830s is based on Preradovich's figure for 1829 and statistics in Koselleck (p. 689) which indicate almost no percentage change between 1831 and 1841.

[97] For a recent work which emphasizes the decline, not the persistence, of the nobility, see William W. Hagen, "Seventeeth-Century Crisis in Brandenburg: The Thirty Years' War, The Destabilization of Serfdom, and the Rise of Absolutism," *American Historical Review* 94 (April 1989): 302–35. Also see Manfred Botzenhart, "Verfassungsproblematik und Ständepolitik in der preussischen Reformzeit," in Peter Baumgart (ed.), *Ständetum und Staatsbildung in Brandenburg-Preussen: Ergebnisse einer Internationalen Fachtagung* (Berlin: Walter de Gruyter, 1983), 450.

hand, was intrigued with medical instruments, optical telegraphy, ballooning, and steam-powered riverboats, investing 20,000 thaler of his personal wealth in the latter. Similarly, he could express hatred for the artillery, yet praise attempts to mass-produce small arms. He opposed extreme antiguild measures as unjust and unwise, but favored elimination of guild "abuses" which restricted the productive capacities of the nation.[98] If the king's ambivalence inclined him to stop short of the ultras' economic demands, political prudence demanded it. Behind the proposals for provincial diets and guild reform was the old Junker agenda of hegemony in Prussia. The all-important political balance could easily tip away from the crown.

Wanting to steer a middle course, Frederick William ordered the bureaucracy to accelerate work on guild reform,[99] then in November 1824 warned Crown Prince Frederick William against pushing too far in this direction. While it was true that Hardenberg's reform measures of 1810/11 were too theoretical and "no longer appropriate" in some areas, "one must still take into consideration how a rapid, forcible transition to a different legal order will only lead to new disturbances and destroy lawful relationships and procedures which have more or less put down roots." The state would modify existing laws in accordance with "the reasoned wishes which the diets may lay before us," but Prussia's financial situation did not allow it "to take back institutions on which it relies for revenue without replacing them with others which guarantee the same financial results."[100]

As is implicit in the king's memorandum, the return of settled, more prosperous times had added public finance to the list of factors softening Frederick William's resolve to alter the deceased chancellor's programs. In 1818 when the king criticized rural industrialization and ordered guild reforms, direct and indirect taxes netted only 30.9 million thaler and the budget was 3.25 million thaler in deficit. By 1824, the state collected 35.4 million thaler and budgets were nearly balanced. The direct tax on business (*Gewerbesteuer*) represented about 4 percent of the total and was rising steadily—1.6 million thaler in 1824; 2.1 million thaler in 1830. Although conservatives like Knesebeck refused to believe it, the economic upturn in evidence throughout the kingdom by the mid-1820s was widely credited to the account of Hardenberg's reforms of the 1810s.[101]

[98] See the sources cited above in n. 66.

[99] Frederick William to the Ministerial Cabinet, 7 February 1824, Rep.90a, J.I.1, Nr.1, Bl.2, ZStA Merseburg.

[100] Frederick William to the Crown Prince, 30 November 1824, Rep.92 Müffling, A.9, Bl.2, ZStA Merseburg.

[101] Ritter, *Role des Staates*, 121; Ohnishi, *Zolltarifpolitik*, 99–100, 111, 113; Varnhagen

The "reasoned wishes" of the estates, however, soon placed pressure on Frederick William to reconsider his stance. In a grudging compromise with the aristocratic opposition during the summer of 1823, the king agreed to the creation of eight provincial diets to advise him.[102] They would meet consecutively beginning in the fall of 1824 and periodically thereafter. The Crown Prince was to head a special liaison committee of like-minded conservatives to smooth communication between the monarch and these restructured estates of the realm. The heartland province of Brandenburg was the first to assemble its diet in November 1824. The body, comprised largely of aristocrats, accepted a resolution "to confront the defects of the legislation originating in the reform period."[103] The delegates advocated compulsory guilds, restrictions on the hiring of unskilled workers, and barriers to the proliferation of small businesses in town and country. And Brandenburg was not alone. Six of the eight diets meeting in the mid-1820s called for a full restoration of the guilds.

By this time Hans von Bülow's Ministry of Business and Commerce had finally begun to formulate the guild legislation which the king had requested in 1818. Housing primarily the Business Department, this division was designed at its inception in 1817 to add ministerial autonomy and prestige to the cause of freedom of trade and enterprise. Small wonder that Bülow, Maassen, Kunth, Beuth, and the ministry's economic liberals delayed guild work as long as politically feasible.[104] By mid-decade, however, influential forces were gathering against Bülow's bureaucratic citadel. His defenses weakened in early 1825 when he lost the support of an important bureaucratic colleague, Kaspar Friedrich von Schuckmann, Minister of the Interior since 1812. A transparent opportunist, the protégé of Hardenberg and former head of the Business Department had abruptly switched political parties after Waterloo, helping the Junkers weaken peasant emancipation in 1816 and intriguing thereafter with Duke Carl and other countrymen in the so-called "Mecklenburg coterie." A member of both the Crown Prince's Commission and

---

Diary I, 1824: 60–61, 1825: 406; Treitschke, *Deutsche Geschichte*, 2:466; Treue, *Wirtschaftszustände*, 223.

[102] Obenaus, *Anfänge*, 202–09, 233–40, exaggerates the significance of the diets and the liaison committee. He also overlooks the king's dislike of these institutions and wariness with regard to the Crown Prince. For examples, see Rochow, *Vom Leben*, 212; and Varnhagen Diary I, 14 December 1820; 20 December 1822; 23 February 1823; 23 March 1825, 1:243, 2:268, 311, 3:254.

[103] Cited in Obenaus, *Anfänge*, 217, and in general for the diets, see 217–18, 486.

[104] For the rationale behind creation of the ministry, see Schön to Hardenberg, 18 June and 13 July 1817, printed in Theodor von Schön, *Aus den Papieren des Ministers und Burggrafen von Marienburg Theodor von Schön* (Berlin: Verlag von Franz Duncker, 1875–76), 4:392, 407. For Bülow's delays on guild reform, see Rohrscheidt, *Vom Zunftzwang*, 565; and Obenaus, *Anfänge*, 487.

new liaison committee with the diets, Schuckmann recommended accommodating their demands for guild revival.[105]

Bülow tried to fight back, complaining that the diets were delaying legislative progress because they "only half-way"[106] accorded with his ideas. But his resistance crumbled when Wittgenstein joined the assault. Chairman of a "budget savings" commission whose hidden agendas included punishing liberals of all sorts by tightening the purse strings around them, he proposed abolition of the Ministry of Business and Commerce and transferal of its departments to Schuckmann in Interior.[107] This was a fully transparent assault on the economic liberalism which the ministry had furthered for the past eight years.

Bülow continued to struggle, arguing that Schuckmann was already too overburdened to manage such a massive new ministry and—what was more important—lacked political views guaranteed to instill trust among traders and manufacturers.[108] Not surprisingly, such objections carried no weight with Wittgenstein. The ax fell on this liberal bastion with royal approval in June 1825. "The [budget] savings are working," observed Varnhagen von Ense, "like a discriminatory law against those out of favor."[109] Ominously for the liberals, the Business Department was placed under Schuckmann and the task of guild reform given primarily to him.

Despite these developments, it was difficult to ascertain in 1825 whether Prussia and her king were shifting to the right or to the left. Friedrich August [von] Stägemann, one of Hardenberg's lieutenants and now a top aide in the Treasury Ministry, was quite alarmed, bemoaning "all the favorable things which had happened for the nobility."[110] Three million thaler in subsidies given to bankrupt aristocratic estate owners; archreactionaries like Knesebeck and Kalkreuth promoted to General; French law in Westphalia abolished; budgets cut in the politically suspect technical branches of the army; destruction of liberal citadels like the Ministry of Business and Commerce; and provincial diets created—everything seemed to point in this feudalistic direction. Yet Varnhagen von Ense, a perceptive diarist who was also worried, took occasional solace in

[105] Rohrscheidt, *Vom Zunftzwang*, 582. On Schuckmann's opportunism, see Varnhagen Diary I, 13 June 1822, 2:137–38. Liberals had noticed that much which emerged from the Interior Ministry was not "ultra," while conservatives noted a certain lack of principle on Schuckmann's part: today a royalist, tomorrow a democrat.

[106] Bülow to Altenstein, 27 May 1825, Rep.90a, J.I.1, Nr.1, Bl.30, ZStA Merseburg.

[107] See excerpts from the commission's report of February 1824 in: Rep.120, A.I.1., Nr.29, Bl.4, 9, ZStA Merseburg; and Varnhagen Diary I, 9 September 1824, 3:130–31.

[108] Bülow to Lottum, n.d. (spring 1825), quoted in Hermann von Petersdorff, *Friedrich von Motz* (Berlin: Verlag von Reimar Hobbing, 1913), 1:240.

[109] Varnhagen to Oelsner, 21 April 1825, printed in Assing, *Briefwechsel*, 3:280.

[110] Quoted in Varnhagen Diary I, 14 August 1827, 4:278.

indications of different trends.[111] Army liberals like Generals Neithardt von Gneisenau and Moritz Ludwig von Schöler had also received promotions—the former to Fieldmarshall—while former political victims like General Grolman and District President Merckel (Silesia) had been reinstated. Moreover, the king refused to sanction a scheme of the aristocratic parties to curtail the status and pay of bureaucratic councillors (*Räte*), those middle-level civil servants, Hardenberg appointees in the main, who were known for their progressivism.[112] It was as if "an invisible force reigned in all things," blocking "extremism of all kinds," and constantly "turning everything back to the middle."[113] Political liberals were also heartened by Frederick William's drab, unpretentious personal lifestyle, his amusement with plays satirizing the nobility, his disregard for fine distinctions of noble rank, and continuing tolerance for the bourgeoisie at court and in public, for these were all sources of anger to ultras who valued social exclusivity highly. Varnhagen worried, however, that these were not the real trends (*Gemeinsames*) in Prussia, only isolated happenings (*Einzelnes*).[114]

The aristocratic parties did not agree that they were "gaining significant ground," as Stägemann worded it. For those convinced that Prussia's entire legal mechanism needed overhauling in accordance with the rhythm of older, superior institutions, no amount of promotions, subsidies, and bureaucratic rearrangements would suffice. Indeed, nothing short of a thorough legal restructuring seemed capable of reinforcing the steadily deteriorating economic position of the nobility. Already before the wars, two-thirds of the aristocracy was without landed property. "Have we not almost become comical title-holders who lack the means?" wondered Princess Adelheid von Carolath.[115] To some extent this condition represented a centuries-old shifting of wealth from petty and moderate aristocratic clans to the families of the grand nobility. By 1800, however, between 10–15 percent of the nobility's estates had also slipped into bourgeois hands. The agricultural crisis of the 1820s drove rates of bank-

[111] For conservative and liberal trends throughout 1825, see ibid., 3:201–433 passim.

[112] Varnhagen Diary I, 12 January 1825, 3:212–13; Treitschke, *Deutsche Geschichte*, 2:183.

[113] Varnhagen to Oelsner, 12 February 1825, printed in Assing, *Briefwechsel*, 3:268. For the king's habit of placing persons with opposing views on commissions in order to force compromises, see Rochow, *Vom Leben*, 95–96.

[114] For the king's continuing personal habits mentioned here, see Varnhagen Diary I, 4 February 1828; 25 March 1828; and 8 February 1830, 5:26–27, 57–58, and 266. See also the sources cited above in n. 93. For the quote, see Varnhagen to Oelsner, 11 November 1825, printed in Assing, *Briefwechsel*, 3:327.

[115] Adelheid von Carolath to Hermann von Pückler-Muskau, 23 March 1831, printed in Ludmilla Assing (ed.), *Briefwechsel und Tagebücher des Fürsten Hermann von Pückler-Muskau* (Berlin, 1873–76), 7:362.

ruptcy and compulsory sales even higher, and as they rose, the middle class share of feudal estates soared to over 20 percent—and as high as 40 percent in badly afflicted provinces like East Prussia.[116] Trends such as these popularized entailment laws designed to keep aristocratic properties intact. The Crown Prince's Commission had graced such demands with its report in 1823, but the king was unmoved, denying the Commission's first specific proposal in June 1825.[117] Neither this archconservative body nor the provincial diets which it had ushered into being seemed "capable of erecting a dam against the onrushing demands of the future,"[118] as Caroline von Rochow described the seemingly inevitable embourgeoisment of Prussia.

If conservatives gained mastery over the hated, bourgeois-infested, semiautonomous *ministries*, on the other hand, then perhaps legal reforms to solidify the nobleman on his land and the craftsman in his shop would follow. The goal did not appear beyond reach. Aristocrats had coopted Schuckmann—or so it seemed—while the new Minister of Finance, Friedrich von Motz, received his influential post in 1825 with the aid of Wittgenstein over Witzleben's recommendation. The Crown Prince also sat on the Ministerial Cabinet, the committee of all the ministers which screened legislative proposals before presentation to the king. Hoping to establish an impregnable position here, Wittgenstein—who, as Minister of the Royal House, possessed a seat—schemed to have Duke Carl of Mecklenburg appointed President of the Council of State *with voice and vote on the Ministerial Cabinet.*[119]

It was indeed a clever scheme. Hardenberg had created the Council in 1817 to review and amend select legislative proposals originating in the ministries.[120] The king approved members and chose whether or not to bring proposals before the Council, but, never convinced of his own infallibility, usually accepted its advice. Throughout his reign, in fact, Frederick William did not oppose a recommendation of the sixty-odd bureaucrats, princes, and notables on the Council of State. This appears to have been the one type of "parliamentary" arrangement acceptable to him. If the ultras could establish majorities on both Council and Cabinet, therefore, the antimodern "dam" would be firmly in place.

[116] Koselleck, *Preussen*, 80–81, 83, 515; Berdahl, *The Politics of the Prussian Nobility*, 277; and Gregory W. Pedlow, *The Survival of the Hessian Nobility 1770–1870* (Princeton, New Jersey: Princeton University Press, 1988), 68.

[117] Obenaus, *Anfänge*, 450–51.

[118] Rochow, *Vom Leben*, 212.

[119] Varnhagen Diary I, 5 September 1825, 3:367. For Motz and Wittgenstein, see Branig, *Fürst Wittgenstein*, 152.

[120] For this paragraph, see Hans Schneider, *Der preussischer Staatsrath 1817–1918* (Munich: C. H. Beck'sche Verlagsbuchhandlung, 1952), 66–80, 153.

It was clearly against the interests of Frederick William to become a cat's paw of the aristocracy. For over two years, however, the brooding, retiring head of state allowed events to drift in this direction to the dismay of liberals. Thus Duke Carl was appointed President of the Council of State in September 1825. During the following year, moreover, "the better party gained the upper hand,"[121] as Marwitz summarized conservative trends on the Council. But this challenge to "the onrushing demands of the future" gained no more momentum. In December 1827 the king granted Duke Carl voice—but not vote—on the Ministerial Cabinet. Receiving only a few more reliable appointments to the Council of State in the late 1820s, the Duke would be bitterly frustrated in his plans to capture a majority there.[122] Advised daily by Witzleben—and aware himself of a duty to preserve Hohenzollern power—Frederick William could allow neither the liberal nor the conservative parties to gain an upper hand in Prussia. The monarch's compromises fell far short of aristocratic expectations and insured a great measure of ministerial autonomy during completion of the many important economic reforms underway at the end of that decade.

It was remarkable, in fact, how quickly the aristocratic position disintegrated. Entailment laws were not introduced and subsidies and loans to noblemen did not prevent the continued influx of middle class property owners into the Junker's domain—the bourgeois share exceeded 33 percent in 1842.[123] These mounting figures call into question the constant historiographical refrain about "the astounding power of assimilation"[124] which enabled "the first estate of the realm" to open its ranks "just wide enough to strengthen its position."[125] Far from voluntarily opening its ranks, in fact, the surviving upper stratum of Junkers felt besieged and acted accordingly. They schemed to block new arrivals from local office and exclude them from provincial diet seats. Desperate blue bloods petitioned the government for grants-in-aid, but the process of deterioration continued. By 1856, the bourgeoisie owned 42 percent of aristocratic farms in Prussia, double the percentage of neighboring Hesse-Kassel, where noblemen preserved their position with state intervention. In-

[121] From Marwitz's memoir of early 1828, printed in Friedrich Meusel (ed.), *Friedrich August Ludwig von der Marwitz: Ein märkischer Edelmann im Zeitalter der Befreiungskriege* (Berlin: E. S. Mittler und Sohn, 1908), 1:702.

[122] Duke Carl to Wittgenstein, 22 February 1830 and 8 July 1831, Rep.192, B-P Hausarchiv, respectively in V.1.17 and V.1.15,16, GStA Berlin.

[123] Koselleck, *Preussen*, 80–81, 83, 515; Rosenberg, *Bureaucracy*, 221.

[124] Wehler, *Gesellschaftsgeschichte*, 2:153.

[125] Koselleck, *Preussen*, 515. But Rosenberg, *Bureaucracy*, 221; Anderson, *Lineages*, 271; and Görlitz, *Die Junker*, 218–20, make the same claim. More recently, the remarkable ability of the Junkers to endure and assimilate is also emphasized throughout Berdahl, *The Politics of the Prussian Nobility*.

deed, by 1856 only 16.4 percent of the landed estates in Prussia were held by the same family which had possessed them in 1807.[126] Thus it seems appropriate to give more serious consideration to Gaetano Mosca's maxim that incorporation of newcomers reinvigorates an aristocracy, whereas a nobility "turns plebs"[127] if the pace quickens uncontrollably. This was certainly the contemporary impression—on both sides of the political fence—of what was happening. Thus Carl von Clausewitz, a moderately liberal army officer, felt the nobility had acquired the character of "a ruin which has been corroded and undermined by time,"[128] while Marwitz went even farther, asserting in 1819 that "the nobility, in truth and in fact, ceased to exist quite some time ago."[129]

The conservatives' goal of blocking a threatening industrialization by restoring the guilds was another setback for a class in decline. Wittgenstein and the enemies of Hardenberg's economic liberalization undoubtedly took some consolation after Duke Carl's failure to control Cabinet and Council in the fact that Schuckmann's influence with fellow ministers guaranteed a safe completion of guild restoration. But they were badly mistaken. Three of the five commissioners appointed by the Cabinet to draft this legislation came from Interior on Schuckmann's recommendation.[130] These included Kunth and Hoffmann, two longtime liberal dynamos from the Business Department. The Minister of Finance, Friedrich von Motz, commissioned Georg Maassen, an economic liberal and former head of the same department. Their presence virtually assured the liberal nature of the commission's report.

Schuckmann's motive for stacking the committee with outspoken opponents of the old guilds only months after endorsing the diets' proguild petitions unfortunately remains a mystery. As suggested above, it is likely that he had been an opportunist all along, never genuinely abandoning his commitment to Hardenberg's reforms. There is also evidence that he quickly perceived Motz as a rising star and wanted to ally with

[126] For Hesse-Kassel, see Pedlow, *The Survival*, 74–75, 98. The bourgeois share of Rittergüter rose to 19 percent in 1808, then leveled off, reaching only 20.3 percent in 1895. For the turnover of estates in Prussia, see Helmuth Croon, "Die Provinziallandtage im Vormärz unter besonderer Berücksichtigung der Rheinprovinz und der Provinz Westfalen," in Baumgart, *Ständetum*, 460, 479.

[127] Gaetano Mosca, *The Ruling Class*, edited and revised, with an introduction, by Arthur Livingston (New York: McGraw-Hill, 1939), 425.

[128] See Clausewitz's essay, "Politische Umtriebe," n.d. (1820s), cited in K. Schwarz, *Leben des Generals Carl von Clausewitz und der Frau Marie von Clausewitz* (Berlin: Ferd. Dümmlers, 1878), 2:203.

[129] See Marwitz's remark of 1819 in Meusel, *Marwitz*, 1:627.

[130] Schuckmann to Altenstein, 28 August 1825, Rep.90a, J.I.1, Nr.1, Bl.36–37, ZStA Merseburg.

him, thereby establishing a formidable bloc on the Ministerial Cabinet.[131] Liberal economic policies would have been one method of impressing him, for Motz was not one to look backward.

Indeed the Minister of Finance proved to be another shock to conservative Berlin. As District President in Erfurt (1816–1821) and Provincial Governor in Magdeburg (1821–1825), Motz had shown signs of economic progressivism, promoting road construction, textiles, and the scientific sheep-raising techniques of Albrecht Thaer, a family friend. But as his biographer, Hermann von Petersdorff, tells us, Motz belonged to no party when he assumed ministerial duties in 1825. Yet by 1827/28, the man Wittgenstein promoted was mocking the medieval romanticism of the pietists, defending the economic reforms of Stein and Hardenberg, and praising the social, financial, and military importance of industrial advance before the king. He also spoke defiantly about military struggle against Austria, the bastion of conservatism in central Europe.[132]

The liberals had quickly incorporated Berlin's newest minister into their network. Georg Maassen's appointment to the guild reform commission in the autumn of 1825 is some confirmation of Petersdorff's claim that Motz began to appreciate Maassen "very quickly" and that "a heartfelt friendship soon bound the two selfless patriots."[133] Ludwig Kühne, a friend of Motz's wife and another zealous advocate of freedom of enterprise and trade, also drew into "a steadily closer working relationship" with him during 1825/26, "such that Motz without Kühne is scarcely imaginable."[134] It was Kühne, in fact, who wrote Motz's biannual report to the king in 1828 with its praise for Hardenberg's "emancipation of manufacturing technology [*fabrikativen Gewerbefleisses*] and trade from the institutional shackles which bound them."[135] Joining Maassen and Kühne in the minister's official entourage by this time were Christian Kunth, whom Motz came to regard as a "hero,"[136] and Peter Beuth. "How happy [Motz] felt," writes Treitschke, "to cooperate with this genial technologist who was so certain that a completely transformed time, a new epoch of discoveries and inventions, had come, who, so confident in the future, swam in the current of this great century."[137] Motz's early

[131] Schön to Stägemann, 22 December 1825, printed in Rühl, *Briefe*, 3:237–38. Schuckmann had written Schön that he (Schuckmann) and Motz, not Count Lottum, were now the real political forces with which to reckon.

[132] Varnhagen Diary I, 16 November 1825, 21 March 1828, 3:406, 5:55; Petersdorff, *Friedrich von Motz*, 2:20–21, 33–40, 42–43, 71–73, 206, 244, 312–15, 333, 374, 384; Treitschke, *Deutsche Geschichte*, 3:462–63.

[133] Petersdorff, *Friedrich von Motz*, 2:72.

[134] Ibid., 2:20.

[135] Ibid., 2:43.

[136] Ibid., 2:333.

[137] Treitschke, *Deutsche Geschichte*, 3:463.

advocacy of vocational education and grandiose railway lines were indications that his vision of industry's future was clearer and sometimes more far-sighted than Beuth's (see chap. 3); but in a sense, Treitschke was right, for it was usually Beuth who pointed the way, preaching the indispensability of newer machines, better factories, and more road connections.[138]

A trip to the western provinces during the summer of 1827 reinforced the lessons taught so diligently by liberal compatriots. Motz marveled in Aachen and Seraing (Belgium) at the machine-making factories of Maassen's son-in-law, William Cockerill, and nodded with approval at the prosperity that industry was bringing to the western provinces. Budget surpluses underscored the wisdom of promoting industry—Motz had amassed surpluses of 15 million thaler by early 1828—and solidified a political position which grew more controversial the farther he moved to the left. It was a measure of Motz's growing prestige and influence that Frederick William agreed in March 1829 to transfer the Business Department and Mining Corps to the increasingly powerful Ministry of Finance.[139] As damaging as the conservatives' earlier victory over Bülow's Ministry of Business and Commerce had been, it was now totally reversed.

During Motz's brief tenure in office (1825–1830), prospects brightened for spreading consumer-oriented manufacturing to cities, towns, and country sites throughout the kingdom. The minister's influence—and that of his liberal allies—was most pronounced in three crucial areas. The commercial treaties begun by Motz and completed by his successor, Georg Maassen, with the opening of the German Zollverein (Customs Union) boosted sales potential and enhanced the attractiveness of new investments. "With these treaties," wrote a normally standoffish David Hansemann of Aachen, "the government has gained much love and trust in the country."[140] Contrary to the reigning interpretation, in fact, Motz and the economic liberals who surrounded him were interested in more than fiscal improvements and diplomatic leverage against Austria. They also saw the large German market as an agent of rural industrial growth and technological advance. As Motz put it in 1829, an extended free trade

[138] It was Beuth, for example, who pressed Motz over modernization of the Eichsfeld worsted industry. See the Beuth-Motz exchange of 29 September and 8 October 1829, D.IV.8, Nr.22, Bd.1, Bl.128–29, ZStA Merseburg. For Beuth's enthusiasm for road construction, see *Verhandlungen* 9 (1830): 242–50, GB/TU Berlin; and Petersdorff, *Friedrich von Motz*, 2:298–99.

[139] For the surpluses, see Stägemann to Olfers, 7 February 1828, printed in Rühl, *Briefe*, 3:391–92; for the aquisition of the two departments, see Petersdorff, *Friedrich von Motz*, 2:21.

[140] Hansemann to Maassen, n.d. (November 1829), cited in Bergengrün, *David Hansemann*, 99.

zone would favor "bigger industry" and "superior fabrication," and purge factories which were "on a weaker footing" or "incorrectly laid out."[141]

Transportation improvements served the same purpose by effectively expanding markets for farmers, traders, and manufacturers. Under Motz, government invested about 15.2 million thaler in over 2,800 kilometers of new roads. The funds increased Prussia's existing road network by one-third. Pointing to a 17 percent decline in the transportation cost of coals between 1825 and 1830, Josua Hasenclever, a prosperous merchant-industrialist from Remscheid, exclaimed that "these cheaper prices arose only because of the roads." One had to live through it, he wrote, "to know how much we would have missed if the blessing of these connections had not been bestowed upon us." The government allocated an additional 1.8 million thaler for canal construction and river improvement schemes. This was a huge total, representing 20.3 percent of net non-agricultural investment in Prussia from 1824 to 1830.[142]

In a conscious effort to provide future businessmen with appropriate schooling, moreover, Motz backed Kunth and other educational reformers who wanted secondary schools in Prussia which emphasized science, mathematics, and modern languages. "In the future," wrote Hansemann, "mathematics must become a major subject of instruction—and in this way we give the death blow to stupidity and its consequences." The king supported this orientation, ordering the Ministry of Education and Ecclesiastical Affairs in September 1829 to place more emphasis on vocational gymnasiums (*Realschulen*). Over the next two decades, more than three of the latter were founded for every new classical gymnasium—51 to 16.[143]

Together with earlier liberal reforms (1807–1817), Motz's programs contributed significantly to the infrastructural and institutional framework of the expanding Prussian economy. As the educational example demonstrates, however, we should not stress the contribution of liberal

[141] J. F. Cotta relayed Motz's arguments to Armansperg in a letter of 31 January 1829, printed in H. Oncken and F.E.M. Saemisch, *Vorgeschichte und Begründung des Deutschen Zollvereins 1815–1834: Akten der Staaten des Deutschen Bundes und der Europäischen Mächte* (Berlin: Reimar Hobbing, 1934), 3:452–53. Also see Motz to Eichhorn, 23 February 1829 (3:462). For authors who discount the industrial motivation, see Tom Kemp, *Industrialization in Nineteenth-Century Europe* (Essex: Longman, 1969), 84; Ohnishi, *Zolltarifpolitik*, 227; and Martin Kitchen, *The Political Economy of Germany 1815–1914* (London: McGill-Queen's University Press, 1978), 40–43.

[142] Tilly, "Capital Formation in Germany," in Mathias and Postan, *Cambridge Economic History*, 7 (1): 413, 436; Jacobs and Richter, *Grosshandelspreise*, 78. I have converted Tilly's net investment figures for 1913 using the index for construction materials (*Baustoffe*) in Jacobs and Richter. For the Hasenclever quote, see his "Memoirs," written in 1841, printed in Adolf Hasenclever, *Josua Hasenclever aus Remscheid-Ehringhausen: Erinnerungen und Briefe* (Halle: Karras, Kröber & Nietschmann, 1922), 33.

[143] Petersdorff, *Friedrich von Motz*, 2:198–210, 350–53. For the Hansemann quote, see Hansemann to Vogt, 23 December 1825, printed in Bergengrün, *David Hansemann*, 48.

bureaucrats to the point of underestimating the role of Frederick William III. Prussia faced clear alternatives and critical decisions in the two decades after her humiliation at Jena. For every liberal proposal there was an opposing argument put forward by very influential elites. That Prussia embarked upon the course she did was in no way inevitable. And from peasant emancipation and abolition of guild privileges before 1811, to patent laws, tariffs, the defeat of guild restoration, the founding of the Zollverein, new roads, and business-oriented schools, ultimate responsibility for the direction of change rested with the monarch. That his ministers and private advisers convinced him repeatedly of the need to move forward in these areas was less a sign of his personal weakness than an indication of his moderation and practicality.[144] In the cases of tariffs, the Mining Corps, and Seehandlung, in fact, it was the king who initiated moderation and compromise. "Guard yourself against unpractical theories and the zeal for innovation that is gaining so much ground," he warned the Crown Prince in 1827, "but be on guard as well against a preference for the old which, driven too far, is almost equally harmful." Only by avoiding these two extremes could he promote "genuinely useful improvements."[145] As we shall see, Frederick William had good reason to worry about his son's yearning for the past. As for the father, a rollback of industry and technology was no more in his makeup than a doctrinaire, ideologically consistent means to scientific and technological advancement. It was under such a "bourgeois king," divorced from academic as well as aristocratic pretensions, prejudices, and predilections, that Prussia embarked upon its industrial journey.

. . . . .

Our story has advanced chronologically to the eve of the revolutionary year of 1830. Subsequent chapters will return on other thematic planes to the early years of the century, then begin to press the analysis forward into the 1830s and 1840s. Before concluding this chapter, however, it would be appropriate to briefly pursue the question of guild reform into these later decades, for this episode highlights the complex, nondoctrinaire nature of the economic modernity which was coming to Frederick William's Prussia.

[144] For the king's involvement in the areas of roads, Zollverein, and education, see ibid., 2:298–99, 351–52; Eylert, *Charakter-Züge*, 3A:203–7; and Karl Bruhns (ed.), *Life of Alexander von Humboldt* (London: Longman's, Green & Co., 1873), 2:100–106. Frederick William would eventually react more positively to military modernization (see below, chap. 5) and mechanized manufacturing (see below, chap. 6).

[145] Frederick William to the Crown Prince, 1 December 1827, printed in Klöden, *Friedrich Wilhelm*, 388–89.

Johann Gottfried Hoffmann and his committee would have had it differently. Their report to the Ministerial Cabinet of February 1835, contained very few concessions for the anti-industrial champions of guild restoration. The document recommended guilds which would serve important social and educational functions and provide handicraftsmen with revitalized organizational solidarity. But the new corporations would be powerless in the business world, *exerting no restrictive influences on hiring, production, pricing, quality, or technology.*[146] Only when society was endangered by the practice of trades by untrained, unreliable persons did the committee recommend that practitioners acquire a license. Careful lest the enemies of industry seize upon the issue of proof of skill, however, Hoffmann's committee refused to specify the trades which required licensing. It was preferable to let changing attitudes and the progress of technology, not rigid, inflexible legislation, determine such matters.

Had the report come a few years earlier, the Ministerial Cabinet might have been more amenable to the committee's laissez- faire recommendations. By early 1835, however, the "heroes" of economic liberalism were gone. Within a span of four years, Kunth, Motz, and Maassen had followed Hardenberg to the grave, leaving Kühne and Beuth without ministerial patrons as the Mining Corps, the Seehandlung, the Prussian Army, and other economic conservatives mounted offensives.[147] Hoffmann's report fell victim to these statist trends. The Cabinet specified scores of businesses which would require a license, including most of the medical, transportation, and construction trades. Local authorities were also given an important role in the testing process.[148]

What the ministers started in motion, the guilds, provincial diets, and conservatives on the Council of State tried to accelerate, calling for guild-controlled licensing of all businesses during the late 1830s. These demands were effectively parried by a Cabinet of Ministers determined to defend its compromise legislation. The Industrial Code of 1845 (*Gewerbeordnung*) remained an essentially moderate piece of legislation.[149] By this time, however, Prussia could no longer be termed a state guided by the principles of economic liberalism.

[146] See Hoffmann's interim report of 30 November 1830, and the final report of 24 February 1835, in (respectively) Rep.90a, J.I.1., Nr.1, Bd.1, Bl.116–25, and Bd.2, Bl.128–159, ZStA Merseburg.

[147] Albert Lotz, *Geschichte des Deutschen Beamtentums* (Berlin: R. v. Decker's Verlag, 1906), 392.

[148] The lengthy ministerial deliberations of 1835/37 are located in Rep.90a, J.I.1, Nr.1, Bd.2–4, passim, ZStA Merseburg.

[149] For the reactions of the provincial diets (1837–38) and the Council of State (1842–44), see the materials in Rep.90a, J.I.1, Adhib B, Bd.1–2, passim, ZStA Merseburg.

There was also less basis of hope for the surviving advocates of rural industrialization. The goal of Hardenberg, Scharnweber, Kunth, and Beuth was only partially realized before railroads ushered in a different era. Their sovereign of forty years was also upset over the development of things. In the late 1830s, for instance, we find Frederick William III complaining about the fast pace of modern life. "Our age loves steam. Everything is supposed to gallop, but peace and quiet suffer because of it."[150] The old king was no different in this respect from the men who had encouraged him to industrialize a quarter of a century earlier. All were uncomfortable with a modernity which had deviated from their visions. It is a theme to which we will return.

[150] Eylert, *Charakter-Züge*, 3A:205.

II

Liberal Brothers in Arms

JOHANN BRAUN left the university and walked slowly westward along Unter den Linden. The classical literature of Greece and Rome held a particular attraction for the chief of Prussia's Experimental Artillery Department, and today, like so many other days since the peace, Braun had sat attentively in the rear of the lecture hall. Soon he neared the neoclassical facade of Number 74, a Schinkel masterpiece which for the past seven years had housed the Artillery and Engineering School. As usual, he was the first member of the army's new Commission on Science and Technology to arrive.

Within minutes, however, the others began to appear. Braun's longtime artillery superior, Prince August, brought his two leading staff officers, Joseph Maria von Radowitz and Carl von Clausewitz. The Corps of Engineers sent its head, Gustav von Rauch, and the General Staff its new chief, Johann [von] Krauseneck. Completing the group were Georg von Valentini, the Inspector of Military Schools, and Ernst von Pfuel, a military theorist and presently commander of the Seventh Landwehr Brigade. The commission usually considered every idea that appeared promising, from the crushing strength of axles to the needlegun.[1] With the meeting about to begin, ancient times would have to be put aside.

The commissioners realized that advances in human knowledge could mean victory or defeat for their Prussian Fatherland. Each approached his investigative task with an intensely patriotic motivation. None, however, placed trust exclusively in the firearms which a soldier carried into combat, the projectiles fired by supporting units, or the massive stone fortresses which guarded the country's frontier. In fact, the officers of Prussia's technical establishment were actually quite ambivalent about the "hard" military technology their commissions and institutes sought to develop (see chap. 5). For they believed that warfare in the modern era would remain a basically human affair. Given a rough equality of armaments, the nation which had earned the confidence of its people through

[1] For the composition of the commission in the early years, see Frederick William to Hake, 28 December 1828, Rep.90, Tit.xxxv, Nr.195, Bl.3–4, ZStA Merseburg; Varnhagen Diary I, 5 December 1828, 5:145; Kurt von Priesdorff (ed.), *Soldatisches Führertum* (Hamburg: Hanseatische Verlagsanstalt, n.d.), 6:222; and Peter Paret, *Clausewitz and the State* (New York: Oxford University Press, 1976), 326. For Braun's pastime at the University of Berlin, see Priesdorff, *Soldatisches Führertum*, 4:132. Also see below, chap. 5.

"soft" technology like enlightened reforms would be able to translate this enthusiasm into an invincible excitement in hand-to-hand fighting.[2] Thus Hermann von Boyen, the ex-Minister of War, valued the "manly sense of independence" which a people brought to its battlefields. More decisive than *individual* bravery or courage, this "warlike spirit" derived from "the lifestyle of a folk, its customs and constitution."[3] Neithardt von Gneisenau had written Clausewitz in a similar vein. The best military technology "will not be of much use if we cannot count on a good spirit in the provinces." And nothing would do more to bolster good feeling, the free-thinking hero of the Napoleonic Wars concluded, "than the granting of a constitution for the entire monarchy."[4]

Our opening scene introduces the men who belonged to the Commission on Science and Technology in early 1830. Not surprisingly or coincidentally, every one of them belonged to the military's silent, yet still very sanguine fraternity of reformers.[5] Many of them were closely connected as well to liberal friendship circles in the civilian bureaucracy. In this chapter we shall discuss their collective failure to advance "soft" technology by 1832. The opening act of a political crisis which progressed steadily in the waning Pre-March, this set of political events is not unrelated to the technical developments discussed in later chapters. For it was the money-making success of political liberals in Prussia's feuding economic and technological agencies—ironically—which permitted the state to avoid massive borrowing and the convening of a parliament.

. . . . .

At the center of the army's political brotherhood were four overlapping circles of talented friends. The first of these tight-knit cliques were "the court liberals"—those advocates of moderate political liberalization who possessed access to the king. Prince August, the monarch's debonair, progressive cousin; Prince William, the shy, liberal-minded brother of

[2] Hermann von Boyen, "Die Geistigen und Körperlichen Eigenschaften des Kriegers," n.d. (1820s), an unpublished essay in I HA Rep.92 Boyen, Nr.442, GStA Berlin.

[3] See his unpublished essay: "Über kriegischen Geist," n.d. (1820s), I HA Rep.92 Boyen, Nr.439, GStA Berlin.

[4] Gneisenau to Clausewitz, 20 August 1818, printed in Hans Delbrück (ed.), *Das Leben des Feldmarschalls Grafen Neithardt von Gneisenau* (Berlin: G. Reimer, 1880), 5:333.

[5] Manfred Messerschmidt, "Die politische Geschichte der preussisch-deutschen Armee," in Gerhard Papke, *Handbuch der deutschen Militärgeschichte 1648–1939* (Munich: Bernard & Graefe Verlag für Wehrwesen, 1975), 4 (1): 121, errs in asserting that reformism was so shallow in the Prussian officer corps that it disappeared soon after Boyen's resignation in 1819. Rather, these ideas survived throughout the *Vormärz*—and beyond. For Boyen's political optimism in particular, see Varnhagen's note on the general's conversation with Johann Eichhorn, Varnhagen Diary I, 20 August 1824, 3:120.

the king; and Oldwig von Natzmer, the latter's dashing friend since childhood, had all stood by Gerhard Scharnhorst, the slain military reformer whose crusading spirit still infused the army Left. Inseparable from them—and usually more influential—was Job von Witzleben, one of the few men near the throne.[6] Wilhelm von Humboldt also shared an intimate personal friendship with these men. Never completely out of favor after resigning from the ministries in 1819, Humboldt was a significant "linking" personality in the network of civilian and military cliques which made up liberal Berlin. He was "Witzleben's truest political and also intimate [*gemütlicher*] friend,"[7] recalled Alexander von Humboldt.

Humboldt's contacts and friendships also spread across the other army circles under discussion here.[8] Probably the best-known of these is the union between Clausewitz and Gneisenau, the great theorist's bosom companion. Like most of their fellow reformers, the two had come up under Scharnhorst,[9] intrigued with "the virtuous ones" against Napoleon,

[6] For the "court liberals," see the friendly correspondence between Prince August and Prince William, Prince William Papers, D22, Nr. 12–43, HeStA Darmstadt; Prince William and Witzleben, Prince William Papers, D22, Nr. 13–46, HeStA Darmstadt and Rep. 92 Witzleben, Nr. 100, Bl. 6–8, ZStA Merseburg; Prince William, Natzmer, and Witzleben in Oldwig von Natzmer, *Unter den Hohenzollern: Denkwürdigkeiten des Generals Oldwig von Natzmer* (Gotha: Friedrich Andreas Perthes, 1887), 1:28, 107, 270; and Prince August and Witzleben, in E. von Conrady, *Leben und Wirken des Generals der Infanterie und kommandierenden Generals des V. Armeekorps Carl von Grolman* (Berlin: Ernst Siegfried Mittler und Sohn, 1894), 3:91–100. Also see Rochow, *Vom Leben*, 80–81, 84–86.

Varnhagen describes Prince August as "one of the very dissatisfied" (Varnhagen Diary I, 26 December 1821, 1:386), and later, as one taking "the constitutional side" with Clausewitz, Willisen, and Alexander von Humboldt in discussions over Portugal (5 June 1828, 5:86). There are many other indications of his liberalism. He corresponded with Wilhelm von Humboldt, fought hard for Grolman, and surrounded himself at luncheons with liberals like Varnhagen von Ense, Friedrich von Raumer, Alexander von Humboldt, Peter Beuth, and Friedrich Adolf von Willisen (see Varnhagen Diary I, 7 January 1820; 29 June 1826; 14 November 1827; 26 January 1828, 1:47, 4:87, 152, 335, 5:14. Theodor von Schön, moreover, considered Prince August an important contact (Schön, *Aus den Papieren*, 6:402), while Ancillon felt the need to lecture the Prince about the pitfalls of parliamentarism (Delbrück, *Gneisenaus Leben*, 5:77–78).

[7] The remark, probably made to Wilhelm Dorow in April 1835, is cited in Hanno Beck, *Gespräche Alexander von Humboldts* (Berlin: Akademie-Verlag, 1959), 142. For Humboldt and the other court liberals, see Varnhagen Diary I, 13 January 1820, 1:50; Natzmer, *Unter den Hohenzollern*, 1:143; Dorow, *Job von Witzleben*, 42; and Prince William Papers, D22, Nr. 12–80, HeStA Darmstadt.

[8] For connections with the other three circles, see Varnhagen Diary I, 25 November 1819, 1:3.

[9] For the brotherhood's devotion to Scharnhorst, see Udo von Bonin, *Geschichte des Ingenieurkorps und der Pioniere in Preussen* (Wiesbaden: LTR Verlag, 1981), 2:191–93; Paret, *Clausewitz and the State*, 67, 69–70; Priesdorff, *Soldatisches Führertum*, 3:279, 4:128; Friedrich Meinecke, *Das Leben des Generalfeldmarschalls Hermann von Boyen* (Stuttgart: J.G. Gotta'sche Buchhandlung Nachfolger GmbH, 1899), 1:112–13; Colmar

and ridden to glory when Europe's revolt began. With Scharnhorst's unfortunate death, Gneisenau assumed the mantle of liberal leadership in the army, drawing so many progressive comrades to his postwar entourage in Coblenz that Tsar Alexander felt he might "have to help the King of Prussia fight his own army."[10]

Early 1819 saw Clausewitz and Gneisenau together in Berlin again. Their presence afforded many opportunities to visit with Prince William, an extremely close friend despite disagreements over the Prince's indecision at Waterloo.[11] Acquaintances were also renewed with comrades from "Wallenstein's Camp"—the name conservatives gave to the Fieldmarshall's politicized Rhenish headquarters—but regular contacts were rare.[12] Of the old circle stationed in Berlin, Gneisenau and Clausewitz saw only three on a regular basis: Franz August O'etzel, a major with Clausewitz at the War School; Karl von der Gröben, one of the few liberal adjutants around Crown Prince Frederick William; and Wilhelm von Scharnhorst, the great reformer's son, now Gneisenau's son-in-law.[13]

Relations remained very cordial, however, with another clique of officers who had frequented Wallenstein's Camp. This group included Pfuel; Valentini; Karl von Müffling, Chief of the General Staff until 1829; and Otto August Rühle von Lilienstern, the chief of the Military Curriculum Commission, affectionately called "Rühl" by the others. These men were so determined to liberalize Prussia that Hardenberg referred to them—misleadingly—as "the Teutonic Jacobins."[14] Their real distinction was a

---

Freiherr von der Goltz, *Kriegsgeschichte Deutschlands im Neunzehnten Jahrhundert* (Berlin: Georg Bondi, 1914), 2:20–21; "General-Lieutenant Rühle von Lilienstern," *Beiheft zum Militair-Wochenblatt*, (October–December, 1847), 126–27, 130–31 (hereafter cited as "Rühle von Lilienstern," MgFa Freiburg); and more recently, Charles Edward White, *The Enlightened Soldier: Scharnhorst and the Militärische Gesellschaft in Berlin, 1801–1805* (New York: Praeger, 1989), 100, 111, 131.

[10] Cited in H. Ulmann, "Die Anklage des Jakobinismus in Preussen im Jahre 1815," *Historische Zeitschrift* 95 (1905): 436. For Gneisenau's entourage in Coblenz, see Schwartz, *Leben des Generals*, 2:172–89.

[11] Paret, *Clausewitz and the State*, 232, 307; Delbrück, *Gneisenaus Leben*, 5:33; Prince William to Natzmer, 2 January 1832, printed in Natzmer, *Unter den Hohenzollern*, 1:270; and Priesdorf, *Soldatisches Führertum*, 3:270–71; and Varnhagen Diary I, 7 January 1820, 1:47. See also Marie von Clausewitz to Prince William, 1 October 1830 and 22 August 1832, Prince William Papers, D22, Nr.13–37; Gneisenau to Prince William, 26 April 1831, Nr.12–118; and Wilhelm von Scharnhorst to Prince William, 5 September 1831, D-22, Nr. 13–63, HeStA Darmstadt.

[12] For the Coblenz circle, see Varrentrap, *Johannes Schulze*, 178–83.

[13] Paret, *Clausewitz and the State*, 213, 265, 310; Karl Linnebach, *Karl und Marie von Clausewitz: Ein Lebensbild in Briefen und Tagebuchblättern* (Berlin: Martin Warneck, 1925), 38; and the correspondence between Gneisenau and Scharnhorst (the son), in Albert Pick, "Briefe des Feldmarschalls Grafen Neithardt v. Gneisenau an seinen Schwiegersohn Wilhelm von Scharnhorst," *Historische Zeitschrift* 77 (1896): 67–85, 234–56, 448–60.

[14] Cited in Treitschke, *Deutsche Geschichte*, 2:605.

degree of artistic, literary, and intellectual sophistication unusual in any military establishment. Rühle and Müffling had even impressed Goethe. Rounding out the circle were: Braun, whose gymnasial education and intense study under Scharnhorst had earned the artillerist his learned friends' respect; Radowitz, who had come into this circle as the protégé of Rühle von Lilienstern in 1826; and Johannes Schulze, Head of Secondary Education in the Ministry of Education and Ecclesiastical Affairs—and another important link to liberals in the civilian bureaucracy.[15]

If there was a real Jacobin in the Prussian Army in the late 1820s, it was Karl Wilhelm von Grolman, Commander of the Fifth Army Corps in Breslau. He was known for his brash manner, antimonarchist statements, and desire to establish a Spartan Republic on the Spree.[16] It had taken "the court liberals" years to convince the king to give this former Scharnhorst devotee and Chief of the General Staff another command. When it finally came in 1825, Frederick William sent him to Silesia, far away from the capital.

Grolman shared fast bonds of friendship with Hermann von Boyen, the ex-Minister of War, and Johann Krauseneck, Müffling's successor as Chief of the General Staff in 1829. To these three more radical proponents of political change, Hardenberg had been a mere "half-revolutionary."[17] But the "Triumvirate"—as Friedrich Meinecke called it—was actually a wider association. Gustav von Rauch of the Corps of Engineers belonged, as did his engineering colleague, Ernst Ludwig Aster, Commandant of Ehrenbreitstein. Another common friend was August von Hedemann, a Husar Brigadier, loving son-in-law of Wilhelm von Humboldt, and one of the few members of the Spanish Society, a radical discussion club, to wear Prussian blue.[18]

[15] For Braun, see his letters to Müffling, 28 February and 31 March 1831, Rep.92 Müffling, zu A.14, ZStA Merseburg; Priesdorff, *Soldatisches Führertum*, 4:128; and White, *Enlightened Soldier*, 111, 131. For the circle, see Varrentrapp, *Johannes Schulze*, 183, 219, 558–59, 566–67; Erich Weniger, *Goethe und die Generale* (Leipzig: Im Insel Verlag, 1942), 92, 102–3, 168; and Heinrich von Brandt (ed.), *Aus dem Leben des Generals der Infanterie z.D. Dr. Heinrich von Brandt* (Berlin: E. S. Mittler, 1869), 2:2–3, 5–6, 10, 17, 55. And for the constitutionalism of this circle, see Jeismann, *Das preussische Gymnasium*, 262; Rühl, *Briefe*, 2:62–67; Treitschke, *Deutsche Geschichte*, 3:364–65; and Varnhagen Diary I, 4:186, 9 February, 1827. Paret, *Clausewitz and the State*, 192, describes Valentini as a moderate conservative.

[16] Brandt, *Aus dem Leben*, 2:40; and Varnhagen Diary I, 8 November 1822, 2:242.

[17] Clausewitz to Gneisenau, 12 October 1816, cited in Delbrück, *Gneisenaus Leben*, 5:151–52.

[18] For this entire clique, see Meinecke, *Das Leben*, 1:392–93, 2:489, 521; Gerlach, *Aufzeichnungen*, 95–96; Varnhagen Diary I, 2 July 1825, 3:320; Diary of Karl August Varnhagen von Ense, 12 January 1841, printed in Ludmilla Assing (ed.), *Tagebücher von K. A. Varnhagen von Ense* (Bern: Herbert Lang, 1972), 1:264, originally published in 1861 (hereafter cited as Varnhagen Diary II); Bonin, *Geschichte des Ingenieurkorps*, 2:190–93; and

None of the triumvirs was so far to the left, however, that ties with the other reformers were jeopardized. For all had fought together and were in basic agreement that intelligent political institutions had to stand in the front line of the nation's defense.[19] Schulze and Braun befriended Aster, for instance, during Gneisenau's sojourn in Coblenz. Pfuel met regularly in postwar Berlin with Grolman, Hedemann, and Wilhelm von Lützow, the former free corps leader, to discuss combat experiences and domestic politics.[20] Clausewitz and Gneisenau found Boyen and Grolman irritating at times, but had no doubt that they were "two of ours," while Prince William looked fondly upon Hedemann, his longtime adjutant. Prince August and Witzleben, finally, were the court liberals who secured the reinstatement of Grolman, a common friend of theirs, with the Fifth Corps. It was this overlapping, interlocking nature of the cliques—or as Varnhagen put it, the "great interrelation"—which drew them closer together into a larger, overtly political brotherhood.[21]

This is not to say that Prussia's liberal fraternity did not have its running feuds. Thus Valentini had been unimpressed with Scharnhorst's lectures before Jena, nearly fought a duel with Gneisenau during the wars, and did not like Clausewitz. The feeling was mutual. Similarly, Witzleben, Gneisenau, Scharnhorst, and Grolman did not like Radowitz, eventually affecting his exile away from the Crown Prince to another post in Frankfurt. Witzleben also arranged Müffling's transfer to Münster in 1829 after the General Staff Chief had strayed opportunistically into the political camp of Duke Carl and Count Wittgenstein.[22] No fraternity is

Natzmer, *Unter den Hohenzollern*, 1:143, 2:78, 3:110, 129. Also see Boyen to Hedemann, 10 October 1820, Prince William Papers, D22, Nr.13–32, HeStA Darmstadt; and Rauch to Boyen, 18 August 1816 and 27 June 1817, I HA Rep.92 Boyen, Nr.1, GStA Berlin.

[19] See notes 2–18 for documentation on the group's liberalism. For Müffling, see below, n. 22. Clausewitz also warrants a qualifying remark. After a decade or more of espousing liberal constitutionalism, Clausewitz wrote a conservative tract (around 1822/23), but it is debatable whether this reflected his genuine political feelings. See Paret, *Clausewitz and the State*, 298–306.

[20] Varrentrap, *Johannes Schulze*, 178–83; Gerlach, *Aufzeichnungen*, 95–96.

[21] For "two of ours," see Clausewitz to Gneisenau, 14 November 1816, printed in Delbrück, *Gneisenaus Leben*, 5:162. Clausewitz disagreed with Boyen and Grolman about details of the *Landwehr* (see letters in 5:151–52, 216), but they remained friends (see 5:237). For "great interrelation," see Varnhagen Diary I, 25 November 1819, 1:3. Also see the Prince William-Hedemann correspondence, Prince William Papers, D22, Nr.13–103, HeStA Darmstadt; and Conrady, *Carl von Grolman*, 3:91–100.

[22] ADB 27:143; Varnhagen Diary I, 12 May 1827 and 30 November 1829, 4:230, 5:251–52. Varnhagen's reports about Müffling's disingenuous conservatism were probably true. See Prince William (the son) to Natzmer, 2 July 1825, printed in *Politische Correspondenz Kaiser Wilhelms I* (Berlin: Hugo Steinitz, 1890), 39; Schneider, *Der preussische Staatsrat*, 92; and Ludwig Dehio, "Wittgenstein und das letzte Jahrzehnt Freidrich Wilhelms III," *Forschungen zur brandenburgischen und preussischen Geschichte*, 35 (1923): 216–19.

without its emotional quarrels, but these were more than insignificant spats. In the highly personalized political culture of a semiabsolute monarchy, political effectiveness often depended on one's proximity to the head of state. No longer trusted, Radowitz and Müffling were banished from the central arena.

Indeed politics had become an extremely potent force in high society, holding together, breaking up, or reshuffling many relationships between friends as the 1810s drew on to the 1820s. The Saturday poetry gatherings of Amalie von Helvig, wife of a retired artillery technician, were torn asunder by political passions, while other more famous salons like that of Rahel von Varnhagen became highly politicized, attracting an increasingly one-sided, liberal following. Former radicals like Karl von dem Knesebeck "became the total opposite" after Napoleon's early victories, evoking feelings of disgust from erstwhile brothers and deep sympathy from new, essentially political compatriots like Gustav von Rochow. The nationalistic demonstrations and disruptions after 1817, moreover, tore the three Gerlach brothers away from "the Germanic clique" around Grolman, Hedemann, and Lützow, driving them rightward to Kottwitz and the pietists. Some friendships, of course, were stronger than politics. Thus Ludwig Gustav von Thile, whose conservatism bordered on mysticism, remained good friends with Boyen, while both Thile and Radowitz bridged the gap between the Gerlachs and the open-minded circle around Valentini and his brother-in-law, Rühle von Lilienstern.[23] These open bridges were still very common in the 1820s—but in the developing pattern of things, were becoming the exceptions.

Friendship cliques were an important part of Berlin's political infrastructure. In the private company of old friends one could feel at ease, simply relax, or join in congenial talk about "war, peace, history, politics, astronomy and God knows what else,"[24] as one young officer remembered a dinner at Valentini's in January 1828. Prayer meetings with Kottwitz and intimate gatherings at the Rochows', at Monbijou Palace with Duke Carl, or at Wittgenstein's card table were also quite innocuous at times. But the small, closed party sometimes served as an exchange of information collected at "semipublic" (*private-öffentliche*) gatherings like

[23] Varnhagen Diary I, 2 January 1821, 14 December 1822, 1:251, 2:264; Brandt, *Aus dem Leben*, 2:10, 17, 51, 55; and Bernhard von Gersdorff, *Ernst von Pfuel* (Berlin: Stopp Verlag, 1981), 45; Gerlach, *Aufzeichnungen*, 101–4; Boyen's "total opposite" remark is cited in White, *Enlightened Soldier*, 63; and Rochow, *Vom Leben*, 116–18, where she observes that Knesebeck and Rochow had "a friendly relationship based on the same political views"; Clausewitz to Gneisenau, 12 October 1816, printed in Delbrück, *Gneisenaus Leben*, 5:151–52; and Treitschke, *Deutsche Geschichte*, 5:19.

[24] Brandt, *Aus dem Leben*, 2:2–3. The passage in Brandt (pp. 1–57) which begins with this quote offers much insight to the world of political "cliques and coteries" (p. 2).

Rahel's salon, Spanish Society evenings, political luncheons at Prince August's suite (see below), and at dinner engagements with the neutral Lawless Society or the more conservative Christian-German Dining Club.[25]

Whether liberal or conservative, however, extended friendship circles became important staging grounds for intrigue and party-political action during critical moments. We have already observed in chapter 1 how the pietist and aristocratic cliques mobilized against Hardenberg's followers. An instance of liberal action occurred in the winter following Hardenberg's death. As Frederick William procrastinated, chose two successors who rapidly passed away, then fell into renewed indecision about the appointment, Berlin's leading liberals in and out of uniform orchestrated a campaign to bring trusted compatriots into the king's privy cabinets. "Herr von Witzleben is working against Duke Carl and Prince Wittgenstein," reported Varnhagen in February 1823, and was "finding encouragement from all sides." This backing came in the form of memoranda presented before the king and crown prince that were "not without effect." It was primarily "the military" along with "many respected civil servants" who were at work. But "the real instigators," his informant relayed, "are entirely in the dark, people who do not come forward."[26] Apparantly emboldened by friendly pressure from his own support network, Witzleben went even farther in early March and recommended that the king choose Wilhelm von Humboldt as chancellor. Afraid that the ex-minister would stray "too far along the devious paths of the modern Zeitgeist,"[27] sensitive about reactions in Vienna, and unwilling to create unnecessary parliamentary challenges to his personal rule, Frederick William finally opted in April to rule without a chancellor. The loyal, harmless Minister of the Treasury, Gustav von Wylich und Lottum, was named "cabinet minister," a personal liaison between monarch and ministers. Friedrich August Stägemann, a Hardenberg appointee who had

[25] For a general discussion of "private Öffentlichkeit" in the Pre-March, a term coined by Jürgen Habermas, see Thomas Nipperdey, "Verein als soziale Struktur in Deutschland im späten 18. und frühen 19. Jahrhundert," in H. Boockmann et al. (eds.), *Geschichtswissenschaft und Vereinswesen im 19. Jahrhundert* (Göttingen, 1972), 29–42. For the Lawless Society and the Christian-German dining Club, see Ludwig Metzel, "Die 'Zwanglose' 1806 bis 1906," *Mitteilungen für die Geschichte Berlins* 23 (1906): 57–60; Gersdorff, *Ernst von Pfuel*, 45; and Barbara Vogel, "Beamtenkonservatismus: Sozial- und verfassungsgeschichtliche Voraussetzungen der Parteien in Preussen im frühen 19. Jahrhundert," in Dirk Stegmann et al., *Deutscher Konservatismus im 19. und 20. Jahrhundert* (Bonn: Verlag Neue Gesellschaft, 1983), 17–18.

[26] Varnhagen Diary I, 21 February 1823, 2:305. Varnhagen's "usually well-informed man," clearly a civilian and a liberal, was probably Stägemann.

[27] Cited in Sweet, *Wilhelm von Humboldt*, 2:388.

figured prominently in the liberals' early conspiracy, was assigned to Lottum's staff.[28]

Thus as mid-decade approached, it remained politically risky to speak or write about constitutions in Prussia. Most army moderates kept prudently silent, waiting for better political opportunities. Some, like Müffling, deserted to the other side. "In Monbijou Palace, Duke Carl's residence, Kamptz and Müffling held forth," writes Treitschke. "Haller's holy teachings were preached there with even greater emphasis than they were in the palace along the Wilhelmstrasse where the crown prince assembled his romantic entourage."[29] To his credit, Müffling merely cloaked his liberal views, proposing surreptitiously in December 1825, that the king establish a bicameral legislature with real influence before worse was forced upon him from below during wartime. But Müffling registered no more success than Witzleben two years earlier.[30] Meanwhile, the privy councillor who would punish Müffling went into a less noble hiding with his views. Thus Witzleben criticized the liberalism of his civilian friend, Alexander von Humboldt, during a social gathering following one of Humboldt's "Cosmos" lectures in December 1827. "But do you really think otherwise, Herr General," asked one of the guests, "even if you must act differently on occasion?"[31] As Witzleben blushed, another guest mercifully changed the subject. The episode demonstrates how difficult it was on occasion for Witzleben to contain his liberalism and function as the king's adviser, not the leader of a party, the role which Hardenberg had played so frequently as chancellor.

As long as the peace in Europe was secure, Prussia's budget was secure. And with the coffers full, liberal parliamentarism remained a theoretical subject in Prussia. As the decade drew to a close, however, the protracted Greek revolt against Turkey upset the equilibrium in Europe. Austria had opposed the rebellion from its outset in 1821, refusing to distinguish the latest challenge to legitimacy from earlier uprisings in the Mediterranean. Vienna was pleased, therefore, when the Sultan's Egyptian forces began to rout a disorganized Greek army in 1825. After his brush with revolution that year, Tsar Nicholas retained a visceral dislike of subversives, but Russian interests dictated continued expansion to the southwest, renewed conflict with the Ottoman Empire, and thus implicit support for Greece. French policies supported Russia, and with a pro-Greek Canning Government in London, a new diplomatic alignment took shape. Austria watched with mounting alarm as naval units of this

[28] Varnhagen Diary I, 1823: 8 February; 13 February; 20 February; 22 April, 2:293, 297, 302, 304, 336.

[29] Treitschke, *Deutsche Geschichte*, 3:363.

[30] Dehio, "Das letzte Jahrzehnt," 216–17.

[31] Cited in Varnhagen Diary I, 31 December 1827, 4:356.

"triple alliance" decimated a Turkish-Egyptian fleet at Navarino in October 1827, and with rage as a Russian army crossed the River Pruth in April 1828. A European conflagration nudged closer to reality during the autumn of 1828 when Metternich convinced the Duke of Wellington's more anti-Russian cabinet to join Vienna in "negotiating" a Russian withdrawal. But Tsar Nicholas objected vehemently. By early 1829, Europe hovered near the brink.[32]

The crisis sparked a debate in Prussia which was only temporarily resolved by the Treaty of Adrianople ending the Russo-Turkish War in September 1829. At the center of the controversy lay the unavoidable question of Prussia's diplomatic and military options should the war spread. Duke Carl, Count Wittgenstein, the Mecklenburgers, and the aristocratic parties stood steadfastly by Austria as the true standard-bearer of conservatism, legitimacy, and continental order. For them, the mere suggestion of deserting the Danubian Monarchy and the German Confederation was a suspicious, near-revolutionary act.

Others had grown weary of Prussia's subsidiary role in the Confederation and desired foreign policies which demonstrated the kingdom's status as a genuine Great Power. Alliances with Bavaria, Württemberg, or Russia would break the fraternal Austrian bonds restraining Prussia and probably produce quick territorial concessions in Poland and Bohemia. "Only a war can help us," said Friedrich von Motz, one of the leading anti-Austrians. His good friend Müffling was full of tough, anti-Habsburg rhetoric, going as far as drawing up contingency plans for deploying five army corps against Austria and Saxony. Other hotspurs were Johann Eichhorn, an influential undersecretary in the Foreign Office, and Prince August of the artillery. By 1828 even Job von Witzleben cautiously supported the bellicose ones.[33]

Prince August's neo-Greek residence near Berlin's Tiergarten became the most prestigious arena for irreverent criticism of Prussian foreign policy. In 1823 he had employed Friedrich Schinkel, Berlin's leading archi-

[32] For the Near Eastern Crisis, see Treitschke, *Deutsche Geschichte*, 3:728–46; and more recently, Lawrence J. Baack, *Christian Bernstorff and Prussia: Diplomacy and Reform Conservatism 1818–1832* (New Brunswick, New Jersey: Rutgers University Press, 1980), 147–60.

[33] Varnhagen von Ense was on cordial terms with leaders in both factions. His diaries are therefore a rich source for this debate. See Varnhagen Diary I, 1826: 29 June, 31 July; 1827: 14 January, 18 April, 12 May, 17 July, 14 November, 14 December, 28 December; 1828: 26 January, 5 March, 25 March, 31 March, 7 May; and 1829: 3 June, 11 June, 20 October, 29 October; cited in 4:87–88, 91–92, 172, 218, 230–34, 265, 334–35, 349–50, 354–55; and 5:14, 43, 57, 59, 61, 72, 208, 210, 241, 243. See also Baack, *Christian Bernstorff*, 157–58, 160; Treitschke, *Deutsche Geschichte*, 3:453, 456; and Petersdorff, *Friedrich von Motz*, 2:374. For Motz's remark, see Petersdorff, 2:374; for Müffling's staff plans, see Brandt, *Aus dem Leben*, 2:14–15.

tect and builder, to renovate the upper story of the family palace in order to entertain in physical surroundings befitting his princely station. Adjacent to the main banquet hall he included a smaller dining room for more intimate political and intellectual discussions. By 1829 the marble goddesses which complemented the purple curtains and golden coffered ceiling of the luncheonette had held out their victory wreaths across time to scores of Greek enthusiasts from Berlin's vibrant liberal community. These included the Humboldts and members of the Spanish Society like Johann Eichhorn; Peter Beuth; Friedrich Schleiermacher, a liberal theologian; and Ferdinand Friese, Chief of the State Bank. Many of the most prominent progressives in uniform—Ernst von Pfuel, Wilhelm von Scharnhorst, and Adolf von Willisen—were guests too. Within August's richly decorated walls one heard impassioned defenses of Greek independence, brash talk of Prussian alliances with Russia and South Germany, insulting denunciations of Metternich and the ultra-reactionary Charles X, and even harsh accusations against Frederick William. The king, it was alleged, had neglected the war-readiness of the Prussian Army.[34]

The aging sovereign had no urge for territorial acquisitions, wanting most of all to live his final years in peace.[35] This was a worthy principle whose defense placed him in the influential position of broker between the contending factions at home and abroad—and enabled him to avoid being drawn into hateful party politics. By avoiding war, of course, the king also stepped out from under the Damoclean sword of parliamentarism, for no loans would be necessary. The restraint of Frederick William and his Foreign Minister, Christian von Bernsdorff, was an important factor preserving peace during the critical winter of 1828/29.[36] By ending the war in September 1829, the Treaty of Adrianople extended Frederick William's lease on unrestricted political power in Berlin.

It would take more than a Near Eastern crisis to embolden Prussia's reform-minded soldiers to another constitutional initiative. There were some assignments, however, which soothed frustrated spirits and strengthened brotherly ties. Thus Rühle, Müffling, and Rauch served under Gneisenau on the army's first rocket commission. One historian has speculated that the dedication of some officers to the development of new weaponry was essentially a sublimation of political desires frustrated since the late 1810s.[37] One problem with this thesis is that all forms of

[34] For the luncheons, see the citations from the Varnhagen diary in n. 33. For the decor, see Eve Haas and Herzeleide Henning, *Prinz August von Preussen* (Berlin: Stapp Verlag, 1988), 134.

[35] Varnhagen Diary I, 5 January 1828 and 25 March 1828, 5:2, 59.

[36] Baack, *Christian Bernsdorff*, 160.

[37] See Ernst August Nohn, *Wehrwissenschaften im frühen 19. Jahrhundert unter beson-*

technology were under suspicion after the Carlsbad Decrees, making it doubtful that technical work could become a socially acceptable expression of suppressed impulses. Nevertheless, there was clearly some psychological mechanism at work which facilitated withdrawal from a frustrating political world where liberal opportunities were rare. Thus Rühle's writings avoided military and political themes after 1820, turning to the more impressive world of ancient Egypt and Persia.[38] Eduard [von] Peucker, head of the War Ministry's Artillery Division, sought an elusive "peace and quiet"[39] throughout his military career. Similarly, Clausewitz complained after Boyen's resignation in 1819 that little remained "but to withdraw indignantly into one's innermost being,"[40] This escape from ugly political realities produced his best theoretical writings in the 1820s, and in early 1830, facilitated total emersion in technical artillery studies on Prince August's Commission on Science and Technology.[41] Prussia's chief artilleryman expressed shame himself at the kingdom's foreign and domestic policies while maintaining a schedule of inspection tours legendary in its intensity. "Our master," recalled one adjutant, "drove his entourage nearly to death."[42] Perhaps, like Clausewitz, his lengthy workday served therapeutic needs.

There were political areas, however, where one could surface with his views. Thus Gneisenau, Radowitz, Schulze, and Rühle worked closely with Valentini on educational reforms after the latter's promotion to Inspector of Military Schools in 1828.[43] This struggle, in fact, was an old one. Even before the disasters of 1806, Gerhard Scharnhorst's party had criticized the widespread aristocratic belief that character counted for everything, while intellectual endeavors actually dampened a soldier's warlike spirit. The art of modern warfare, rebutted the reformers, demanded leaders with both natural skill *and* higher education. Royal decrees brought the first substantive victories for this newer viewpoint in 1808.

---

derer Berücksichtigung der Wehrtechnik (Münster: Westfälische Wilhelms-Universität, 1977), 346.

[38] "Rühle von Lilienstern," 177, MgFa Freiburg.

[39] Peucker to his son, Eduard, 5 September 1848, FSg. 1/151 Peucker, BA Aus Frankfurt/Main.

[40] Clausewitz to Gneisenau, 26 December 1819, cited in Paret, *Clausewitz and the State*, 281.

[41] For Clausewitz's intense study, see Clausewitz to Gneisenau, 31 July 1830, printed in Delbrück, *Gneisenaus Leben*, 5:593–94.

[42] Cited in Priesdorff, *Soldatisches Führertum*, 3:283.

[43] "Rühle von Lilienstern," 168, MgFa Freiburg; Varrentrapp, *Johannes Schulze*, 558; L. von Scharfenort, *Die Königlich Preussische Kriegsakademie* (Berlin: Ernst Siegfried Mittler und Sohn, 1910), 155; Priesdorff, *Soldatisches Führertum*, 4:58; B. Poten, *Geschichte des Militär-Erziehungs-und Bildungswesens in den Landen deutscher Zunge* (Berlin: A. Hoffmann, 1896), 217; W. Nottebohm, *Hundert Jahre militärischen Prüfungsverfahrens* (Berlin: Ober-Militär-Prüfungskommission, 1908), 23–30.

Peacetime officer candidates had to demonstrate knowledge of history, geography, foreign languages, mathematics, and a host of technical subjects before a formal examining committee. The founding of the War School in 1810, the Artillery and Engineering School in 1816, and special divisional schools in 1818 provided further cultural momentum.[44]

But a conservative backlash set in after the Carlsbad Decrees in 1819. The party around Duke Carl, himself a former member of Scharnhorst's Military Society, managed to exclude the Military Curriculum Commission, headed by Rühle von Lilienstern, and the Examinations Committee, chaired by Gneisenau, from any practical control over the course of study in divisional schools or the officers' exams administered there. The consequence was a partial return to the old system. Where high academic standards were honored—as in the War School, the Artillery and Engineering School, and some of the divisional schools—officers received an education; where aristocratic bearing was valued above all else—as in most of the army's eighteen divisional schools—almost anyone could pass.[45]

Frederick William had little sympathy with these conditions. The bourgeois king placed a much higher value on education than rustic Junkers, and this opinion received added weight after Witzleben took over the military cabinet in 1817. Only peace, fiscal restraint, and the monarch's desire to be left alone allowed this blatant disregard of the reformers' educational criteria to affect policy.[46] After the first significant budget surplus in 1827—and a beckoning European war in 1828—Frederick William's small reserve of patience expired.

Rühle, Schulze, and Gneisenau now prevailed with a drastic restructuring of the divisional schools. Stiff entrance tests were required and the remedial curricula typical of most of these institutes abolished. Demanding course work would prepare officer candidates for a grueling set of qualifying exams. The history-politics requirements alone were humbling: Ancient Civilizations of the World; Ancient Mediterranean to Augustus Caesar; Medieval Europe; Modern Europe; France after Louis XIV; Prussia 1415–1701; Prussia 1701–Present; Wars of Recent Times;

[44] Demeter, *German Officer-Corps*, 63–73; Nottebaum, *Hundert Jahre*, 1–23; Poten, *Geschichte*, 219; Bernhard Schwertfeger, *Die grossen Erzieher des deutschen Heeres: Aus der Geschichte der Kriegsakademie* (Potsdam: Rütten & Loening, 1936), 13–25; and Nitschke, *Preussische Militärreformen*, 123–27.

[45] White, *Enlightened Soldier*, 39; Nottebaum, *Hundert Jahre*, 24–27; Poten, *Geschichte*, 210–14.

[46] Dorow, *Job von Witzleben*, 68; Nottebaum, *Hundert Jahre*, 29; Poten, *Geschichte*, 211. Also see Eduard [von] Peucker, *Denkschrift über den geschichtlichen Verlauf welchen die Vorschriften über das von den Offizier-Aspiranten darzulegende Mass an formaler Bildung und die Erfolge dieser Vorschriften seit der Reorganisation des Heeres im Jahre 1808 genommen haben* (Berlin: R. Decker, 1861), 27–28.

and Political and Cultural Geography of the Modern World. Cadets also had to master mathematics, German, French, and military science. The king approved the new system in June 1828 and Georg von Valentini administered its implementation a year later.[47]

Behind the reform lay an even more radical design conceived by Rühle and Schulze in 1828. Prussia's divisional schools were sorely unprepared for the pedagogical task suddenly thrust upon them. In fact, both men regarded the new curriculum as merely a preparatory measure for another project first discussed in Scharnhorst's day. If Prussia were to survive in modern times, its army must become one with its people—a nationalized military in a militarized nation. Boyen's national militia (*Landwehr*) remained one promising means of achieving this goal. A merger of military and civilian schools was another. Thus Rühle and Schulze proposed creation of a special army teaching faculty which would supplement instruction in the gymnasiums and universities. The divisional schools, even the elite war schools themselves, would become superfluous. As armed conflagration in Europe neared, Valentini and the Minister of War, Georg Leopold von Hake, supported the "Teutonic Jacobins."[48]

Hake made no formal proposal in the Ministerial Cabinet, but may have broached the issue to Witzleben for discussion with the king. We know only that "opposition of a varied sort"[49] thwarted the plan. The War Minister and his compatriots undoubtedly became embroiled in a complicated controversy. Duke Carl, Count Wittgenstein, General Knesebeck, and other leading conservatives were as opposed to popularizing military education as they had been to the national militia. Inclusion of Prussia's gymnasiums in the proposal made it doubly objectionable to these aristocrats, for many liberal schoolmasters used the study of Greek mythology and reading of Polibius and Livy to preach about the superiority of shared power and inevitable death which came to tyrants. Even advocates of constitutionalism like Clausewitz objected to this politicizing in the classroom, mocking professors "who knock about with a few Greek and Latin authors and have their heads full of ancient freedom and an-

[47] Poten, *Geschichte*, 214–18; Nottebaum, *Hundert Jahre*, 30.

[48] Schwertfeger, *Die grossen Erzieher*, 65; Varrentrapp, *Johannes Schulze*, 558–59; "Rühle von Lilienstern," 169, MgFa Freiburg.

[49] "Rühle von Lilienstern," 169, MgFa Freiburg. The protocols of the Ministerial Cabinet for the late 1820s (Rep.90a, B.III.2b, Nr.6, Bd.15–18, ZStA Merseburg) show no discussions of this proposal. In general for the present paragraph and the next, see Varrentrapp, *Johannes Schulze*, 323–32, 410–15, 429–30; Karl-Ernst Jeismann, *Das preussische Gymnasium in Staat und Gesellschaft* (Stuttgart: Ernst Klett, 1974), 258–63; Friedrich Paulsen, *Geschichte des gelehrten Unterrichts* (Berlin: Walter de Gruyter, 1921), 318–23; Petersdorff, *Friedrich von Motz*, 2:350–52.

cient constitutions which they do not understand."[50] Conservatives preferred practical, unpolitical, unphilosophical subjects like math and science to the classics—provided that the technical instruction was combined with strict religious training.

Such conclusions led to an unusual alliance against the exponents of classical education. As the Ministry of Education and Ecclesiastical Affairs' foremost advocate of classical secondary education, Johannes Schulze had made very few concessions to adherents of vocational education (*Realbildung*) like Motz, Kunth, and Leopold von Bärensprung, the Mayor of Berlin. These men objected to the study of classical languages as an unnecessary and useless curriculum for aspiring businessmen, proposing that the state found special "vocational gymnasiums" (*Realschulen*) to teach math, science, and modern languages. The political scales began to tip against Schulze, however, when the court conservatives spoke out in favor of vocational secondary curricula. The allies received prestigeous support in 1827 with the return to Berlin of Alexander von Humboldt. For the country's most famous scientist mounted a concerted campaign to convince Frederick William to accept modern schooling. The king had his own political suspicions of the classicists and was impressed with the seeming practicality of the new educational approach. In 1829, therefore, he ordered the Ministry to accelerate the growth of vocational schools. So it was that Hake's overtures about militarizing the classical gymnasiums were destined to get a cold reception.

A modified version of the plan was accepted, however, by the Artillery and Engineering School. Under the proposed reorganization, all connection with the substandard divisional schools would be discontinued and applicants required to certify completion of the second form of a gymnasium. Prince August and his adjutant, Radowitz, were sensitive to Rühle von Lilienstern's argument that Prussian officers needed a classical, humanistic education to ensure the respect of subordinates or facilitate a later career change.[51] The two also shared Rühle's liberal political agenda.[52] However, like Peter Beuth (see chap. 3), Prince August and Radowitz tempered their enthusiasm for the classics with a practical appreciation of technical subjects. Along with history, geography, and foreign languages, cadets would take a battery of advanced courses in math, physics, and chemistry during a tour of study which was extended from

[50] Carl von Clausewitz, "Umtriebe," n.d. (1822/23), printed in Carl von Clausewitz, *Politische Schriften und Briefe* (Munich: Drei Masken Verlag, 1922), 166.

[51] See Poten, *Geschichte*, 400–403, where Radowitz and Prince August paraphrase an earlier statement by Rühle, printed in "Rühle von Lilienstern," 168, MgFa Freiburg.

[52] For Radowitz's liberalism, see Priesdorff, *Soldatisches Führertum*, 6:225. For Prince August, see above, n. 6.

two to three years. The essential portions of the plan were approved in the winter of 1829/30.[53]

Constitutionalism and humanistic curricula were linked very closely to a third cause dear to the kingdom's military reformers. With few exceptions, these men had fought for Prussia and remained loyal to her. But they were also devoted to the concept of "Germany" and disliked the reactionary, Austrian-led confederation created at Vienna in 1815. Coupled with an enlightened constitution, Prussia's military might, economic strength, and excellent school system would offer South Germans an attractive alternative. For if it came to a fraternal struggle with Austria or another Punic War with France, Prussia would need an alliance with Russia as well as the added security of a military league of friendly German states.[54] However, with the king very reluctant to provoke Austria and adamantly opposed to any policy which threatened the peace, these matters could only proceed slowly and indirectly.

Such an opportunity arose in the tense winter of 1828/29. Württemberg and Bavaria had expressed the desire for a tariff union with Prussia unofficially during the 1820s, but Motz, Prussia's principal trade minister, had been unresponsive. Somewhat of a gambler, the Minister of Finance felt that Prussia should consolidate its position in the north by forging treaties with Nassau, Electoral Hesse, and the Grand Duchy of Hesse, then wait for a more formal offer from the south. When the Grand Duchy of Hesse signed a trade agreement with Prussia in February 1828, Bavaria fumed and Württemberg contented itself with semiofficial visits to Berlin by Johann Friedrich Cotta von Cottendorf in September and November of 1828.[55]

The newspaper mogul and vice-president of the Württembergian second chamber found Motz and Witzleben more accommodating, preoccupied as they were with the escalating Near Eastern Crisis. Motz noted that some units of the army were already on war alert, that Saxony was leaning toward Austria, and suggested that Bavaria and Württemberg befriend Prussia. "The most suitable [means] to this [friendship] would be

[53] Poten, *Geschichte*, 397–405.

[54] For expressions of anti-Austrian and pro-Russian sentiment, see Witzleben to Natzmer, n.d. (1821), cited in Natzmer, *Unter den Hohenzollern*, 2:88–89; and Varnhagen Diary I, 26 January 1828; 5 March 1828; 25 March 1828, 31 March 1828, 5:14, 43, 57, 59, 61. For closer ties with South Germany, see "Rühle von Lilienstern," 170, MgFa Freiburg; Treitschke, *Deutsche Geschichte*, 4:216; Dorow, *Job von Witzleben*, 73–74; Meinecke, *Das Leben*, 2:458; Dehio, "Wittgenstein und das letzte Jahrzehnt," 217; and Baack, *Christian Bernstorff*, 134–35.

[55] For the tariff talks of 1828/29, see Petersdorff, *Friedrich von Motz*, 198–202; Baack, *Christian Bernstorff*, 128–33; and Daniel Moran's more recent *Toward the Century of Words: Johann Cotta and the Politics of the Public Realm in Germany, 1795–1832* (Berkeley: University of California Press, 1990), 227–37.

a closer understanding on trade policies."[56] Witzleben was even more specific. Prussia wanted no territorial acquisitions in Germany, but was committed to the defense of the Fatherland. This role rightly fell to Prussia with its respect for "justice and public opinion" and belief in man's "free moral and intellectual development."[57] If Bavaria and Württemberg entered into an economic agreement with their sovereignty unattenuated, the foundation would be laid for closer ties. Negotiations began that winter, and in May 1829, south joined north in a tariff union. Rumors circulated that a military pact would soon follow.

A Prussian–South German–Russian alignment lost some urgency with the Treaty of Adrianople in September 1829—but gained it back quickly after Charles X appointed Jules de Polignac premier. Nervous eyes had trained on France for two years as Charles X moved from one ugly confrontation to another with angry liberals in the National Assembly.[58] Fear of French revolution and the wars that usually followed mounted throughout the winter of 1829/30. For it seemed doubtful that domestic peace could endure in a country where the king was an absolutist, the premier believed he could converse with the Virgin Mary, and parliament was packed with liberal anticlerics. The domestic confrontation in Paris turned bloody in July 1830.

In Metternich's memorable phrase, Europe "caught cold" in the seven months following the July Days. Emboldened by the contagious success of revolution in France, patriots rebelled in Belgium, Hanover, Brunswick, Electoral Hesse, Saxony, Poland, and Italy. European armies were mobilized, demobilized, and mobilized again in the two years which followed. There was no shortage of bellicose rhetoric in France about pushing its natural frontier to the Rhine, or in eastern Europe about saddling horses to suppress pernicious western revolution. On two occasions, in fact, local fighting nearly escalated into all-out war. One occurred during the spring of 1831 when Russian forces invaded Poland—and the French king ignored angry demonstrations for intervention. The other took place in the fall of 1832 when a French army besieged Antwerp—and the eastern sovereigns acquiesced. Europe's crisis did not pass until March 1833.

During the early months of unpredictability and rising tension, Berlin pulsated with political emotion. Friedrich August Stägemann was taken aback by the "loud cries of opposition" which evoked shrill "counter-

[56] Cited in Petersdorff, *Friedrich von Motz*, 2:204.

[57] Cited in Baack, *Christian Bernsdorff*, 134.

[58] For examples of Prussian nervousness over French developments, see Gneisenau to Clausewitz, 11 November 1828, and Clausewitz to Gneisenau, 20 August 1829, printed in Delbrück, *Gneisenaus Leben*, 5:553–54, 567. In both letters references are to the general state of opinion in Berlin.

cries"[59] for a legitimist crusade. Heinrich von Brandt, a young protégé of Valentini who taught at the War School, recalled similarly that 1830 began a "total transformation"[60] to the midcentury style of party politics. Varnhagen von Ense had begun to trace this politicization, polarization, and party factionalism as soon as he came to Berlin in 1820, but he, too, was sensitive to a new spirit. "Everyone ran after the latest news," he wrote, and "few felt the need to hold back their opinions. Admittedly," he added, "one sought out mostly only those with a like mind."[61] For Varnhagen this meant the progressives at Steheli's Café or at home with a "dear circle of like-minded friends"[62] which included Friedrich Hegel, Eduard Gans, and Alexander von Humboldt. Brandt's memoirs also testify to the intimate connection between friendship circles and the emerging party-oriented politics. During the anxious winter of 1830/31 he avoided "the salons of the upper estates"[63] for smaller get-togethers with liberal friends like Louis Blesson and Franz von Cyriaci, editors of a successful military journal. Brandt also kept frequent company with friendly cliques around Valentini, the Helvigs, Beuth, and Schinkel, Beuth's dear friend.

For Brandt, these social evenings with close friends seem to have been a welcome haven from the heavier atmosphere of intrigue which prevailed at some of the salons and cafés.[64] Small private gatherings, however, were themselves actively political affairs on occasion. Thus Hermann von Boyen hosted numerous dinners to discuss his political agenda with other generals. Boyen had always regarded parliamentary reforms

[59] Stägemann to Cramer, 20 September 1830, printed in C. A. Varnhagen von Ense, *Briefe von Chamisso, Gneisenau, Haugwitz, W. von Humboldt, Prinz Louis Ferdinand, Rahel, Rückert, L. Tieck u.a.*, edited by Ludmilla Assing (Leipzig: F. A. Brockhaus, 1867), 2:198.

[60] Brandt, *Aus dem Leben*, 2:54.

[61] Varnhagen Diary I, August, 1830, 5:298.

[62] Varnhagen to Heine, 16 February 1832, and to Prince Pückler-Muskau, 29 May 1836, printed in K. A. Varnhagen von Ense, *Kommentare zum Zeitgeschehen: Publizistik, Briefe, Dokumente, 1813–1858*, edited by Werner Greiling (Leipzig: Reclam, 1984), 81, 117. For Steheli's Cafe, see Treitschke, *Deutsche Geschichte*, 4:207; and Brandt, *Aus dem Leben*, 2:37, 56.

[63] Brandt, *Aus dem Leben*, 2:55.

[64] See Brandt's remarks in *Aus dem Leben*, 2:42, 51. See also Gordon A. Craig, *The Politics of the Prussian Army 1640–1945* (New York: Oxford University, 1955), 80. Craig cites an 1821 essay by Brandt's friend, Louis Blesson, who wrote that professional soldiers had no right to share in the politics of the day because "the deliberating soldier is really no longer a soldier but a mutineer." The fact remains, however, that Blesson and Brandt were liberal politically and moved in these circles. See Brandt (pp. 15–16, 17, 45); and the criticism by Baron de la Motte Fouque, an ultraconservative, of Blesson's proud assertion—made in the same work cited by Craig—that the Romans were "in favor of freedom" [*Zeitschrift für Kunst, Wissenschaft und Geschichte des Kriegs* 63 (1856)] (hereafter cited as Zeitschrift-Krieges, MgFa Freiburg).

as one of the most important weapons in Prussia's arsenal and believed adamantly that political changes should be made before war broke out. Having made little or no impression on the king with a private memorandum in August 1830, the aging "Frondeur"—as Brandt dubbed the old friend of the reform party—was probably attempting to enlist the support of more influential compatriots like Krauseneck and Rühle von Lilienstern who agreed that Europe's fragile peace was the moment to exploit liberalism's first good opportunity in a decade.[65]

Whether they were sanguine about the chances of parliamentary reform or not, consensus reigned among Prussia's reformist brothers that the nation should maintain a defensive posture and look to South Germany and Russia as principal comrades-in-arms. Such a position was consistent with the belief that the country would be weaker in wartime without the benefit of this "soft" military technology. As revolutionary fires spread that first fall, Grolman and Prince August were the only real hotspurs. Both expressed shame that Prussia's army had not marched westward, even though each had a different enemy in mind. Thus Grolman disagreed vehemently with Krauseneck and Valentini at a small dinner which Gneisenau hosted in late October. The arrogant triumvir blurted out his desire to fight for the Belgian revolutionaries against Holland. A few weeks later, Prince August joined the other princes in criticizing Witzleben before the king at dinner. Ever mindful of the need to keep the king's trust,[66] Witzleben calmly defended Frederick William's peaceful policies toward France and recognition of the new French regime. Nor did he harbor grudges against his outspoken friends. If confederate troops took the field, Witzleben wanted Prince August in command of a Prussian–South German force on the middle Rhine with Grolman as the liberal artillerist's chief-of-staff.[67]

Clearly the insurrectionary nature of the situation made it wise to place advocates of change in key positions. Witzleben, chief of the Personnel Department of the War Ministry since 1825, made good use of his second influential position to accomplish this goal. It was no coincidence that the controversial Police Chief of Cologne was sacked or that Ernst von Pfuel was named Commandant of Cologne-Deutz. Political appearances also

[65] For Boyen's dinners and agenda, see, respectively, Brandt, *Aus dem Leben*, 2:42; and Meinecke, *Das Leben*, 2:437, 476. For Krauseneck and Rühle in 1830, see Paret, *Clausewitz and the State*, 401.

[66] See Varnhagen Diary I, 31 March 1828, 5:61; and Clausewitz's observations in a memoir describing the winter of 1830/31, printed in Schwarz, *Leben*, 2:303, 308, 313.

[67] Brandt, *Aus dem Leben*, 2:42; Gneisenau to Clausewitz, 18 August, 30 October, 17 November, and 22 November 1830, printed in Delbrück, *Gneisenaus Leben*, 5:604, 613, 621, 627; Clausewitz to his wife, Marie, 9 June, 23 June, and 11 September 1831, printed in Linnebach, *Karl und Marie von Clausewitz*, 446–47, 455–56, 494; Paret, *Clausewitz and the State*, 400–405; Baack, *Christian Bernsdorff*, 188.

explain why Prince William (the king's brother) was hurried to the same city as military governor of Rhineland-Westphalia. The post was "just as important as it was an honor," wrote Witzleben to William, for simply the news of the popular prince's appointment would make "a most blessed impression."[68] In December 1830, moreover, the king approved Witzleben's recommendation that Gneisenau and Clausewitz receive command of the four army corps assigned to guard Prussia's eastern frontier against Polish rebels. Finally, the delicate task of negotiating South German military participation along the western borders went to Rühle von Lilienstern, a man whose liberalism was well regarded in Munich, Stuttgart, Karlsruhe, and Darmstadt.[69]

Like Boyen, however, Prussia's brothers in arms wanted more than liberal appearances. Indeed, the tense winter of 1830/31 offers an excellent opportunity to observe the workings of Prussia's "autocratic-democratic" party structure. In January, the fourth estate of the Westphalian Diet, led by Friedrich Harkort, owner of a machine-making factory in Wetter, raised the issue of Frederick William's parliamentary promise of 1815. The first estate, led by Count Landsberg, recoiled from the idea of publicly pressuring the king. Eager to advance aristocratic causes in Berlin, however, Landsberg rallied the Diet around a motion to ask Prince William, the new military governor, to broach the matter of a parliament privately with his brother. Karl von Müffling, Commander of the Westphalian Corps, was similarly opposed to the "scandalous"[70] initiative of Harkort's party, but the old Teutonic Jacobin had always favored the same goal. So he wrote to Prince William, the court liberal, encouraging him to accept the Diet's offer. Müffling also wrote to his new party friends in Berlin, chastizing them for denying a provincial diet's competence to raise national issues during an emergency and for mocking Harkort as a "Pumpernickel Lafayette."[71] Prince William, like Müffling, had

[68] Witzleben to Prince Wilhelm (the king's brother), 24 September 1830, Prince William Papers, D-22, Nr.29–1, Bl.6–7, HeStA Darmstadt.

[69] For the Police Chief's sacking and Pfuel's appointment, see Theodor von Rochow-Briest, Prince William's adjutant, to Prince William, 12 September 1830, Prince William Papers D-22, Nr.13–123, Bl. 89, HeStA Darmstadt; and Gersdorff, *Ernst von Pfuel*, 85; for the eastern command, see Gneisenau to Clausewitz, 4 December 1830, printed in Delbrück, *Gneisenaus Leben*, 5:631; for Rühle's mission, see "Rühle von Lilienstern," 170, MgFa Freiburg, and especially Baack, *Christian Bernsdorff*, 265–74; for Witzleben and army personnel, see Rudolf Schmidt-Bückeburg, *Das Militärkabinett der preussischen Könige und deutschen Kaiser: Seine geschichtliche Entwicklung und staatsrechtliche Stellung 1787–1918* (Berlin: E. S. Mittler, 1933), 32.

[70] Müffling to Prince William, 9 January 1831, Prince William Papers, D-22, Nr.29–2, Bl.22–23, HeStA Darmstadt. For Harkort and developments on the Diet, see Bergengrün, *David Hansemann*, 118–19.

[71] Müffling to Prince William, 18 February 1831, Prince William Papers, D-22, Nr.29–3,

no party ties to Harkort. While still two days' ride from Cologne in late December, for instance, he met with a deputation from the Diet and lectured them about "how thoroughly inappropriate it is to remind the king of a promise which he would also fulfill, because he made it."[72] After hearing from Müffling, however, the prince asked Witzleben to find a propitious moment to inform the king about the Diet's actions. Frederick William knew all too well, of course, that his brother felt a monarch *should* fulfill a promise once he had made it. Witzleben quickly accommodated his friend,[73] for such incidents provided the crafty general with another excellent opportunity to finesse the king. Indeed, there is a great deal of evidence which suggests that Witzleben was already deeply involved in a constitutional intrigue of the riskiest sort.[74] "One cannot demand of Witzleben, with his thoroughly liberal schooling," wrote Leopold von Gerlach, "that he should resist such masters as the two Humboldts, Grolman and Krauseneck."[75] From Münster to Berlin parties were working across the divide of state and society toward a common end goal.

But the man closest to the top had to be careful. Thus Clausewitz and Krauseneck met over dinner frequently that first winter to arrange for Gneisenau to be given near-dictatorial powers.[76] Clausewitz eventually came to the realization, however, that neither the king nor Witzleben "felt so strongly pressed by the times [that they had] to place themselves

---

Bl.49, HeStA Darmstadt. The "Pumpernickel Lafayette" remark was made by the Crown Prince, quoted in Bergengrün, *David Hansemann*, 119.

[72] Prince William to Witzleben, 12 January 1831, Rep.92 Witzleben, Nr.100, Bl.6–7, ZStA Merseburg.

[73] Witzleben to Prince William, 20 January 1831, Prince William Papers, D-22, Nr.29–1, HeStA Darmstadt.

[74] Dorow, *Job von Witzleben*, 67, claims that Witzleben, always approached because of his influence with the king, became embroiled politically after the Belgian Revolution. Schöning's letter to Natzmer (n.d. [early 1834], printed in Natzmer, *Unter den Hohenzollern*, 2:77) makes a similarly general assertion. More specifically, it is known that Witzleben and Wilhelm von Humboldt worked together in the winter of 1830/31 to soften the proposed revisions of Stein's Municipal Ordinance of 1808 (see Duke Carl to Wittgenstein, 26 January 1831, B-P Hausarchiv, Rep.192 Wittgenstein, V.1.17, GStA Berlin). Clausewitz also observed the close political relationship between the two men (see his memoir of the winter of 1831, printed in Schwartz, *Leben*, 2:308, 311). Moreover, Leopold von Gerlach heard a rumor (letter to Ludwig von Gerlach, 27 December 1830, printed in Gerlach, *Aufzeichnungen*, 195) that Witzleben and Humboldt were working on a draft constitution. Ludmilla Assing (Varnhagen von Ense, *Briefe von Chamisso*, 1:14) notes the very widespread nature of the rumor, but correctly rejects the notion that Humboldt had a charge from the king.

[75] Leopold von Gerlach to Ludwig von Gerlach, 27 December 1830, printed in Gerlach, *Aufzeichnungen*, 195.

[76] For Krauseneck and Clausewitz, see Clausewitz's memoir of 1830/31, cited in Schwartz, *Leben*, 2:303–15.

under the tutelage of such a guardian" (*um sich einen solchen Vormund zu setzen*). Clausewitz was particularly struck by Witzleben's reluctance to let others "look at the cards" and displace him as the one "at the middle of all military and political negotiations."[77] If Witzleben wanted to protect his position as adviser to the king, it was imperative that he refrain from joining blatantly party-oriented intrigues. With good reason, for Frederick William was growing wary of everything around him. "All truth and faith has disappeared from the world," he said in December. "The times of the Faustrecht have returned—one has to fortify himself in order to be safe."[78]

The timing could not have been worse for a second initiative which emanated from the western provinces. David Hansemann, a leading member of the Chamber of Commerce and City Council in Aachen, penned a lengthy memorandum for the king in December 1830 which advocated political reform as the wisest measure in a time of crisis. Although he was ostensibly writing for himself alone, Hansemann, like Harkort, was highly respected by local businessmen and had previously stepped forward as their spokesman. As his biographer writes, the young entrepreneur intended to convey to Frederick William the political wishes of "the patriotic-minded, educated and independent middle class . . . of the Rhineland."[79] His work reached the king with the aid of Stägemann, "one of the last genial spirits from the great time of Hardenberg, Stein and Scharnhorst," as Hansemann recalled later.[80] Frederick William replied in February 1831 that Minister Schuckmann would take Hansemann's proposals under consideration. That the man from Aachen could report this outcome so enthusiastically to Georg Maassen, a very close party ally in the bureaucracy, is some indication, perhaps, that both knew how quickly Schuckmann's ministry was falling into the hands of a common political friend, Peter Beuth.[81] Indeed, the Chief of the Busi-

[77] For the quotes, see ibid., 2:313, 308, respectively.

[78] The king's statement appears in the Diary of Leopold von Gerlach, 4 December 1830, cited in Leopold von Gerlach, *Denkwürdigkeiten aus dem Leben Leopold von Gerlachs* (Berlin: Wilhelm Hertz, n.d.), 1:59–60.

[79] Bergengrün, *David Hansemann*, 106. For his activity on the Chamber of Commerce and City Council, see pp. 75–81.

[80] Ibid., 118.

[81] For Hansemann's friendly political relations with Maassen and Beuth, see ibid., 59, 70, 81–82, 89, 99, 118. It is also interesting in this respect that Beuth had interceded successfully with Schuckmann on behalf of Hansemann in January 1829, while the latter was presenting the Chamber of Commerce's petition to establish a wool marketing fair (*Wollmarkt*). The decision was made in 1830 to pension Schuckmann (see Rochow-Briest to Prince William, 12 September 1830, Prince William Papers, D-22, Nr. 13–123, HeStA Darmstadt), but for unknown reasons he stayed on as head of a new Ministry of the Interior for Business and Commerce. Beuth, however, was the driving force here.

ness Department was taking every opportunity to convince ministerial colleagues who looked askance at Rhenish developments of "how patriotically Prussian the Rhinelanders were."[82] Hansemann's effort provides another useful glimpse at the cooperation of small cliques or parties bridging state and society in Prussia's authoritarian political culture.

But another example illustrates how doubtful it was that Hansemann had succeeded in influencing the king. That same February Prince William dropped all previous subtlety in his dealings with Frederick William and submitted a lengthy memorandum of his own. In order to assure the loyalty of Rhinelanders and bolster the spirit of Prussian troops during a French invasion which he projected for April at the latest, the people's wishes for a parliament should be granted.[83] The king's reply was discouraging. "No representative constitution will help in the least," he wrote, against "the crazy drive to topple everything which exists." The "most complete proof of this" was supplied by parliamentary countries like England, France, Belgium, and Switzerland, where "things were really the worst [*(wo) es am ärgsten geht*]." Thus "[even] if one could come to say heartfelt things about [adopting] the same [institutions] for here, that which really happens [in the world] fully suffices to bring one back away from this."[84] From Duke Carl to Metternich, conservatives feared that Frederick William was "entirely in the hands of the liberals."[85] They were obviously wrong, for only financial exigencies requiring public borrowing would force this illiberal king to convene a representative assembly.

Such an emergency loomed increasingly closer, however, as crisis followed crisis in 1830 and 1831. Friedrich von Motz had amassed a war chest of 15 million thaler from his annual surpluses of the late 1820s, but with army troop strength raised from 130,000 to nearly 200,000, these funds were quickly exhausted.[86] Only by reducing troop levels in numerous units during the spring of 1831 did Prussia avoid "financial operations which have to be reserved for an actual war," explained Witzleben to Prince William.[87] But with the continued drain of eastern and western mobilizations that summer and fall—troop strength stood at 210,500 in October 1831—high officials began to resign themselves to the seemingly

[82] Wissmann to Stägemann, 29 June 1831, printed in Rühl, *Briefe*, 3:488.

[83] Prince William to Frederick William, 23 February 1831, Prince William Papers, D-22, Nr.29–6, Bl.18–19, 24–27, HeStA Darmstadt.

[84] Frederick William to Prince William, 7 March 1831, Prince William Papers, D-22, Nr.11–1, Bl.133, HeStA Darmstadt.

[85] Duke Carl to Wittgenstein, 28 February 1831, B-P Hausarchiv, Rep.192 Wittgenstein, V.1.17, GStA Berlin.

[86] Stägemann to Olfers, 7 February 1828, printed in Rühl, *Briefe*, 3:391–92; Witzleben to Prince William, 26 May 1831, Prince William Papers, D-22, 29–2, Bl.20, HeStA Darmstadt. For troop levels in this passage, see Wehler, *Deutsche Gesellschaftsgeschichte*, 2:385.

[87] Witzleben to Prince William, 26 May 1831, as cited in n. 86.

inevitable necessity of convening a popular assembly. Christian Rother and Friedrich Stägemann approached Count Lottum, the cabinet minister, with separate plans in August for structuring estates of the realm from the provincial diets and the Council of State, while later that winter Müffling presented him with similar proposals. Not a liberal, Lottum nevertheless could see no other way out of the situation.[88] These conditions help to explain why Peter Beuth, Georg Maassen, and Ludewig Gerhard, Chief Captain of the Mining Corps, proposed in early 1832 that the government borrow money to construct a short railway in the Ruhr.[89] The liberals must have assumed that borrowing was becoming an acceptable means of financing such projects, for no one would have dared to suggest such a thing before the Belgian crisis.

Unfortunately for the adherents of parliamentarism, the liberal hour was quickly passing. The Polish and Italian uprisings had been scourged by early 1832, the German revolutions quarantined, and the new French regime proven less virulent than the Jacobin and Napoleonic contagions which had previously swept Paris. Accordingly, Prussian army strength declined to 155,500 by May 1832. Peace was not yet assured, but war no longer appeared inevitable. Metternich probably sensed this first. During the fall of 1831 the wily Austrian genius began to encourage German statesmen to adopt stringent anti-parliamentary safeguards. The adoption of the Six Articles at Frankfurt in June 1832 confirmed the principles of Carlsbad approved by the Confederation in 1819.

Prussia devised a nonparliamentary solution to its budget woes that same summer. Rother's Seehandlung raised 12.6 million thaler from interest-bearing lottery tickets, lent it to the government, then held 4.6 million back in its own coffers as credit against the loan.[90] Unfortunately, we have no evidence about the origins of this public financial coup or Rother's thoughts about it. As discussed in chapter 6, the fund enabled him to undertake a massive industrial expansion and realize personal goals for the economy which had interested him since the early 1820s. Over these same years, however, Rother had believed that the estates of the realm were the only appropriate answer to Prussia's public financial dilemma. Only months earlier he had come forward with his own parliamentary proposals.[91] He must have been troubled that the Seehandlung lottery had extended the life of absolutism in Prussia.

[88] Rother to Stägemann, 3 August 1831, printed in Rühl, *Briefe*, 3:489; Dehio, "Das letzte Jahrzehnt," 216–19.

[89] Gleim, "Zum dritten November 1888," *Archiv für Eisenbahnwesen* 11 (1888): 800–801.

[90] Schrader, *Die Geschichte*, 10; Radtke, *Die preussische Seehandlung*, 80–81. Radtke adds to the detail provided by Schrader, but it is difficult to accept his rejection of Schrader's claim that the lottery was related to the state's public financial crisis.

[91] Varnhagen Diary I, 22 April 1821, 1:288; see also Rother to Stägemann, 3 August 1831, cited above in n. 88.

But Rother was not the only liberal technocrat whose economic success worked at cross purposes with his stated political goals. In the 1820s, for instance, the surpluses from Mining Corps operations amounted to about 12 million thaler, while the business tax (*Gewerbesteuer*) netted roughly 18 million thaler. Without these monies, Motz would never have been able to amass his war chest of 15 million thaler. This fund, together with Seehandlung loans and continued surpluses from the Mining Corps and the business tax, accounted for around 32 million of the 35.2 million thaler of extraordinary military expenditures during the crisis years of 1830/32.[92] When the French besieged Antwerp in November 1832, the state possessed adequate funds to weather the last war scare of the persistent Belgian crisis.

By the autumn of 1832, in fact, Prussia's liberals were in complete disarray. Rühle von Lilienstern's mission to South Germany figured prominently in this party-political collapse. A copy of Foreign Minister Bernstorff's memorandum to the king of January 1831 had reached Metternich a few months after Rühle's trip to the southern courts. The document, drafted by the minister's top aide, Johann Eichhorn, advocated liberal press laws, economic unity in Germany, and north-south military cooperation. Metternich promptly informed Wittgenstein of the impropriety of such a document circulating for liberal effect in the South. Bernstorff was soon forced to resign, while Eichhorn narrowly escaped his monarch's wrath. Suspicion for the source of the leak centered on Rühle, who, like Eichhorn, managed to preserve his career, but not his chances for wielding influence under Frederick William III.[93]

More significant was the fall of Job von Witzleben. It is not certain whether Frederick William cooled toward his privy councillor for recommending undesirable constitutional reforms in the midst of a dangerous crisis. Witzleben himself seemed to sense his political vulnerability in early 1833, for he retracted his support for such schemes.[94] But it may have been too late. In October 1833 he was reassigned to head the War Ministry. "Many say Witzleben's influence has lessened, that he was shown he was not almighty," wrote one of Oldwig von Natzmer's contacts at court. "His hasty, liberal dealings in important matters of state were

[92] For the extraordinary expenditures (35.2 million thaler), see Wehler, *Gesellschaftsgeschichte*, 2:385. Business tax proceeds rose from 1.6 million thaler in 1824 to 2.1 million in 1830 (Treitschke, *Deutsche Geschichte*, 2:466) and probably totaled around 5 million in the more depressed period 1830–32. For Mining Corps surpluses in 1826 and 1827 (1.5 million thaler per year), see chap. 4, n. 26. Corps surpluses in 1830–32 were probably somewhat less (1.33 million thaler per year, or a total of 4 million). For the Seehandlung loan (8 million thaler) and Motz's war chest (15 million thaler), respectively, see above, notes 90 and 86 (Stägemann letter).

[93] Dehio, "Das letzte Jahrzehnt," 220–26.

[94] Ibid., 226.

censured by the discreet conservatives, Lottum and Wittgenstein. Others . . . say the king did not want to let [Witzleben] get far away."[95] Witzleben's appointment as War Minister may therefore have been a compromise solution so typical of Frederick William. But there can be no doubt that Wittgenstein emerged as the king's closest adviser after the early 1830s—and that Witzleben's influence quickly ebbed.[96]

The only real consolation for these cliques of civilian and military reformers was the founding of the Zollverein in January 1834. Here, at last, was substantial progress toward German unity. Beyond the political-diplomatic-military plane, however, the Zollverein was cause for divisiveness. While appreciating its many noneconomic advantages, those around Motz also saw the customs union as an agent of commercial expansion, industrial growth, and technological advance (see chap. 1). To Ludwig Gustav von Thile, on the other hand, the low tariffs of the Zollverein were a Godsend because they helped agriculture and hurt industry—that "cancerous growth on the land."[97] Thile's liberal brothers in arms placed much more trust in constitutions than he did. As we shall see in chapter 5, however, there was somewhat more common ground when issues of capitalism, industry, and technology were under discussion.

The consolation of the Zollverein could not counterbalance the other political setbacks, of course, nor could it soothe a fraternity debilitated by personal tragedies. Cholera, so often an ugly camp follower in wartime, spread north from the Russo-Turkish conflagration to Poland and Posen in 1831. There it claimed the lives of two inseparable friends, Gneisenau and Clausewitz. A few years later Valentini and Braun died, beginning the depressing process of natural attrition for a generation which had reached manhood before the turn of the century. Just as devastating were the losses in civilian ranks during the first half of the decade. Motz, Kunth, and Maassen passed away, as did Schleiermacher and Wilhelm von Humboldt. "Friend Humboldt has died," wrote Beuth sadly to Rother.[98] He knew that no replacements could be—or were being—found.

[95] Schöning to Natzmer, cited in Natzmer, *Unter den Hohenzollern*, 2:77.

[96] Dehio, "Das letzte Jahrzehnt," 228, argues convincingly that Wittgenstein's star rose after Witzleben's appointment. On Wittgenstein's influence with the king in the late 1830s, see also Rochow, *Vom Leben*, 263, 278.

[97] Cited in Alphons Thun, *Die Industrie am Niederrhein und ihre Arbeiter* (Leipzig: Duncker & Humblot, 1879), 190. See also ADB 38:28–32.

[98] Beuth to Rother, 8 April 1835, Rep.92 Rother, Ca Nr.17, Bl.20, ZStA Merseburg.

III

The Aesthetic View from Pegasus

THE SKIES OF LONDON were accommodating on August 4, 1826. Presented with an uncharacteristically dry day in England, two Prussians left their rooms at St. Paul's Coffee House for one last walk through the city. The promenade took them to Temple Bar and Lincoln's Inn Fields to scrutinize the Palladian architecture of Inigo Jones. The first visitor was Carl Friedrich Schinkel, the leading architect of Prussia, whose grand creations had already transformed Berlin from its prewar drabness. The second was Peter Beuth, Schinkel's bosom friend and kindred spirit. The business trip about to end had taken them to France and Great Britain, the industrial powers of western Europe. Beuth and Schinkel would board the steamer "Lord Wollesly" in the morning for Calais, and thence, after a fortnight's coach ride through the developing regions of Belgium and western Germany, to Berlin. To be fresh for this arduous trek, they turned, strolled back to their guest house, and retired early.[1]

The homeward leg of the journey afforded ample time to exchange impressions about their four months abroad—and perhaps also to dream aloud. Schinkel loved to experiment with new lines and styles, but western influences would not deter him from completing his latest and most ambitious project, the Royal Museum of Art, as a neo-Greek edifice. Beuth, a lover of art himself, looked to his own special future for Prussian industry.[2] Schinkel was so captivated by his friend's fantastic descriptions that he later put it down on canvas. Prussia's great builder painted a prosperous industrial colony serviced by roads, navigable rivers, and quiet canals, all tucked away somewhere in the German foothills. Soaring far above the business world below was an Olympian Beuth, seated atop the winged Pegasus, blowing mirthful puffs of pipe smoke into the clouds.

According to mythology, Pegasus had filled the fountain of the Muses on Helicon with one stroke of his hoof. Schinkel knew that Beuth longed

[1] For the details of the last day of their trip, see the Diary of Carl Friedrich Schinkel, 30 July and 4 August 1826, printed in Alfred von Wolzogen (ed.), *Aus Schinkels Nachlass* (Berlin: Verlag R. Decker, 1863), 3:130,134, and for the entire journey, 2:139–51 and 3:1–138. It is also important to note that the factory visits in 1826 were arranged by Beuth, for Schinkel was traveling in another capacity as state architect to observe museum architecture in Paris and London.

[2] The painting of Beuth on Pegasus discussed in this paragraph is reproduced in Matschoss, *Preussens Gewerbeförderung*, 72.

Figure 2. Friedrich Schinkel's painting of Peter Beuth riding Pegasus above a factory town, 1837.

to bring forth a wellspring of industrialization that would be socially, politically, and aesthetically superior to that of the pioneering industrial countries. Manufacturing should be widely dispersed to provincial towns and sites in the countryside, for, unlike the "chamber pots"[3] which some English cities had become, the ideal society possessed an organic harmony between agriculture and industry, country and town. Accordingly,

[3] Beuth to Schinkel, July 1823, printed in Wolzogen, *Aus Schinkels Nachlass*, 3:139.

Beuth's itinerary in 1826 included scores of country factories located in cloisters, castles, and quaint, romantic settings. As far as the industrial worker of the future was concerned, he could do well to follow the lead of disciplined English laborers, but required education to appreciate the improvements which modernity afforded over the harsher, peasant way of life. He would also need training to withstand the dehumanizing effects of factory labor and avoid the "machine wars"[4] which had rocked the first industrial nation. The social misery of England's lower classes bothered Beuth too, but only self-help, not the dole, offered a way out.[5] When Beuth pictured the ideal industrialist, he saw a culturally refined Christian who took an artistic pride in factory buildings, machinery, and products, but had the business acumen and self-confidence to survive in a competitive world without heavy doses of state aid. The ideal state, finally, was a parliamentary technocracy which joined this entrepreneurial elite with highly trained, technologically competent bureaucrats who would lead the way in all things.

Like other visions, the view from Pegasus was never fully realized. As their coach rocked across the Brandenburg plain ever closer to the sober realities of Berlin, doubts must have begun to encroach on dreams—first, and perhaps most irritatingly of all, on parliamentary dreams. Like so many of the young bureaucrats who came up with Hardenberg in 1810, Beuth had relished the coming of a new constitutional order. Then in 1813/14, he served with Wilhelm von Lützow's black-bloused, rough-riding volunteers, an experience which heightened his sense of nationalism, political radicalism, and anti-Semitism. Beuth wanted to open government to the upper middle class into which he was born, divide the nation's tax burden more equally between the bourgeoisie and aristocracy, purge state and economy of members of the Hebrew faith, and incorporate these changes in a written constitution that would enable Prussia to fulfill the Germanic promise of the wars of liberation.[6] When expectations for a constitution dwindled after the Carlsbad Decrees and Hardenberg's death, Beuth appears to have joined many of the chancellor's younger appointees in encouraging the king's close advisor, Job von Witzleben, to take up the cudgel of parliamentary reform.[7]

[4] Beuth to Kramsta, 28 January 1831, Rep.120, D.V.2c, Nr.9, Bd.1, Bl.48, ZStA Merseburg.

[5] For the ideal worker and the dispersion of industry, see Beuth to Baumann, 5 January 1829, Beuth-Baumann Correspondence (N 7/1), sww Dortmund; Beuth to Bülow, 6 December 1823, Rep.120, D.I.1, Nr.12, Bl.25–34, ZStA Merseburg; and Schnabel, *Deutsche Geschichte*, 3:299–302, 353–54, 436.

[6] Varnhagen Diary I, 14 January 1820, 1:52–53; Schneider, *Preussischer Staatsrat*, 191–92; and Beuth to Vincke, 15 June 1815, Vincke Papers A.III.2, StA Münster. For Lützow, Vater Jahn, and anti-Semitism in these circles in general, see ADB 13:662–64, 19:720–22; and Wehler, *Deutsche Gesellschaftsgeschichte*, 2:335.

[7] For the hopes placed in Witzleben by younger bureaucrats, see Varnhagen Diary I, 21

In the meantime, hopes were kept alive—and frustrations vented—in irreverent political clubs like the Spanish Society. There gathered the overlapping circles of friends who were least satisfied with the political status quo: Friedrich Schleiermacher; Georg Reimer; Johann Eichhorn and the other "Germanic patriots"—Karl Friese; Friedrich Link; Christian Rother; Friedrich Stägemann; and fellow "Freemasonic liberals"—Georg Maassen; Peter Beuth; Heinrich Spiker; and like-minded "Anglophiles."[8] At the time of his return from London, Beuth was encouraging another friend, the popular Governor of Westphalia, Ludwig von Vincke, to advance the liberal cause. Thus Stägemann, who placed his own hopes in another common friend, Wilhelm von Humboldt, wrote that "Vincke, aroused, I think, by Beuth, writes a lot, but has little credibility because he sets off his powder too often without real need."[9] Such speculation over likely champions illustrates that despite years of repression, the liberals' spirit remained uncrushed. And with some justification, for Berlin was somewhat freer by mid-decade and, in their eyes, Prussia was only a financial crisis away from parliamentarism. Hadn't Humboldt's close friend, Rother, who knew as much about the status of public finances as anyone in Berlin, doubted their soundness openly in parlor society, assuring his sympathetic listeners that "the estates [of the realm]"[10] were the only answer?

It was more difficult for Schinkel and Beuth to dream about a technocratic order as their coach passed through the gates of Berlin. For nearly twenty years the men of the Business Department had striven to usher the bourgeoisie into a government where technologists already wielded tremendous influence. Hardenberg's instinct for preserving his own authority undermined initial efforts. Fearing a sort of tryanny of expertise,

February 1823, 2:305; that Beuth had close contacts with Witzleben, see Schinkel to his wife, 25 April 1826, printed in Wolzogen, *Aus Schinkels Nachlass*, 2:146.

[8] Friedrich Schleiermacher founded the Spanish Society (see Johannes Bachmann, *Ernst Wilhelm Hengstenberg: Sein Leben und Wirken nach gedruckten und ungedruckten Quellen* [Güttersloh: C. Bertelsmann, 1876–92] 175–78). For the appearance of the persons listed here at meetings of the Spanish Society, see Varnhagen Diary I, 14 January 1820 and 31 December 1823, 1:52–53, 2:459. For the three friendship circles listed here, see, respectively, (1) ADB 5:737–41, 27:709–12, 31:439–48; (2) Adolf Trende, *Im Schatten des Freimaurer- und Judenthums: Ausgewählte Stücke aus dem Briefwechsel des Ministers und Chefs der preussischen Bankinstitute Christian von Rother* (Berlin: Verlag der Deutschen Arbeitsfront, 1938), 21, 42; Friedrich Kleisner, *Geschichte der deutschen Freimaurerei* (Berlin: Alfred Unger, 1912), 178, 180; and Radtke, *Seehandlung*, 19, 22–23 (Friese and Link belonged to the Grand Lodge of the Royal York Friendship, Rother and Stägemann to the Grand National Lodge of German Freemasons, both English versions of Freemasonry. See also below, chap. 6); and (3) Beuth to Schinkel, 1 August 1823, printed in Wolzogen, *Aus Schinkels Nachlass*, 3:148. See also Stägemann to Olfers, 31 December 1825 and 24 January 1826, printed in Rühl, *Briefe*, 3:240, 244.

[9] Stägemann to Olfers, 23 September 1826, printed in Rühl, *Briefe*, 3:267.

[10] Varnhagen Diary I, 22 April 1821, 1:288.

the chancellor squelched a proposal in 1811 that envisioned a plenary body of technical councillors to guide the ministries. The pared-back result was the Technical Deputation, a low-paid body of adjunct bureaucrats who reported to the head of the department.[11]

Undeterred, Beuth returned to the technocratic concept at the end of 1817.[12] Central to his proposal were higher salaries, larger budgets, and an elevated status for the Technical Deputation. These improvements would facilitate the Deputation's mission of accelerating technical transfer and enhance the staff's powers of persuasion in the ministries which affected industrial policy. With austerity budgets becoming the order of the day—and technophobic reactionaries like Kottwitz and Müller gaining credibility in the capital—Beuth clearly wanted to strengthen his hand in the ministerial citadels.

But the district governments (*Regierungen*) were important too. In order to ensure even greater influence there, it was necessary to train a cadre of technological experts in Berlin and systematically promote them through the lower rungs of the state. Aspiring technocrats could supplement their university study of cameralism with one or two years of specialized instruction under the Technical Deputation. Trainees would gain valuable insights into the mechanical principles behind the new technologies, the complexity of patent decisions and the difficulty of selecting the most appropriate technologies for Prussia's needs. The completion of these studies would lead to an apprenticeship in a district government. There the young bureaucratic apprentices could probe deeper into the world of machinery, business, and government before promotion to district councillor for industry (*Gewerberat*).

The industrial councillor's job required a constant canvassing of the district and listening to the experiences, advice, and complaints of entrepreneurs, while offering information and counsel in return on the latest scientific and technological advances. Beuth recommended that the industrial councillor be drawn into the work of other departments "whenever matters of national welfare, industry, and trade"[13] were at stake, thus enabling him to exert the same controlling influence on industrial policy in the district government that the Technical Deputation would wield in Berlin. In the most quickly industrializing districts, he proposed that industrial councillors advance rapidly to presidential rank. In short,

[11] An attempt by the Ministry of the Interior in 1809/10 to establish a plenary body of technical experts which would function at the ministerial level was rejected for these reasons. See the proposal of Geheimer Staatsrat Bose, n.d. (1810), and Hardenberg's response of 15 March 1811, in Rep.74 H.XV, Nr.2, Bl.10–23, 29, ZStA Merseburg.

[12] Beuth to Bülow, n.d. (December 1817), Rep.120 A.II.1, Nr.1, Bd.2, Bl.33–36, ZStA Merseburg.

[13] Ibid.

Beuth envisioned a bureaucracy permeated with technocrats who would steer the nation slowly to industrial manhood.

"It is always highly useful," he wrote a few years later, "to lay a solid foundation in the affairs one pursues."[14] Beuth was undoubtedly on *terra firma* within the new Ministry of Business and Commerce, for its chief, Hans von Bülow, quickly recommended the proposals as "reasonable"[15] to the king in January 1818. That the minister waited more than a year before receiving a reply was good indication, however, that Beuth's ability to manipulate affairs did not extend beyond Bülow. Indeed, Frederick William, and behind him, Hardenberg, were still clearly opposed to the technocratic state: the Technical Deputation could serve in an "instructive, advisory" capacity "but never participate in the administration [of the nation]."[16] Nor could district governments comply with Beuth's recommendations without creating artificial divisions between departments—or worse, placing industrial councillors in a "privileged position"[17] over their colleagues. A budget increase for the Deputation and some key personnel changes were the only substantive results.

The reasons for this setback were more complicated, however, than the logical power instincts of king and chancellor. For Beuth's campaign to construct a technocratic state was part of a developing cultural struggle within the bureaucracy itself. Beginning in the late 1700s, young Prussians started flocking to German gymnasiums and universities in the hopes of securing lucrative civil service positions in their rapidly expanding kingdom. The surge to institutions of higher learning resumed after 1815 as a result of new territorial acquisitions in western Germany and stiffer educational requirements. Between 1820 and 1830, the number of students acquiring "cultural currency" (Collins) in Prussian universities rose from 2,450 to 4,949. The largest group studied theology (41.3%), hoping for respectable posts in seminaries, parish ministries, or the church administration. Law students (33.4%) eagerly paid much higher premiums for the more influential positions in Prussia's courts as well as in the prestigeous central bureaucracy. Opting for legal studies was a rational decision as long as bureaucratic entrance exams tested in law. Challenging law's privileged position at the top of the state, however, were the newer, rising credentials of medicine (11.4%) and philosophy

[14] Beuth to Baumann, 9 November 1830, Beuth-Baumann correspondence (N7/1), sww Dortmund.

[15] Bülow to Frederick William, 24 January 1818, Rep.120 A.II.1, Nr.1, Bd.2, Bl.37–47, ZStA Merseburg.

[16] Frederick William to Bülow, 4 March 1819, Rep.120 A.II.2, Nr.1, Bd.2, Bl.51–52, ZStA Merseburg. For Hardenberg's views on technocracy, see Hardenberg to Schuckmann, 15 March 1811, as cited in n. 11.

[17] Frederick William to Bülow, 4 March 1819, as cited in n. 16.

(13.8%)—including philology, history, natural science, and Beuth's subdiscipline of cameralism. But his proposals for expanding the cameralist's role in government were destined to founder against the objections of the arrogant and far more numerous lawyers.[18] Technological advice was regarded as helpful to those whose training in jurisprudence gave them the broad, global perspective required to interpret and make intelligent use of expert advice, but was not deemed worthy of the special importance which the technologists themselves attached to it.

Nevertheless, Beuth persisted in trying to shape a new bureaucratic tradition. After repeated complaints by district presidents in the early 1820s that their councillors knew the law, but were ignorant of factories and technology, Beuth again proposed that young officials study in Berlin under the Technical Deputation. Unable, and probably unwilling, to finance such a costly educational program, the conservative Minister of the Interior, Friedrich von Schuckmann, notified the governors of Brandenburg, Upper Silesia, Westphalia, and the Rhineland in April 1826, that each aspirant's year in Berlin would have to be self-financed.[19] Given the low salaries of most young bureaucrats, Beuth could not have been very surprised to learn after his return from England that none had yet applied.

. . . . .

By the autumn of 1826, Schuckmann and the "Mecklenburg Coterie" were another source of unpleasantness for the men of the Business De-

[18] For the percentages, see Wehler, *Gesellschaftsgeschichte*, 2:214, 220. For works which stress the fragmentation of the cultivated bourgeoisie, see Lenore O' Boyle, "Klassische Bildung und soziale Structur in Deutschland zwischen 1800 und 1848," *Historische Zeitschrift* 207 (December 1968), 584–608; Gillis, *Prussian Bureaucracy*, 22–39; Robert M. Bigler, "The Social Status and Political Role of the Protestant Clergy in Pre-March Prussia," in Hans Ulrich Wehler (ed.), *Sozialgeschichte Heute: Festschrift für Hans Rosenberg zum 70. Geburtstag* (Göttingen: Vandenhoeck & Ruprecht, 1974), 175–90; Charles E. McClelland, *State, Society, and University in Germany 1700–1914* (Cambridge: Cambridge University Press, 1980), 114–17; Claudia Huerkamp, "Die preussisch-deutsche Ärzteschaft als Teil des Bildungsbürgertums: Wandel in Lage und Selbstverständnis vom ausgehenden 18. Jahrhundert bis zum Kaiserreich," and Christoph Führ, "Gelehrter Schulmann—Oberlehrer—Studienrat: Zum sozialen Aufstieg der Philologen," in Werner Conze and Jürgen Kocka (eds.), *Bildungsbürgertum im 19. Jahrhundert. Teil I: Bildungssystem und Professionalisierung in internationalen Vergleichen* (Stuttgart: Ernst Klett, 1985), respectively, 358–87, 417–57; Rudolf Vierhaus, "Der Aufstieg des Bürgertums vom späten 18. Jahrhundert bis 1848/49," in Kocka, *Bürger und Bürgerlichkeit*, 69–75 (and other sources cited there); Lepsius, "Zur Soziologie," in Kocka, *Bürger und Bürgerlichkeit*, 86–88; and Wehler, *Gesellschaftsgeschichte*, 2:211–34.

[19] Schuckmann to the Oberpräsidenten in Coblence, Münster, Potsdam, and Breslau, 10 April 1826, Rep.120 A.II.5b, Nr.9, Bl.1, ZStA Merseburg.

partment. While the work of "guild reform" seemed safe in the hands of Kunth, Maassen, and Hoffmann (see chap. 1), there were nevertheless many signs that economic reactionism was the order of the day. For one thing, the directorship of the department, vacant since 1818, was not filled when the Ministry of Commerce and Business was abolished in 1825, leaving all "reporting councillors" (*Vortragende Räte*) at the strained mercy of Schuckmann. Thus when Privy Councillor Semler, the Business Department's expert on trade and commerce, proposed publication of a business journal to justify state policies before the public, Schuckmann slapped down the idea, blurting that he "hated the public."[20] Even more alarming, reactionary politicians like Gustav von Rochow were finding support in unexpected places for their campaign to restore guild privileges—namely, from Baron vom Stein. [21] The unfolding plot worried Gneisenau, the leading army liberal, who feared that "a great phalanx of zealous confessors will jettison all the new industrial legislation."[22] When Duke Carl was appointed President of the Council of State in 1825, moreover, Beuth "and a few other defenders of industrial interests"[23] were sacked from their Council committees. Privy Councillor von Graevenitz, one of Duke Carl's reliably conservative appointees to the Council of State, also appears on the lists of the Business Department at this time. His function in a politically suspicious department, apparently, was to keep an eye on the liberals.[24]

As we observed in chapter 1, however, different winds were blowing in Berlin by the late 1820s. Indeed, with Friedrich von Motz occupying the Ministry of Finance, the Business Department's fortunes started to rise. Peter Beuth was promoted to head it in January 1828 and his yearly budget was raised from 65,000 to 100,000 thaler. This led to major changes in the internal workings of this department. While the directorship was vacant, a functional division of labor had been observed and decisions were made collegially. Christian Kunth headed a team of factory inspectors; Friedrich Skalley handled legal matters; Oberlandesbaurat Eytelwein supervised construction, and architectural and engineering education; while matters of trade and statistics, respectively, went to Privy Councillors Semler and Ferber. But Peter Beuth was un-

[20] Varnhagen Diary I, 27 January 1828, 5:18.

[21] See Stein's "Über Entwerfung eines zweckmässigen Gewerbepolizei-Gesetzes" (2 January 1826), and Stein to Rochow, 24 May 1826, printed, respectively, in Botzenhart, *Briefe*, 6:923–30, 7:3.

[22] Gneisenau to Schön, 24 February 1827, cited in Treitschke, *Deutsche Geschichte*, 3:377.

[23] Schleiermacher to Gass, 19 November 1825, printed in Wilhelm Dilthey (ed.), *Aus Schleiermacher's Leben* (Berlin: Georg Reimer, 1863), 4:210.

[24] Varnhagen Diary I, 24 December 1828, 5:151; *Handbuch über den Königlichen Preussischen Hof und Staat*, 1828.

doubtedly the department's most dynamic member, revitalizing the Technical Deputation in 1819, opening the Technical Institute in 1821, and founding the Association for the Promotion of Technical Activity in Prussia that same year. More and more influential as the decade unfolded, Beuth was the logical choice in 1828. Now *he* approved all plans, thereby ending years of collegial decisions. The previous division of labor yielded increasingly, moreover, to one-man domination of departmental business. Thus it was Beuth, not Semler, who helped Motz negotiate commercial treaties with South Germany, and Beuth, not Eytelwein, who soon controlled construction, and architectural and engineering education. When Motz acquired the Business Department for his own ministry in March 1829, Schuckmann was no longer an important obstacle and Graevenitz was left isolated. After Motz's death in July 1830, the establishment of a new Ministry of the Interior for Commerce and Business allowed Beuth to operate autonomously, for an aging, ailing, and ineffective Schuckmann could only go through the motions of presiding over the Business Department and Mining Corps.[25]

Technological priorities in the department changed under Beuth. Like Kunth, Scharnweber, Raumer, and Maassen, the young director had been part of Hardenberg's revolution from above with its emphasis on comprehensive freedom of enterprise, low tariffs, and rural industries. Unlike former superiors, however, he had doubted whether linen could meet cotton's challenge well before funds were denied for this purpose in 1820.[26] Consequently, we find the ambitious technocrat advocating a shift of departmental resources in the years thereafter. The main purpose of Beuth's English trip, in fact, was to investigate the "worsted" industries of Yorkshire which spun combed, "long-staple" wool into compactly twisted yarn. He also wanted to buy some of England's best long-haired sheep, purchase of which had been legalized by parliament only the year before.[27] Just prior to his western sojourn, moreover, the Business De-

[25] Handbuch (cited in ibid.), 1828; Treitschke, *Deutsche Geschichte*, 3:466; Petersdorff, *Friedrich von Motz*, 2:244, 326; Mieck, *Gewerbe-politik*, 29, 32.

[26] For Beuth's early association with and support for Hardenberg, see Beuth to Vincke, 10 September 1811, Vincke Papers A.III.23, and Beuth's memorandum on guild reform, 8 February 1810, Vincke Papers A.V.35, StA Münster; Beuth, Raumer, and Scharnweber to Hardenberg, 30 July 1810, Rep.92 Schöll 29, Bl.122–140, ZStA Merseburg; Niebuhr to Stein, 29 June 1810, and Niebuhr to Hardenberg, 6 July 1810, printed in Gerhard and Norwin, *Die Briefe*, 2:124–45, 130. For Beuth's early doubts about linen, see his draft report for Bülow, n.d. (early 1818), Rep.120, D.V.2c, Nr.3, Bd.1, Bl.97–103, ZStA Merseburg.

[27] For the motivation behind Beuth's trip, see Beuth to Frederick William, 25 February 1826, and Beuth to Bülow, 25 June 1826, Rep.120, D.I.1, Nr.13, Bl.77, 84, ZStA Merseburg. For the legalization of English exports, see John James, *History of the Worsted Manufacture in England* (London: Frank Cass & Co., 1968), 399–400 (originally published in 1857).

partment spent over 42,000 thaler on an impressive set of French machines for spinning fine combed wool. The machines were constructed in the machine shop of Cockerill's textile factory in Berlin. The goal was to spin worsted yarn from the best wool and sell it to manufacturers in the Eichsfeld region of Prussian Saxony and in the lower Rhine around Eupen and Viersen. Impressed by the product, Prussian entrepreneurs would modernize and prevent English, French, and Saxon firms from dominating the growing worsted trade.[28]

Beginning in February 1826, the Business Department yielded even more to consumer preferences by extending its promotional efforts for the first time to cotton. During the 1810s, cotton had been castigated by department officials as a textile which required expensive, imported raw materials and enjoyed no comparative advantage in world trade. By the 1820s, however, consumption of cotton goods was nearly ten times what it had been at the turn of the century. Promotion of cotton could now be defended as an import substitute. Beuth's goal was to establish a cotton "carding" and spinning bastion around Elberfeld. He wanted the local machine-making firm of Harkort and Kamp to reproduce machinery—chosen by Berlin—and sell it at competitive prices to a select group of promising textile manufacturers in neighboring counties.[29]

As Beuth's star rose, he grew impatient with other departmental traditions. It seemed to him that one of his predecessors' original goals—modernization of existing, small-scale operations—was slipping farther and farther away. He grew frustrated with Silesian flax farmers who completely ignored Belgian methods, country weavers who "knew" at a glance that "fast-feeder" looms were worse than traditional hand methods, and Berlin handicraftsmen who were "too distinguished" to adopt machinery or even appear on the street "in attire less than a prince with cloak and fur collar."[30] Beuth took increasingly little consolation, more-

[28] May to Beuth, 16 September 1825, Severin to Schuckmann, 18 August 1826, Rep.120, D.IV.1, Nr.15, Bd.1, and Beuth to Cockerill, 19 January 1830, Bd.2, Bl.35; Kunth to Beuth, 13 March 1828 and Beuth to Kunth, 4 April 1828, D.IV.8, Nr.22, Bd.1, Bl.43–51, ZStA Merseburg. For the efforts to sell the department's wool, see the documents in Rep.120, D.IV.1, Nr.15, Bd.1, Bl.175–82, ZStA Merseburg. For German combed wool, see Horst Blumberg, *Die Deutsche Textilindustrie in der industriellen Revolution* (Berlin: Akademie-Verlag, 1965), 74–79, 84–85, 102–3, 384, 387, 405.

[29] For the first efforts to promote cotton, see Beuth to Friedrich Werth, 26 February 1826, Rep.120, D.VI.1, Nr.4, Bd.1, Bl.2, Kunth to Schuckmann, 21 July 1826, Rep. 120, D.XIV.1, Nr.17, Bl.87–91, and Beuth to Sutterroth, 2 April 1827, Rep.120TD, Abt.I/Fach 5, Nr.101, Bl.28, ZStA Merseburg. Detailed documentation on the construction and distribution of the machinery is found in Rep.120, D.VI.1, Nr.4, Bd.1; D.VI.2, Nr.11, Bd.1; and Rep.120TD, Abt.I/Fach 5, Nr.101. For a list of cotton factories in the Rhineland in 1829, see 403/190, LHA Koblenz.

[30] For the prejudiced weavers, see Regierung Liegnitz to Schuckmann, 29 March 1832, Rep.120, D.V.1, Nr.4, Bl.28, and Schuckmann (Beuth) to Regierung Breslau, 10 August

over, in the bulk of the "model" factories identified by Berlin to demonstrate new machinery, for he believed that most of these "showmaster" entrepreneurs overestimated the degree to which they had narrowed the technological gap with England.[31] Hence the great care taken in screening model factories for the Elberfeld project.

Beuth's pessimism concerning existing factories warmed him to the concept of joint-stock companies (*Aktiengesellschaften*). Small groups of daring entrepreneurs should form partnerships and construct the new factories which would enable Prussia to compete. The Business Department should be prepared to do its part by expending great sums on some of the necessary machinery. As oldtimers like Kunth and Maassen watched with disapproval, their friend and protégé moved up in the bureaucracy and began to turn his back on the small man. Thus he wanted *new* industrial ventures to revitalize the Eichsfeld worsted industries. Beuth also proposed joint-stock schemes at this time for a cotton finishing center (*Appretur*) in Viersen, an integrated, "combing to dyeing" woolen operation in Luckenwald and a machine-building complex in Breslau.[32]

We can discern a fairly consistent pattern to Beuth's aspirations. The idea of establishing a parliament open to only the most successful members of the Christian upper classes was an early sign of an elitism that grew as he advanced through the state. This penchant for exclusivity was also evident in his repeated attempts to permeate government with technological experts. Beuth's impatience with cautious factory owners and his preference for bold new entrepreneurs who would advance "out of the old rubble"[33] were further indications of elitist designs. In sum, the physician's son from Cleve was trying to create and elevate a new bourgeois elite bridging state and society.

Such notions probably began to germinate at the turn of the century in the lecture rooms of the University of Halle. Beuth arrived there in 1799 as a student of cameralism just as the eighteenth century traditions of this discipline were beginning to blend with the antimercantilistic, in-

---

1829, D.V.2c, Nr.8, Bl.136, ZStA Merseburg; for Berlin, see Beuth to Baumann, 5 January 1829, Beuth-Baumann Correspondence (N 7/1), sww Dortmund.

[31] Beuth to Boehme and Schlief, 29 June 1828, Rep.120, D.IV.1, Nr.26, Bl.62–63, ZStA Merseburg.

[32] Beuth to Rother, 19 January 1828, Rep.120, D.IV.1, Nr.26, Bl.11–14; Beuth to Kunth, 4 April 1828, D.IV.8, Nr.22, Bd.1, Bl.50–51; Beuth to Diergardt, 10 December 1828, D.I.1, Nr.20, Bl.115–17; and Beuth to Merckel, 6 February 1830, D.XIV.1, Nr.35, Bd.1, Bl.1, ZStA Merseburg.

[33] Beuth to Delius, 14 November 1836, Rep.120, D.V.2d, Nr.12, Bd.1, Bl.221, ZStA Merseburg. For Beuth's impatience with timid owners, see Beuth to Büschgen, 12 October 1831, Rep.120, D.VIII.2, Nr.23, Bl.10; Beuth to Farthmann and Dörner, 28 November 1837, Rep.120, D.VIII.1, Nr.14, Bd.1, Bl.146, 150; and Beuth to Quentin, n.d. (summer 1838), Rep.120, D.VIII.2, Nr.14, Bd.1, Bl.61–62, ZStA Merseburg.

dividualistic economics of Adam Smith. The intellectual merger found expression in the teaching of Ludwig Heinrich Jakob, one of Beuth's professors and Halle's premier "cameralist." The individual entered civil society, he taught, in order to acquire the wealth, property, and security which were the foundation stones of happiness. The means to this end consisted "partly in the private powers of individuals [and] partly in the public and combined powers of the state." When the former were weak, the latter were required to foster "those principles which have to be set in motion in a people so that the means for the satisfaction of need are produced, increased and also distributed and appropriately used."[34] One cannot find a more succinct statement of the moderately laissez-faire guidelines which Beuth would employ as head of a Business Department dedicated to helping businessmen help themselves.

Long before his rise to the top of this department, in fact, Beuth had held out a friendly hand to Berlin's premier entrepreneurs. As early as 1817 he started to hold informal, weekend gatherings of manufacturers at his home in Schönhausen to exchange information on technological issues and discuss common political interests. Soon the new circle of friendship was extended to Beuth's older network of friends in the civil service, university, army, and artistic community. After four years of these Sunday discussions, the friends of industrial and political progress united in the Association for the Promotion of Technical Activity in Prussia. Functioning by early 1821, this associational brotherhood boasted 146 members, four standing committees, regular general meetings, and a bimonthly journal which defended the modernist cause and added to Prussia's stream of technical information from foreign lands.[35]

The Association never attracted great numbers of state servants and industrialists. Hampered by a high turnover rate of 60 percent, membership rose slowly and steadily, eventually peaking at about nine hundred in the late 1830s.[36] But Chairman Beuth was less interested in numbers than in key personalities who could lend prestige and political weight to his mission. By the mid-1820s, he had recruited a glittering contingent of friends and dignitaries: Prince August and Prince Karl of the royal family; Fieldmarshall Neidhardt von Gneisenau; Wilhelm von Scharnhorst,

[34] Keith Tribe, *Governing Economy: The Reformation of German Economic Discourse 1750–1840* (Cambridge: Cambridge University Press, 1988), 171 (text and n. 70), cites Jakob's lecture texts. For cameralist studies within the College of Philosophy at Halle, see Johann Christoph Hoffbauer, *Geschichte der Universität zu Halle bis zum Jahre 1805* (Halle: Schimmelpfennig und Co., 1805), 412–14, 439–40.

[35] Matschoss, *Preussens Gewerbeförderung*, 33–37.

[36] For peak membership statistics, see *Verhandlungen* 20 (1841), GB/TU Berlin. Turnover was too high to be explained by deaths. Thus in 1832 actual membership stood at 700, while membership card numbers extended as high as 1208 (see Hotho to Vincke, 31 December 1832, Oberpräsidium 2778, Bl.111, StA Münster).

son of the great military reformer; the brothers Alexander and Wilhelm von Humboldt; Christian Daniel Rauch, Berlin's leading sculptor; and—as impressive as anyone—Schinkel. Joining these standouts were scores of technical and scientific experts from every governmental agency responsible for these affairs: Rother of the Seehandlung; Gerhard of the Mining Corps; Rauch of the Army Corps of Engineers; Kunth from Commerce and Business; Motz and Maassen from Finance; Könen of the Medical Corps and many more.

Beuth could also be pleased with the quality of business membership. For his list of pioneering entrepreneurs read impressively like a "Who's Who" of early Prussian industry: iron producers Dinnendahl, Krupp, Jacobi, Haniel, and Thyssen; machine producers Harkort, Kamp, Freund, Hummel, Borsig, and Egells; and textile manufacturers Gropius, Dannenberger, Tappert, Cockerill, Alberti, Ruffer, and Diergardt. As we shall see, Beuth and the businessmen often disagreed over economic and technological policy; nevertheless, most of them were personal friends whose membership he valued very highly.[37]

As gratifying as these successes were, Beuth recognized the magnitude of work which lay ahead. He knew that it was not this generation of entrepreneurs, but in all likelihood the next which would fulfill his dreams for the Prussian bourgeoisie. This explains why so much of the Privy Councillor's energy in the 1820s was absorbed by the politics of educational reform.[38] Like many of the technocrats, Beuth placed great store in mastering practical subjects, complaining that Berliners were becoming "so rhetorical and overeducated that everything simply human [like work] seems too vulgar for them."[39] Accordingly, the changes which he implemented between 1817 and 1821 featured a nationwide system of one-year business schools (*Gewerbeschulen*) in each of Prussia's twenty-five governing districts. These were crowned by the exclusive Technical Institute in Berlin, a two-year school for the most accomplished provincial students. Departing from the "trade school" (*Handwerkerschule*) concept which "rejected [scientific] subjects out of hand as not belonging there,"[40] Beuth's schools offered physics, chemistry, and mathematics in

[37] For membership, see *Verhandlungen* 5 (1826): 4–11, GB/TU Berlin. For Beuth's friendship with many western industrialists, see Beuth to Quentin, 15 March 1838, Rep. 120, D.VIII.2, Nr. 14, Bd. 1, Bl. 55, ZStA Merseburg; Matschoss, *Preussens Gewerbeförderung*, 34; and Treitschke, *Deutsche Geschichte*, 3:467.

[38] For this paragraph, see Goldschmidt and Goldschmidt, *Das Leben*, 141–56; Lundgreen, *Techniker in Preussen*, 57; and Christiane Schiersmann, *Zur Sozialgeschichte der preussischen Provinzial-Gewerbeschulen im 19. Jahrhundert* (Weinheim: Beltz Verlag, 1979), 34–56, 123–24, 128.

[39] Beuth to Baumann, 5 January 1829, Beuth-Baumann Correspondence (N 7/1), SWW Dortmund. The letter bemoans the denigrating attitude of Berliners toward work.

[40] Beuth to Steineccius, 17 July 1817, printed in Straube, *Die Gewerbeförderung*, 14.

order to provide future shop or factory owners with the wide range of practical knowledge needed in the new era of free enterprise. He insisted, furthermore, that technical schools draw Seniors from "the most excellent school in the town,"[41] whether it be a city school or a gymnasium. In this way aspiring entrepreneurs could receive a broad general education before specializing in business-related subjects. He fought consistently and successfully for his educational approach, frustrating the efforts of Kunth and Leopold von Bärensprung, the mayor of Berlin, who wanted to found a new system of vocational gymnasiums (*Realschulen*) which would emphasize science, math, and modern languages at the expense of Latin, Greek, and the classics taught in city schools and gymnasiums. Beuth's system appears to have been a compromise between an excessive emphasis on ancient languages, which he believed was "too time-consuming" for the businessman, and their total elimination, which smacked of the view that boys should not study beyond "what they will need when they become men."[42]

Certainly, Beuth did not consider knowledge of ancient times irrelevant for those who would make contributions in the modern world. For instance, he believed it was necessary for the Technical Deputation "to bring the influence of refined taste and knowledge of antiquity to bear [on the business world]."[43] Similarly, Beuth wanted his business schools to include the practice of architectural drawing in order "to cultivate a refined taste based on the study of the ancients." Drawing projects of this sort, worked and reworked by his Seniors during the school year, would be "the student's traveling companion throughout life."[44] As evidenced by these proposals, Beuth wanted to impart a social status or pedigree to manufacturers which, in his opinion, they could not acquire without exposure to classical aesthetics. This was an obvious personal preference, for Beuth revered the "applied art" (see below) of the ancients, amassing a private collection of replicas and applauding enthusiastically as his

[41] Beuth to the Regierung Trier, 8 February 1829, quoted in Schiersmann, *Zur Sozialgeschichte*, 69.

[42] See Beuth's report of 17 July 1822, printed in *Verhandlungen* 1 (1822):133, GB/TU Berlin. Lundgreen, *Techniker*, 30, errs—and contradicts his own evidence—in drawing the conclusion that Beuth was uninterested in "a higher pre-schooling" (*eine höhere Vorbildung*). For the more formal schooling actually acquired by students entering Beuth's schools, see Lundgreen, 100–101, 106–7, 110–11. Herbert Blankertz, *Bildung im Zeitalter der grossen Industrie: Pädagogik, Schule und Berufsbildung im 19. Jahrhundert* (Hanover: Hermann Schroedel, 1969), 81–84, also errs in claiming that Beuth's goal was a simple, unpretentious education. Schiersmann, *Sozialgeschichte*, is more reliable.

[43] Beuth to Bülow, 27 June 1817, cited in Matschoss, "Geschichte der Königlich Preussischen Technischen Deputation," 253.

[44] Both quotes here from Beuth to Steineccius, 17 July 1817, are printed in Straube, *Die Gewerbeförderung*, 15.

friends Schinkel, Rauch, Tieck, and Schadow brought the world of the heroes to life again—not through study of extinct languages, but in the buildings and statues of neoclassical Berlin.[45]

This semipolitical use of art combined, in all likelihood, with other, more overtly political reasons for attaining "knowledge of antiquity." Beuth had studied cameralism in the College of Philosophy at Halle while its great philologist, Friedrich August Wolf, was at the height of his teaching career. During Beuth's first Berlin years, moreover, he befriended one of Wolf's greatest admirers, Wilhelm von Humboldt, working closely with the ex-minister to advance Schinkel's revival of neoclassical art. Given Beuth's liberal politics and love for antiquity, it seems certain that he agreed with Wolf and Humboldt that the political lessons of antiquity were indispensable in the waning age of absolutism. Prussia, like Athens and Rome, would rise to greatness under just institutions, not tyranny. Schinkel, too, wanted his exquisite Greek structures to remind official Berlin of the political ideals of Periclean Athens and serve, in the meantime, as an inspiration "which helps us to persevere through the injustices of time."[46]

There was a darker side, however, to neoclassicism. As argued persuasively by Martin Bernal, Germany's enthrallment with ancient Greece was part of a liberal political agenda with strong *racial* overtones. Rising pride in the German people as a homogeneous "kind" or "race" induced many scholars of the late 1700s and early 1800s to search for legitimating role models from antiquity. "Knowledge of the Greeks is not merely pleasant, useful or necessary to us!" exclaimed Humboldt. "No, in the Greeks alone we find the ideal of that which we should like to be and produce."[47] The result was a largely imagined or fabricated Classical Greece purged of its obvious Egyptian (African) and Phoenician (Semitic) influences. Thus Wolf interpreted the *Iliad* as a product of the original European race in the purity of its childhood, while Humboldt believed

[45] Matschoss, *Preussens Gewerbeförderung*, 71–72, 75–76; Rudolf von Delbrück, *Lebenserinnerungen* (Leipzig: Verlag von Duncker & Humblot, 1905), 1:135.

[46] For Wolf and his college's curriculum at Halle, see Hoffbauer, *Geschichte*, 412–14, 439–40. For Beuth's friendship with Humboldt, see Beuth to Rother, 8 April 1835, Rep.92 Rother, CaNr.17, ZStA Merseburg; for the joint promotion of neoclassical art, see Varnhagen Diary I, 20 September 1825, 3:379; and Bernsdorff, *Aufzeichnungen*, 1:262. And for the political lessons of antiquity, see Varrentrapp, *Johannes Schulze*, 380; Jeismann, *Das preussische Gymnasium*, 258–63; and Martin Bernal, *Black Athena: The Afroasiatic Roots of Classical Civilization* (New Brunswick, New Jersey: Rutgers University Press, 1987), 283–90. The Schinkel quote is cited in Gerd-H. Zuchold, "Bemerkungen zu Schinkels Verhältnis zur Kunst der Antike," in Eberhard Bartke et al., *Karl Friedrich Schinkel 1781–1841* (Berlin: Verlag Das Europäische Buch, 1981), 384.

[47] Cited in Bernal, *Black Athena*, 287. For Bernal's general argument, see pp. 281–316, 338–40. For a similar remark by Humboldt, see Varrentrapp, *Johannes Schulze*, 243.

that Greek was a language uncontaminated by foreign elements. Bernal's thesis helps to explain how Beuth could mock the pyramids as the product of backward slave labor in one issue of the proceedings of the Association for the Promotion of Technical Activity, then praise a drawing of Greek temple architecture in another. This new work also makes sense of the paradoxical, seemingly inconsistent image of Beuth, a sophisticated aesthete and devotee of classical art, denouncing the Jews and urging the Council of State to put constraints on a race which was allegedly undermining the productive forces of Germans.[48] While Beuth's neoclassical aesthetics and anti-Semitism were quite compatible, it is impossible to determine which was cause and which the effect. He probably imbibed the xenophobic hatred of Jews which swept the Rhineland after the first French invasions, then reinforced these prejudices during his studies at Halle and stint with Lützow. As argued more extensively below, the ex-cavalryman was a different and more complex person than the benignly progressive caricature found so often in the accounts of nationalist historians and modernization theorists.

Indeed the man Schinkel would paint atop Pegasus had ambitious designs for his industrialists. Traveling in the woolen district of Leeds on an earlier trip in 1823, he found a businessman of high culture whom any Prussian manufacturer would do well to emulate:

> I have some fever too [he wrote to Schinkel], thus also the fantasies that come with fever. *Nota bene*: knights or friars or robbers are not what my dreams are made of. Mine have spinners [and] weavers who, rich in inventiveness, create, bask in millions, build villas on the hills of their land, practice the arts, and are hospitable. It is always something new and pleasant to me when I see how the factory owner hangs the factory owner [in him] on the hook at five o'clock and goes to his villa to be with himself, his family, [and] his hobbies. . . . I have spent pleasant evenings with a cloth manufacturer in Leeds. His park and villa lie on a hill a half German mile from town. The view from there onto two lovely water-blessed valleys and the ruins of Kirkstall Abbey is very beautiful. Inside [one finds] beautiful paintings: a Paolo Veronese, Poussin, [and] Claude, plus beautiful things from Athens that a deceased son sent him.[49]

The author of these lines obviously saw more in rural industrialization than cheap costs of production and easy access to agricultural raw materials and waterpower. There was an aesthetic aim here also. Manchester's "forest of tall smoke stacks"[50] did not revolt Beuth as much as it did

[48] *Verhandlungen* 1 (1822): 165–66, 2 (1823): 212–13.
[49] Beuth to Schinkel, 30 July 1823, printed in Wolzogen, *Aus Schinkels Nachlass*, 3:144.
[50] Beuth to Schinkel, July 1823, printed in ibid., 3:141.

Schinkel, but he preferred country plants built at the end of tree-sheltered lanes, hidden in picturesque cloisters and castles, or, like this one, near a "lovely water-blessed valley."

The letter also says something about Beuth's view of art. The technocrat was apparently influenced by the prevailing romantic school which had removed art entirely from the world of mundane, practical concerns, setting it apart on a different plane as something to contemplate and enjoy for its own sake. There was another artistic tradition, however, which weighed more heavily with him—as witnessed most convincingly by his own private art collection. It was a tradition which he had encountered through formal schooling and a voracious reading of ancient history. This older school, largely *passé* by the 1820s, looked at crafts and skills that created something practical, useful, and beautiful as art forms.[51] Thus when Rauch's Bavarian colleague, Dinger, fell ill, Beuth grew fearful that Germany would "lose the art of casting [statues] thinly and beautifully on a large scale."[52]

Like rural industry, "*Kunstgewerbe* " (applied/industrial art) was central to Beuth's vision of a refined industrialization in Prussia. He gathered around his person the practitioners of ceramics (Feilner), enameling (Wagenmann), gilding (Hossauer), tapestry (Hotho), and stood very close, as we know, to Berlin's architects and sculptors.[53] It was in the area of applied art, in fact, where Beuth grew excited about integrating modern technology and the art of the ancients. "We lack neither the means nor the opportunity," he wrote in 1824, "to reproduce the artistic treasures of antiquity in plasters, prints, and castings, then use them in a practical way as decorations. But how seldom this happens! Few of my countrymen are such dull characters that they would not react with pleasure in finding the head of Ceres on a medallion from Syracuse." Such works should be produced in quantity and widely displayed, he urged, not "hidden away in cabinets."[54]

Like his steps to spread "knowledge of antiquity," Beuth undertook concrete measures to inject art into the business world. From the beginning of Beuth's headship, the Technical Deputation was encouraged to emphasize "exquisite forms."[55] It did so in interesting ways. Thus an

[51] M. H. Abrams, "Art-as-Such: The Sociology of Modern Aesthetics," *The American Academy of Arts and Sciences Bulletin* 38/6 (March 1985), 8–33.

[52] Beuth to Baumann, 9 September 1832, Beuth-Baumann correspondence (N 7/1), SWW Dortmund.

[53] Treitschke, *Deutsche Geschichte*, 3:468; Matschoss, *Preussens Gewerbeförderung*, 32, 36.

[54] Verhandlungen 3 (1824): 170, GB/TU Berlin.

[55] Beuth to Bülow, 27 June 1817, cited in Matschoss, "Geschichte der Königlich Preussischen Deputation," 253.

early manual of the Deputation coauthored by Beuth and Schinkel contained copper engravings and lithographic prints of buildings, fabrics, rugs, tapestries, curtains, vases, and other interior decorations, all with ancient motifs. While other manuals were designed specifically to facilitate technological modernization, this one was "to have a refining influence" (*geschmackbildend wirken*) on manufacturers who were urged repeatedly "not to be misled into composing on your own, but rather to duplicate [the manual's designs] faithfully with industriousness and good taste."[56] The Berlin Technical Institute—whose original goal was to spawn a cadre of business leaders, not engineers, for the next generation—pursued much the same end. Whether specializing in chemistry, metal-working, or building—the Institute's three majors—students were required to learn "free-hand drawing with special reference to architecture and the decorative arts" (*Verzierungskünste*).[57] Metal-working majors, who comprised about 40 percent of the approximately one thousand students who studied at the Institute before 1848, had to demonstrate "outstanding artistic talent" (*hervorstechendes Kunsttalent*)[58] before advancing to an internship in the machine shop.

The artistry refined there included forging, casting, molding, turning, shaping, and model-making. As demonstrated by these workshop requirements, Beuth did not limit his concept of "art as craft" to the decorative or fine industrial arts like gilding or sculpturing, but extended it to include the well-crafted mechanism. Thus after Tappert constructed a series of wool-spinning models for the Technical Deputation, Beuth praised him as an "artist" (*Artist*) in the sense of "a connoisseur of art" (*Kunstverständiger*).[59] The technocrat clearly found the semiautomatic motion of machines and systematic interaction of precisely built parts aesthetically pleasing. The diversity of captivating machinery and products of "special precision and artistry" displayed at the Business Department's technological exhibitions of 1822, 1827, and 1844 were also part of this overall campaign, as Rudolph von Delbrück observed, "to integrate art with industry" (*die Kunst in das Gewerbe einzuführen*).[60]

To many of Prussia's older economic experts, the elitist inclinations of

[56] For the first quote, see Beuth to Bülow, 17 July 1822, cited in Straube, *Die Gewerbeförderung Preussens*, 38; the second quote, from the manual itself, is cited in Bartke et al., *Karl Friedrich Schinkel*, 259.

[57] See Beuth's instructional regulations of 14 February 1832, printed in Straube, *Die Gewerbeförderung Preussens*, 23.

[58] Ibid.

[59] Cited in Matschoss, "Geschichte der Königlich Preussischen Technischen Deputation," 258.

[60] Delbrück, *Lebenserinnerungen*, 1:135, served briefly under Beuth in the 1840s; for the "precision" quote, see Viebahn (of the Exhibit's organizing committee) to the Regierung Köln, 5 July 1844, Abt.1, Nr.28, Fasz.1, R-W WA Köln.

Beuth, Schinkel, and the coterie of artists around them were impossible to accept. Heinrich Weber, a member of the Technical Deputation since its inception in 1811, objected to the sophisticated, high-brow content and style of the annual yearbook of the Association. Arguing that the proceedings catered only to "the well-informed businessman, educated artist, experienced mechanic or learned technologist," Weber decided to edit his own journal in 1828 aimed at the "less educated class of business people," and designed to convey "the essence of the matter."[61] There was also frequent discord between Beuth and his erstwhile boss, Kunth, who was uncomfortable with the exclusiveness of the Technical Institute. The institution "works toward the future," he observed in 1826, by educating "artisans of a higher order."[62] In so doing "it limits too much the means for the present, the practical example on a broad scale, especially in the provinces . . . [where] merchants, farmers, foresters, builders, soldiers, etc."[63] were in need of basic technical education. It is difficult to imagine that the nation's prestigious statistician, Johann Gottfried Hoffmann, felt any different. The Business Department would consider it a "national loss," he had written in 1812, if schools became "mere institutions for accustoming the eye to beautiful forms."[64] Technological improvements had to serve "the requirements of the great masses of the nation" not just "endulge the ostentatiousness of the rich." Only a curriculum "that led the artisan to think about his trade"[65] could assure such advances. There was much here that Beuth could agree with, but his greater respect for classical curricula and belief that modern industrial technology was the vernacular art form of the *haute bourgeoisie*, helping to refine this new elite and enhance its chances of acquiring a prominent place in the old political order, certainly distinguished him from the Webers, Kunths, and Hoffmanns in his own camp.

Beuth's and Schinkel's dream of aesthetic industrialization was not as completely unrealistic as critics charged. Thus Schinkel affected a veritable revolution in taste among upper class consumers during the 1820s. "The smallest and the largest [objects] took on more refined forms," wrote Theodor Fontane. Ovens and stoves which previously looked like "monsters" turned into enameled "ornaments," while iron grilles and gates "ceased to be a mere collection of rods and bars." Tables, cabinets, glasses, goblets, china, picture frames, and gravestones—all felt Schin-

[61] *Zeitblatt für Gewerbetreibende und Freunde der Gewerbe*, Nr.1, 1 January 1828 (copy in 441/5475, Bl.9–15, LHA Koblenz).

[62] Kunth to Stein, 7 April 1826, printed in Goldschmidt and Goldschmidt, *Das Leben*, 129–30.

[63] Ibid.

[64] Hoffmann to Eytelwein, 21 June 1812, printed in Straube, *Die Gewerbeförderung*, 13.

[65] Ibid.

kel's touch. Beuth's programs also enjoyed a favorable reception in some circles. Having constructed his factory in an old castle overlooking the Ruhr, for example, Friedrich Harkort constantly reminded anxious customers that the completion of orders required time because his machinery was a matter of art, not shopkeeping. The neoclassical style of factory architecture which a few entrepreneurs selected for their factories and train stations were other indications of moderate success, as were the popular art-as-such exhibits displayed at the Industrial Exhibition of 1844. The steady demand for stipendia at the Technical Institute also showed that Beuth's desire to combine the practical and the aesthetic was shared by others. Rugged individualists like Franz Haniel sent their sons to become "artisans of a higher order," while technical perfectionists like Alfred Krupp were full of praise for the Institute. Although Beuth himself came to doubt the success of this campaign, bemoaning the turn away from "the clear forms of antiquity" in the late 1830s toward baroque and rococo decorations with their conservative, absolutist overtones, it is nevertheless clear that some Prussians were not averse to the new definition which he and Schinkel had given to refinement.[66]

There was a certain cogency, however, to the arguments of Beuth's intramural opponents. For a technological revolution pleasing to the eye was something which most businessmen gave low priority. That Beuth was so preoccupied with aesthetics was a clear reflection, in fact, of the crevasse which separated him from the regular world of business. Neither his childhood in the home of a physician, nor a university education, bureaucratic training, or frequent tours of factories and discussions with their owners enabled him to appreciate the microeconomic decision-making process of the entrepreneur like an "insider." As an analysis of the ambitious textile projects in Elberfeld and the Eichsfeld/lower Rhine demonstrates, investment in new plant and machinery was an especially complex and risky business which the "outsider" from Berlin did not fully understand.

The reader will recall that Beuth hoped to develop the area around Elberfeld into a center for cotton carding and spinning able to hold its own with English competition. A stiff tariff of 50 thaler per centner (a

[66] Fontane is cited in ADB 54: 24; Beuth in Bartke et al., *Karl Friedrich Schinkel*, 266 (but see also pp. 382–83). For the remainder of the paragraph, see Schnabel, *Deutsche Geschichte*, 3:280–81; Jürgen Eckhardt and Helmut Geisert, "Industriearchitektur und industrialisierte Bauweise in der Nachfolge Schinkels," in *Karl Friedrich Schinkel: Werke und Wirkungen* (Berlin: Nicolaische Verlagsbuchhandlung Berlin, 1981), 147–57; Delbrück, *Lebenserinnerungen*, 1:135; Lundgreen, *Techniker*, 93–94; Haniel to Beuth, 13 May 1834; Beuth to Haniel, 1 July 1835; and Beuth to Haniel, 11 September 1837, Haniel Papers 210, Haniel Museum Duisburg; and Krupp to Werninghaus, 8 July 1836, F.A.H., II.B.87, Bl.42, HA Krupp Essen.

weight of approximately 110 pounds) was levied on cotton fabrics, pushing the price of English imports about 40 thaler above Rhenish prices in the late 1820s. Behind this wall, 679 weaving, dyeing, and finishing establishments flourished in the Rhineland.[67] In contrast, the tariff on cotton yarn was only 2 thaler per centner, offering no protection for Rhenish spinning factories which faced tough foreign competition. In 1826/27, for instance, Rhenish coarse yarns cost 25 thaler more than English mule-spun yarn which was easily substitutable for Rhenish yarn, that is, *possessed no qualitative disadvantages vis-à-vis the domestic product*. Consequently, only sixty-two entrepreneurs had entered the spinning trade, most of these in the District of Düsseldorf.[68] Nevertheless, Beuth was convinced that it was not free trade, rather "the whole miserable state of [our] machines"[69] which explained why Prussian spinning lagged behind. Therefore he chose the machine-producing partnership of Kamp (Elberfeld) and Harkort (Wetter) to spearhead a drive to push the coarser English yarns from the area.

The Technical Deputation sent the partners copies of the "double speeders" and "eclipse speeders," new American carding engines, and the "throstle spinner," an improved version of the old water frame. Both

[67] For a list of cotton factories in the Rhineland in 1829, see 403/190, LHA Koblenz; for tariff levels, see Dieterici, *Statistische Übersicht*, 1842, 40–73. Thomas Banfield, *Industry of the Rhine* (London, 1846–48), 2:13, cites the price of English cotton goods in Germany in 1846. A full price series for the Pre-March is obtained by allowing this price to float in correlation with the value of exported cotton piece goods given in B. R. Mitchell, *British Historical Statistics* (Cambridge: Cambridge University Press, 1988), 761. Dieterici, *Statistische Übersicht*, 1838, 397, states that the ratio of raw cotton to cotton goods prices in Prussia in 1829–31 was 1/6. A full price series for cotton goods is produced with the Pre-March prices of raw cotton in Germany given in Jacobs and Richter, *Die Grosshandelspreise*, 66.

[68] For cotton factories in the Rhineland and tariffs, see the sources in n. 67. One member of the Elberfeld-Barmen Chamber of Commerce (von Prath memorandum of March 1831, Abt.22, Nr.112, R-WWA Köln) stated that most Rhenish manufacturers imported cotton from England and that English cotton yarn prices were 12–15 percent below the Rhenish in Elberfeld. James A. Mann, *The Cotton Trade of Great Britain: Its Rise, Progress and Present Extent* (London: Frank Cass & Co., 1968), vii, 104, gives the "declared real value" (i.e., the actual value returned to the exporting merchant inclusive of transportation costs) of reexported cotton, 1825–47. Dieterici, *Statistische Übersicht*, 1838, 34–35, figured the price ratio of raw cotton/cotton yarn in 1829–31 at 20/50. Rhenish yarn prices are obtained, therefore, by multiplying the English cotton price by 2.5, while declared real value of exported English yarn is given in Mann, *Cotton Trade*, 95–96. Using these price figures and partial estimates, Rhenish yarn prices averaged 11.6 percent above the English between 1830 and 1835, very close to the 12–15 percent figure of Prath in 1831 (see above). Using the same method for the years 1825–29—the period of Beuth's initiative—Rhenish yarn prices averaged 43.5 percent above the English.

[69] Beuth to Sutterroth, 2 April 1827, Rep.120TD, Abt.I/Fach5, Nr.101, Bl.28, ZStA Merseburg.

devices were designed specifically for carding coarse cotton yarn. Between 1827 and 1829 Harkort and Kamp reproduced these devices and installed them in a number of local factories whose owners had impressed Beuth with their desire to modernize. The Department expected the owners of these select works to appreciate the superiority of the sample machine, purchase more of them, or perhaps completely retool along American lines.

Yet by 1832 Elberfeld factories had placed only a few orders with Harkort and Kamp of the expensive (300–500 thaler) devices. Most of the chosen manufacturers claimed that it went against economic reason to compete against imports. Others added that the Prussian machines did not always work, or were too expensive or less appropriate than other machines available abroad. Kamp, an avid free-trader, pointed to his accusers' allegedly instinctive, irrational resistance to modern improvements.[70] Beuth rebuked district officials for failing to provide the detailed "cost-price" data necessary to determine whether the Kamp-Harkort machine works were as "well-managed"[71] as the British, then resorted to his own laissez-faire instincts in a final report. He castigated the protectionists for ignoring the state's progressive scheme and attributed their charges against Kamp and Harkort in demeaning fashion to political anger over Prussia's low tariffs on yarn. "After all [Kamp and Harkort] are operating the . . . machines, therefore their people must know what they are doing."[72]

The entire complex episode demonstrates Beuth's distance from events and lack of business acumen. The speeders and throstle spinners worked very quickly, but were never adopted on a large scale outside the United States because both wasted large amounts of cotton.[73] Of more importance, the two devices increased overhead costs in a way difficult for the uninitiated to detect. Both worked at higher speeds than the older carding engines, spinning jennies, and water frames. This made necessary a thorough rearrangement of the machinery and a very complicated linkage to the water wheel—unless, that is, the owner was bold enough to replace his entire stock of machines or to purchase an expensive aux-

[70] Peill to Oberpräsident Pestel, 3 October 1831, 403/3345, Bd.1, Bl.221–24, and Peill to Pestel, 18 October 1831, 403/3345, Bd.3, Bl.2–3, LHA Koblenz; Regierung Düsseldorf to Beuth, 27 August 1832, Rep.120, D.VI.2, Nr.11, Bd.1, ZStA Merseburg; Jung to Wittgenstein, n.d. (March 1845), Rep.192 Wittgenstein, BP Hausarchiv V.8.5, GStA Berlin.

[71] Schuckmann (Beuth) to the Ministry of Finance, 1 January 1833, Rep.120, D.VI.2, Nr.11, Bd.1, Bl.87, ZStA Merseburg.

[72] Ibid.

[73] For the two machines, see Julia De L. Mann, "The Textile Industry: Machinery for Cotton, Flax, Wool, 1760–1850," in Charles Singer et al., *A History of Technology* (New York: Oxford University Press, 1958), 4:287, 291.

iliary steam engine for the American machines.[74] The total costs associated with the investment were therefore higher than they appeared.

Entrepreneurs also had to consider the risk associated with these costs. And most were still unwilling to gamble on their ability to compete against cheaper English imports. To be sure, from 1827 to 1833 the price differential between English and Rhenish yarns narrowed steadily from 25 to 5 1/2 thaler per centner. Somewhat emboldened by these trends, textile manufacturers from Elberfeld-Barmen, Hagen, Siegen, Bonn, Cologne, Aachen, Viersen, Rheydt, and Gladbach met in Godesberg in early 1834 to ask for state assistance before embarking upon technological modernization. Their requests included a 4 thaler tariff on yarn and subsidies of 3 thaler per spindle for new machinery. Hoping to assuage exporters of cotton fabrics and piece goods who depended on the lower tariff for cheap materials, the Godesberg conferees proposed tariff rebates of 4 thaler for exporters who used foreign yarn.[75]

Beuth was unmoved, persisting in his belief that manufacturers *with the same technology* should be able to survive with minimal protection. The ministers supported Beuth by rejecting the Godesberg proposals in 1835.[76] "It is very sad," complained J. C. Peill, one of the leading delegates, "that . . . Privy Councillor Beuth has such stubborn, false views regarding a progressive cause [like tariff protection] which he is so intent on belittling."[77] Meanwhile, Britain's comparative advantage increased as its industrialization accelerated. Between 1834 and 1847, the differential between English and Rhenish yarns widened to 7 thaler, a 17 percent gap which the adoption of new Prussian machinery, however aesthetically crafted, could not bridge.[78] Franz Haniel put it quite bluntly in a letter to Beuth: "It is impossible to compete with this country."[79] That the *Geheimrat* always knew better was deeply resented in his native province. For his part, Beuth was hurt that his good friend from Viersen, Friedrich Diergardt, had instigated the Godesberg meetings. In 1835, for instance, we find Beuth referring derogatorily to "the aristocracy of rich factory owners" on the lower Rhine.[80] Deep rifts were beginning to

[74] Swiersen to Vincke, 18 October 1829 and 27 November 1829, Rep.120 D.VI.1, Nr.4, Bd.2, Bl.43–45; and Hofmann to Beuth, 28 January 1829, Rep.120, D.I.1, Nr.20, Bl.153, ZStA Merseburg.

[75] The meeting ran from 27 January to 1 February, 1834. See the protocols in: Abt. 22, Nr. 112, R-W WA Köln.

[76] Oberpräsisent Spiegel to the Elberfeld-Barmen Chamber of Commerce, 11 June 1835, Abt. 22, Nr. 112, R-W WA Köln.

[77] Peill to Pestel, 24 May 1831, 403/3345, Bd.1, Bl.208, LHA Koblenz.

[78] For price trends after 1834, see the sources cited in n. 68.

[79] Haniel to Beuth, n.d. (1836/37), Portfolio 210, Haniel Museum Duisburg.

[80] Beuth to Rother, 24 March 1835, Rep.92 Rother, Ca Nr.17, Bl.18–19, ZStA Merse-

open between the Business Department and those whose interests it claimed to serve.

The negative experience in the western provinces reinforced Beuth's conviction that joint-stock companies were needed to construct the newer, larger factories equipped from the beginning with the latest technology. This was his intention for the worsted industries of the Eichsfeld and lower Rhine around Viersen and Eupen. In early 1826 the Business Department paid Cockerill 42,000 thaler to build an elaborate set of steam-driven "Mule-Jennies" like those of the French entrepreneur, Dobo, and to install some of them in Cockerill's crowded Berlin plant.[81] Because years earlier Berlin's most famous entrepreneur had refused to gamble that limited German sales would cover steep investments in steam-driven machinery, Dobo himself was brought in to rig the test room with steam. The experiment was finally ready for public view in September 1827. During subsequent years, the Business Department offered its worsted yarns to factory owners in the targeted areas in the hope that Dobo's machines would be adopted there on a wide scale. But not until July 1832—when Cockerill himself flirted *temporarily* with the idea of building a new spinning factory in Cottbus for combed wool—did Prussia's entrepreneurs show even the slightest interest in the new technology. Consequently, joint-stock partnerships were never formed.[82]

The problem was not with the machines as such. More familiar with wool than cotton, Beuth's technological instincts were correct this time. Mule-Jennies gradually came to dominate the more highly developed worsted industries of the Saxon duchies during the 1830s and 1840s before finally yielding to self-acting spinners in the 1850s and 1860s.[83]

The barriers to this sort of progress in the Eichsfeld were complex.[84] The combed wool factories of this once prosperous region were small, employing mostly hand-operated spinning jennies. Owners balked at the

---

burg. For Diergardt and the meetings, see Friedrich Otto Dilthey, *Die Geschichte der Niederrheinischen Baumwollindustrie* (Jena: Gustav Fischer, 1908), 18.

[81] See the sources cited above in n. 28.

[82] Karl Lärmer, "Zur Einführung der Dampfkraft in die Berliner Wirtschaft in der ersten Phase der Industriellen Revolution," *Jahrbuch für Wirtschaftsgeschichte* 1977 (Part 4): 114, overlooks the fact that Cockerill had abandoned steam power in the early 1820s. For this, and the remainder of the paragraph, see Kunth to Bülow, 9 May 1821; Severin to Schuckmann, 18 August 1826; Schwendy to Cockerill, 16 June 1827; Beuth to Cockerill, 22 September 1827, Rep.120, D.IV.1, Nr.15, Bd.1, Bl.37, 157, 163; and Cockerill to Beuth, 19 July 1832, Bd.2, Bl.49–50, ZStA Merseburg.

[83] For Mule-Jennies in Germany, see Blumberg, *Deutsche Textilindustrie*, 84–85.

[84] For this paragraph, see Schwendy to Cockerill, 16 June 1827, and Beuth to Cockerill, 19 January 1830, respectively in D.IV.1, Nr.15, Bd.1, Bl.157, and Bd.2, Bl.35, ZStA Merseburg; Beuth's article in *Verhandlungen* 6 (1826): 187–88, GB/TU Berlin; James, *Worsted Manufacture*, 417, 492; and Blumberg, *Deutsche Textilindustrie*, 77, 79.

extra construction costs and mechanics' wages associated with steam engines. They also knew that Dobo's machines were very expensive and required highly paid textile workers to operate. Their shops, on the other hand, produced a cheaper yarn that was preferred by many German weavers because it was considered more durable. Beuth scoffed at what he saw as prejudice, asserting that jennies were "admittedly inexpensive [to operate], but never supply a good, standard [*gleichförmig*] yarn."[85] The marketplace supported his contentions in this case, for the tight texture and regularity of the mule yarn imparted to it a quality which sustained sales *despite a higher price*. Jenny-spun worsted yarn was driven from external markets during the 1830s and Eichsfeld weavers were forced to survive locally on a mere one-seventh of their earlier output.

Of course, Beuth had not placed great hopes in the small factories in the first place, preferring instead to stimulate new joint-stock ventures. Sparse evidence prevents a more rigorous analysis, but among merchant capitalists less wedded to the old ways anxiety must have arisen over the prospect of surviving foreign competition without higher tariffs, especially after such an immense original investment. Prussia taxed woolen yarns, including worsted yarn, at the moderate rate of 6 thaler per centner. English worsted yarn, however, had fallen steadily and precipitously in price from 166 thaler in 1821 to 98 thaler per centner in 1829.[86] That year Beuth offered to sell some of the Business Department's Dobo yarn *at English prices* to B. G. Scheibler, owner of a worsted factory in Eupen which had mechanized along English lines in the early 1820s. Interestingly enough, Scheibler was "completely astounded" by the offer, doubting "its veracity"[87] because the prices Beuth quoted were equal per centner to what he paid his spinners in 1829. And by 1831 English prices had fallen to 78 thaler, a clear reflection of Britain's maturing industrial economy. With no qualitative edge to shield them against the cheaper import, Prussian manufacturers could probably be forgiven for refusing to follow the Thuringian example. Protected by prohibitive tariffs, worsted entrepreneurs from the Saxon duchies formed a syndicate in the late 1820s, installed Mule-Jennies, and began to battle for the German market.[88]

[85] See Beuth's article in *Verhandlungen* 6 (1826): 187–88, GB/TU Berlin.

[86] For tariffs, see Dieterici, *Statistische Übersicht*, 1838, 401; and for English prices, see James, *Worsted Manufacture*, 419, 492, 514.

[87] One of Beuth's agents described Scheibler's reaction. See Hoffmann to Beuth, 16 May 1829, Rep.120, D.I.1, Nr.20, Bl.197–210, ZStA Merseburg. For Scheibler's earlier modernization, see Blumberg, *Deutsche Textilindustrie*, 77.

[88] For the tariff policies of the states surrounding the kingdom of Saxony, see the petition of a Saxon businessman, C.W.E. von Wietersheim, 1 October 1827, printed in Curt Bökelmann, *Das Aufkommen der Grossindustrie im sächsischen Wollgewerbe* (Heidelberg: Universitäts-Verlag J. Horning, 1905), 88.

Beuth grumbled about the "lethargy"[89] of Prussian businessmen and never saw the wisdom of higher tariffs. Rather, he concluded privately to Cockerill that the price tag on the machinery had been too high.[90]

Consistently enough, the man captivated by beautiful machines sought the solution by attempting to produce them less expensively in Prussia. Beuth knew that a group of Silesian businessmen were eager to form a joint-stock company and construct a machine-building plant in Breslau, but lacked the capital. Therefore he emboldened them in 1830 with a gift of lathes, planers, shapers, and other exquisite machine tools worth 22,000 thaler. With the company still undersubscribed in early 1832, Beuth asked Christian Rother's Seehandlung to provide the necessary funds. It will suffice to say at this point that Rother agreed. Discussed at greater length in chapter 6, the Breslau machine-building factory opened its doors in November 1833. Beuth had made an exception in this case, departing from his moderately laissez-faire principles solely in order to supply Prussian entrepreneurs with cheaper machines. The plant's sales expanded slowly but steadily in the 1830s, as did its reputation for inexpensive, fairly reliable products.[91] By themselves, however, machines would not enable Prussia to compete with a land whose textiles benefitted from abundant capital and economies of scale. Construction costs in Prussia were 40 percent higher and operating costs 20 percent above those in England. One can find few historical cases where "infant industry" tariffs were more appropriate. But tariffs remained low with the result that imports, chiefly British, made up 47 percent of worsted yarn, 64 percent of cotton yarn, and 42 percent of linen yarn consumption by the 1840s.[92]

. . . . .

The completion of the Breslau machine tool project occurred at a time of decline for political and economic liberalism. Once high, hopes for par-

[89] For charges of "lethargy," see the correspondence about the Eichsfeld project between Interior and Finance of September-November 1829, in Rep.120, D.IV.8, Nr.22, Bd.1, Bl.102–37, ZStA Merseburg.

[90] Beuth to Cockerill, 19 January 1830, D.IV.1, Nr. 15, Bd.2, Bl.35, ZStA Merseburg.

[91] See the discussion and documentation in chap. 6.

[92] For comparative cost information, see Herbert Kisch, "The Textile Industries in Silesia and the Rhineland: A Comparative Study in Industrialization," in Peter Kriedte et al., *Industrialization before Industrialization* (Cambridge: Cambridge University Press, 1981), 186–87. For the other statistics here, see Blumberg, *Deutsche Textilindustrie*, 405; Dieterici, *Statistische Übersicht*, 1844, 344, 511, 514. The percentage of linen imports is based on consumption figures of one pound—or 4–5 ells (Dieterici, 1838, 410)—per head (Landes, *Unbound Prometheus*, 171; Banfield, *Industry of the Rhine*, 2:22), or 141, 593 centner of total consumption.

liamentarism had plummeted by the early months of 1833 as the Belgian crisis passed finally into history. The death of Georg Maassen in November 1834 marked a similar turning point for the fortunes of laissez-faire. "When one engages them in private conversation," observed a Hessian diplomat in Berlin, "the junior civil servants [*Geschäftsleute*] display signs of much dissatisfaction. Prussia could have lost all of her other ministers and the consequences would not have been as considerable."[93] While liberal bureaucrats slipped into despondency, economic conservatives maneuvered for position. Ludwig Kühne, Maassen's second lieutenant and principal trade negotiator, was too sarcastic, tactless, and politically liberal to ascend to Maassen's seat. The Ministry of Finance went instead to Albrecht von Alvensleben, a royal favorite known less for his liberalism than his lukewarm commitment to it. Thus Christian Rother soon turned Alvensleben against the Zollverein, arguing that Prussia's economic interests were poorly served by the now famous institution. Rother himself received the Business Department in a newly established ministry responsible for trade, road- and canal-building, and all manufacturing outside of the metallurgical sector. Not content with this arrangement, Gustav von Rochow and Postmaster Karl Ferdinand Nagler, two members of Wittgenstein's "younger, stronger party," schemed to bring all of Prussia's economic agencies under conservative leadership.[94] Although this plot miscarried, the now more conservative Ministerial Cabinet made its preferences clear by amending J. G. Hoffmann's guild reform proposals (see chap. 1) and scuttling a planned liberalization of Prussian mining law (see chap. 4).

Given the conservative revival in Berlin, Beuth's interests were better served by loyalty to his old friend and new chief. In February 1835, for instance, he defended Rother's appointment before Rhenish chambers of commerce which had criticized the new ministry because of the Seehandlung's history of competition with private industry. While Rother took the waters, moreover, Beuth kept him abreast of every party intrigue in the capital.[95] These personal and political circumstances help to explain the fact that each man pursued his own separate economic agenda in the mid-1830s without generating obvious friction. In his capacity as Director of the Seehandlung, Rother made a significant start with the vast busi-

[93] Schwedes, the Hessian plenipotentiary, made the remark in February 1835, cited in Lotz, *Geschichte*, 392.

[94] For the "younger, stronger" quote, see Rochow, *Vom Leben*, 263. For the remainder of the paragraph, see Beuth to Rother, 24 March 1835, 8 April 1835, and 1 June 1835, Rep.92 Rother, Ca Nr.17, ZStA Merseburg; Lotz, *Geschichte*, 392; Petersdorff, *Friedrich von Motz*, 2:20–21; and Treitschke, *Deutsche Geschichte*, 4:543–44, 576–78.

[95] Beuth to Rother, 20 February 1835, Rep.92 Rother, Ca Nr.17, ZStA Merseburg. See also the letters to Rother cited in n. 94.

ness empire that would later strain their relationship (see chap. 6). As head of the new ministry, however, Rother held his hand over the Business Department, enabling it to proceed with different means toward Beuth's goal of light country manufacturing in aesthetic surroundings. Rather than encourage large urban concentrations, Beuth continued the traditional strategy of dispersing industry to the provinces through the now familiar practice of disseminating knowledge and sample machines. One of his most ambitious projects at this time, for example, was state aid for a large linen finishing establishment in the countryside outside Bielefeld. Rother's leadership in the ministry may have contributed to this renewed emphasis on linen, for, like earlier advocates of rural industrialization, he valued it highly. Improved flax seeds, more efficient flax-roasting ovens, and a panoply of British and American linen machines were additional hallmarks of this revived program.[96]

Under Rother's congenial administration, Beuth was given a chance to realize another aspect of his vision. In 1836 he returned for a third time to the idea of attracting young bureaucrats to Berlin for a practicum under the Technical Deputation. And this time they received a small allowance to do so. But, as the years drew on to his retirement in 1845, Beuth's last hopes for technocracy gradually faded away. For the provision of stipends was not sufficient inducement for junior bureaucrats to reconsider the realistic prejudices which favored training in jurisprudence. The result, observed the Ministry of Finance, was that "district governments are largely without members who are knowledgeable about industrial technology."[97]

Beuth was not the only senior bureaucrat who failed to supplant legal study from its leading cultural position. Thus in 1839 Karl von Altenstein, the Minister of Education and Ecclesiastical Affairs, recommended that bureaucrats pay far greater attention to science and technology. In order to prepare officials for the new industrial age, he wanted a course of university study which combined two years of law with two years of the liberal arts. Moreover, state entrance exams should test in history, philosophy, political economy, and natural science as well as in the law. His suggestions reflected the prestige which the humanities and physical sciences had gained as a result of the concerted efforts of colleges of philos-

[96] For the Bielefeld project, see Rep.120, D.VIII.2, Nr.2, Bd.1; for the Viersen plant, Rep.120, D.VIII.2, Nr.14, Bd. 1 and 2; and for the other linen promotions, see documents in Rep.120, D.V1, Nr.16 and Nr.21 and Rep.120, D.V.2c, Nr. 8–9, ZStA Merseburg.

[97] Ministry of Finance to the Oberpräsident in Stettin, 15 February 1848, Rep.120, A.II.5b, Nr.9, ZStA Merseburg. Even in the developed industrial district of Düsseldorf, bureaucratic apprentices trained for only three of twelve months in the district industrial department. See the training regulations of March 1844, in Reg.D.Präs.B., Nr.12, HSA Düsseldorf.

ophy to compete with the primary discipline of law. During the 1830s there was a decline in university enrollments from 4,949 to 3,579 as well as a shift away from law to philosophy. While enrollments in law fell from 33.4 percent to 25 percent, enrollments in philosophy rose from 13.8 percent to 22.7 percent. Although Altenstein's proposals represented a significant revision of existing practice, they still remained within the general tradition which posited the bureaucrat as one whose broad, superior education enabled him to manipulate the specific advice of experts. Yet Altenstein was unable to convince his fellow ministers. The commission to which he directed his report finally concluded in 1846 that legal studies fully sufficed as training for higher civil servants. Small wonder that Beuth's far narrower technocratic agenda was unappealing to most Prussian bureaucrats of the Pre-March.[98]

Indeed a mere seven junior officials came to study with the Technical Deputation. The districts of Düsseldorf, Arnsberg, Oppeln, and Liegnitz were the only ones which possessed industrial councillors capable of acting in the spirit of Beuth's programmatic memorandum of 1817. Councillor Quentin of Düsseldorf was probably the most accomplished and energetic graduate of the Technical Deputation, able to blend the goals of the Business Department with the needs of the industrial bourgeoisie. In 1838 we find him coaxing 10,000 thaler from Beuth to support another linen finishing establishment near Viersen, and in 1844 he formed a committee of industrialists to counteract widespread Rhenish reluctance to participate in Beuth's last great industrial exhibition. The young councillor befriended Beuth's friends in the province, provided a reliable stream of information to Düsseldorf and Berlin, and joined enthusiastically in the political activities of socially concerned businessmen. His sympathy for the Frankfurt Parliament ended a promising career in 1849. Truly serving and bridging the interests of bourgeois factions in and out of government, Quentin was unfortunately the exception rather than the rule. And without thirty or forty more like him in Prussia, technocracy was a dead letter.[99]

[98] For Altenstein's initiative and its failure, see his memorandum of 24 September 1837, cited in Hans-Joachim Schoeps (ed.), *Neue Quellen zur Geschichte Preussens im 19. Jahrhundert* (Berlin: Haude & Spenersche Verlagsbuchhandlung, 1968), 241–50; and Schoeps, *Neue Quellen*, 238–39. For the university percentages, see Wehler, *Gesellschaftsgeschichte*, 2:214, 220.

[99] See the applications and records of the seven young bureaucrats interested in the project after 1836 in Rep.120, A.II.5b, Nr.9, ZStA Merseburg. For Quentin's mediation between Beuth and the linen manufacturers of Viersen, see Rep.120, D.VIII.2, Nr.14, Bd.1–2, ZStA Merseburg; for his popularity in and out of government in the district of Düsseldorf, see Viersen manufacturers to the Regierungspräsident in Düsseldorf, 12 January 1842, Reg.D.Präs.B., Nr.1018, and the annual report of the Regierungspräsident, 24 June 1843, Reg.D.Präs.B., Nr.1462, HSA Düsseldorf. For Quentin, the Industrial Exhibit of 1844 and

It seems doubtful, however, that a more universal technical bureaucracy would have been able to eliminate the widening rift between the center and the periphery. In the late 1820s, provincial businessmen had begun to found their own local associations to promote technological change because Berlin seemed out of touch. The acceleration of these trends in the 1830s eventually undercut membership in Beuth's Association for the Promotion of Technical Activity in Prussia, which stagnated from 1838 to 1840 before beginning a rapid descent.[100] "As thankfully as [Hansemann] acknowledged [past favors]," writes his biographer, "as much as he sincerely revered men like Motz, Maassen, Beuth, Kunth, Stägemann, and others, protestations, at first faint, then gradually growing more vehement, against the conceit and self-satisfaction of the bureaucracy as well as its defective expertise in specific areas, mount throughout all of his commentaries on political and economic relations in Prussia."[101] Even in Berlin where early contact with Beuth had been closer and friendlier, and the aura about the technocrat greater, manufacturers started to doubt the technological wisdom of the government. This was evident by 1839 when the capital's leading businessmen formed a polytechnical society as "a complement" (*Ergänzung*) to Beuth's association without inviting him to participate.[102]

For his part, Beuth had lost faith in all but a handful of Prussian businessmen. Thus in 1838 he cautioned Quentin about promising too much government support in Viersen lest weak-kneed industrialists expect Berlin to provide all of the capital.[103] And in 1843 he reacted angrily to the growing demand in business circles and among the provincial diets for a ministry of trade and business separate from the Ministry of Finance

cooperation with businessmen on social issues, see Regierung Köln to the Duisburg Chamber of Commerce, 19 February 1844, and the circular of the Düsseldorf chapter of the Verein für das Wohl der arbeitenden Klassen, signed by Quentin, in Abt.20, Nr.2, fasz.6, R-W WA Köln; and for Quentin in 1849, see the documents in 403/4148, LHA Koblenz.

[100] New associations were founded in Erfurt (1827); Cologne (1829); Schleusingen (1829); Suhl (1834); Düsseldorf (1835); Duisburg (1835); Coblence (1835); Aachen (1836); and Dresden (1837). That Berlin was out of touch, see Regierungspräsident Düsseldorf to Carl Boeninger, 10 September 1835, Abt. 20, Nr.2, Fasz.6, R-W WA Köln; and the materials in Reg.D.Präs.B., Nr.1030, HSA Düsseldorf; and 403/3452, LHA Koblenz. The membership decline of Beuth's Association is presented graphically in Matschoss, *Preussens Gewerbeförderung*, 160.

[101] Bergengrün, *David Hansemann*, 82.

[102] On the feelings of Berlin manufacturers in the 1840s, see Hartmut Kaelble, *Berliner Unternehmer während der frühen Industrialisierung* (Berlin: Walter de Gruyter, 1972), 239–52 (for the polytechnical association in 1839, see p. 238). Dissatisfaction among businessmen with seemingly arbitrary patent decisions is emphasized in Treue, *Wirtschafts- und Technikgeschichte Preussens*, 321ff.

[103] See Beuth's correspondence with Quentin in the summer of 1838, Rep.120, D.VIII.2, Nr.14, Bd.1, Bl.59–64, ZStA Merseburg.

(which had assumed control of the Business Department after the dissolution of Rother's ministry in April 1837). "The distribution of the various bureaucratic departments," Beuth lectured the Provincial Diet in Königsberg, "must be left exclusively in our hands."[104]

The railway mania which seized Prussia after the mid-1830s accelerated Beuth's falling out with the business community. As discussed more graphically in chapter 7, the dream of rural industrialization was beginning to yield during Beuth's years with Rother (1835–1837) to the reality of heavy industrial take-off in the modern era. It was some measure of their frustration with these unforeseen developments that the two waged a lengthy, almost "antimodern" delaying action against railroads and the overheated, dangerously overcapitalized industrialization which they believed would accompany the new transportation technology. But Beuth's anger over the seemingly ineluctable advance of the railways was probably exacerbated by fears and prejudices of another, vulgar sort. A decade earlier he had warned the Council of State that Hardenberg's emancipation of the Jews had been a mistake. "The lazy race of Jews will . . . monopolize such branches of the economy which yield significant profit without physical exertion." Then they would "subdue the productive class of the nation until the unmitigated burden leads the people to take the law into their own hands (*zur Volkshülfe führen wird*)."[105] In order to ward off what he perceived as a calamity, Beuth proposed, among other things, that Jews be forbidden from founding, or continuing to operate, commercial enterprises. That Jewish banking firms like the Mendelssohns of Berlin, Rothschilds of Frankfurt, and Oppenheims of Cologne were among the leading investors in railroad expansion undoubtedly struck a deep chord of revulsion and alarm in the intolerant privy councillor, helping to erode earlier procapitalistic beliefs and evoke anti-Semitic fears of unhealthy, counterproductive investments.[106]

Beuth the elitist, the aesthete, the antimodernist, and the anti-Semite—these are not the images to which we are accustomed. Heinrich von Treitschke described him, for example, as confidently swimming "in the current of this great century." Such "Whig history" can also be found in the work of Conrad Matschoss, who asserted that the head of the Busi-

[104] Beuth to the Provincial Diet of Prussia, 16 May 1843, Rep.120, A.I.1, Nr.23, Bd.6, Bl.43, ZStA Merseburg.

[105] Beuth's largely overlooked address to the Council of State in 1823 is cited in Schneider, *Der preussische Staatsrat*, 191–92.

[106] For the significant participation of Jewish banking families in railroad development in the mid-1830s, see Kaelble, *Berliner Unternehmer*, 154; and Wilhelm Treue, "Das Privatbankwesen im 19. Jahrhundert," and id., "Die Kölner Banken 1835 bis 1871," printed in Wilhelm Treue (ed.), *Unternehmens- und Unternehmergeschichte aus fünf Jahrzehnten* (Wiesbaden: Franz Steiner Verlag, 1989), 541–43, 571–72.

ness Department "unfurled futuristic images of German industry" which included "huge factory cities" that looked forward in time to "the Berlin, Rhineland, Westphalia and Upper Silesia of today."[107] Wilhelm Treue projected a similarly heroic image, claiming that "in practically no other person were the essential traits of a great entrepreneur so genuinely represented as in Beuth."[108] W. O. Henderson fell into the same category by stating that Beuth's real monument was not his statue in the Schinkelplatz, "but [rather] the great achievements of the industrial scientists of the [Second] Reich."[109] Implications that Beuth would have been pleasantly surprised by the urban-industrial world of the 1880s or 1920s, that he epitomized a pioneering industrialist, or that he paved modern industry's route, belie the skewed, refracted perspective of modern generations too caught up in their own times to perceive the reality of a vastly different era. The great technocrat is not to be found along the simple, linear, "past-to-present" progression drawn by many historians—he is off on another plane.[110]

Perhaps the best illustrations of this were a backward-looking love affair with the art of antiquity and a seemingly escapist preoccupation with horses—*both of which grew more intense as modernity's veil lifted.* Beuth drew close to August Borsig during the late 1830s, deriving gratification from the locomotive king's receptivity to the notion of aesthetic industrialization. The former student of the Technical Institute built his engines with "tastefulness" and surrounded both his villa and factory with the Greek pillars that were Schinkel's ideal. Beuth continued to invite industrialists and state servants to his home in Schönhausen on Sundays, but he found it difficult to animate these friendly gatherings with the old excitement unless the art world came under discussion. Then, as one visitor recalled, the conversation "moved along excellently."[111] Indeed, Beuth's artistic entourage, already impressive in the 1820s, expanded during the later years to include young neoclassicists like Friedrich

[107] See Treitschke, *Deutsche Geschichte*, 3:463; and Matschoss, *Preussens Gewerbeförderung*, 71.

[108] Wilhelm Treue, "Deutsche Wirtschaftsführer im 19. Jahrhundert," printed in Treue, *Unternehmens- und Unternehmergeschichte*, 211.

[109] Henderson, *The State*, 118.

[110] See also Otto Büsch and Wolfram Fischer in their introduction to Mieck, *Preussische Gewerbepolitik*, vi; the author (Mieck), 235–39; and Barbara Vogel, "Die 'Allgemeine Gewerbefreiheit' als bürokratische Modernisierungsstrategie in Preussen," in Dirk Stegmann et al. (eds.), *Industrielle Gesellschaft und politisches System* (Bonn: Verlag Neue Gesellschaft GmbH, 1978), 73, 77–78.

[111] For Borsig's villa, see ADB 36:484–85; for Beuth and Borsig (and the "tastefulness" quote), see Treue, "Deutsche Wirtschaftsführer," in Treue, *Unternehmens- und Unternehmergeschichte*, 220; the description of Beuth's gatherings in the early 1840s is found in Delbrück, *Lebenserinnerungen*, 1:136.

Drake, Johann Heinrich Strack, and August Kiss. It was Strack, a pupil of Schinkel, who carried out the work on Borsig's buildings, while Drake, largely unknown, later executed the bronze reliefs on Beuth's monument in the Schinkelplatz. Kiss, a bronze sculptor and foundry instructor at the Technical Institute, won a special place in Beuth's heart with his "Amazon Warrior on Horseback Fighting a Tiger," completed for the Old Museum in 1838. As his fixation with horses deepened, Beuth expanded his library holdings on the subject to 178 volumes and undertook private researches into Greek, Roman, and Arabic horse-breeding culture.[112] Rudolf von Delbrück, a young staffer in the Business Department in 1843, recalled that Beuth could often be found by the ground floor arena of the royal stables and riding school which became a sort of second office in the twilight of a long career. Delbrück also marveled at Beuth's "rich collection of replicas of the most outstanding works of industrial art from remote antiquity" and paid homage to his overall campaign "to integrate art with industry." But Delbrück was saddened that "advance of age and stubbornness of character did not permit [Beuth] to take a new tack."[113] It was the aesthetic alternative—one to which Beuth was more rigidly committed than ever—that greeted visitors to the great Industrial Exhibition of 1844. Artful machinery was displayed there alongside the products of the soil, the country factory, the instrument-maker's bench, and the artist's casting furnace.[114]

The Exhibit opened with a new king of four years, Frederick William IV, on the throne.[115] His first professional biographer, Hermann von Petersdorff, refers to Frederick William's "satisfaction"[116] with the great technological fair. That the reaction was positive may have been a reflection of the monarch's tempered view toward industry. The frustrations of combating industry's march, the practical advice from his father about the importance of Prussia's economy, numerous trips through the industrializing provinces, and a chic Anglophilia had all contributed to an abandonment of earlier efforts to curb the growth of factories. It would be a mistake, however, to describe Frederick William IV as an advocate for Beuth during the latter's waning years in office. As Crown Prince, he had championed railroads over the bitter opposition of Beuth and Rother.

[112] ADB 16:35–36 (for Kiss) and 36:484–85 (for Strack); Matschoss, *Preussens Gewerbeförderung*, 71–72, 75–76.

[113] Delbrück, *Lebenserinnerungen*, 1:137 (for the quotes: 1:135).

[114] Ibid., 1:135; Viebahn (of the Exhibition's organizing committee) to the Regierung Köln, 5 July 1844, Abt.1, Nr.28, Fasz.1, R-W WA Köln.

[115] In general, see Treitschke, *Deutsche Geschichte*, 5: passim; Hermann von Petersdorff, *König Friedrich Wilhelm der Vierte* (Stuttgart, 1900); and Ernst Lewalter, *Friedrich Wilhelm IV: Das Schicksal eines Geistes* (Berlin, 1938). A new study of Frederick William is forthcoming from David Barclay of Kalamazoo College.

[116] Petersdorff, *König Fiedrich Wilhelm*, 58.

Nor did Frederick William sympathize with Beuth's aesthetical engagement with technology. Thus in April 1844, Beuth gave his sovereign a finely cast brass picture frame, beautiful in its metallic simplicity, only to see the king gild it, then give it away to one of the religious "zealots" (*Frömmler*) in his entourage. Frederick William's apparent accommodation with the modern business world, moreover, was a superficial one. The king's real interests lay with religion, literature, and music, while his fascination with railroads did not prevent him from devoting considerable attention to the economic and political rehabilitation of the nobility. Frederick William gave enthusiastic backing to Gustav von Rochow, a former devotee of Adam Müller, in the establishment of the Board of Agriculture (1842), an institution which promoted farming improvements in a "zero-sum" struggle against industry for state resources.[117] The king's advisers were also at work on a new patent for the nobility which was designed to prevent the embourgeoisment of the landholding elite (see the Conclusion). All of this was incompatible with Beuth's grand middle-class agenda.

Court schemes and intrigues threatened to add injury to insult. Theodor von Schön, the ambitious, contentious Governor of East and West Prussia, cautiously championed the business world's demand for an autonomous ministry responsible for trade and industry in 1841. The threat from the old "Jacobin" triggered a counterintrigue in 1842 from two members of Frederick William IV's pietistic circle, Carl von Voss and Leopold von Gerlach. The two "zealots" proposed a member of their party, Friedrich von Rönne, for the new ministry. This can only be interpreted as a clever "retreat to the front" (*Flucht nach vorn*), for Voss and Gerlach were no friends of industry. With the king favoring business representation and often expressing annoyance that the bureaucracy paid no heed to "the mass of experience which exists in the commercial and industrial public,"[118] the plot's chances were good. No friend of the Müllerites and pietists, Prince William, brother of the new king, intervened at the eleventh hour with a compromise which prevented the feudal entourage from wreaking total havoc in Prussia's already divided economic policy-making apparatus. Beuth and his colleagues in the Business Department would keep all of their positions, while Rönne would head a

[117] Varnhagen Diary II, 20 April 1844, 2:287; Petersdorff, *König Friedrich Wilhelm*, 58; Pruns, *Staat und Agrarwirtschaft*, 1:54–59.

[118] For the whole episode, see Voss to Gerlach, 2 April 1841, printed in Schoeps, *Neue Quellen*, 273–75; Schön to Brünneck, 4 and 7 February 1842, and notes of a conversation between Brünneck and the king, 14 January 1842, printed in Schön, *Aus den Papieren*, 3:494, 497, 478; Varnhagen Diary II, 22 May 1844, 2:299; and the king's decree of 7 June 1844, Rep.120, A.I.1, Nr.21, Bl.6, ZStA Merseburg. The king's remarks about the bureaucracy are cited in Petersdorff, *König Friedrich Wilhelm*, 58.

Board of Trade (*Handelsamt*) to advise the king. Prince William had prevented the worst from happening, but, as one of Beuth's younger colleagues lamented, this resolution of the conflict barred the way for any "consistent policy"[119] toward industry.

After mid-decade approached and passed, the longtime Head of the Business Department asked to be relieved of his duties. His division relegated more and more to mundane, unappreciated patent decisions, his plans for parliamentary technocracy a chimera largely forgotten, his vision of rural, aesthetic industrialization rapidly yielding to railroads, heavy industry, and the ugly reality of a modern world, there seemed little reason to carry on.[120] So, after the Exhibition closed its doors, Peter Beuth stepped down from his steed. The view from Pegasus vanished with him.

[119] Delbrück, *Lebenserinnerungen*, 1:146.

[120] For Beuth's bitterness, see Delbrück, *Lebenserinnerungen*, 1:135–36; for the tendency to concentrate solely on patents, see Matschoss, "Geschichte der Königlich Preussischen Technischen Deputation," 262–64.

IV

The Spirit of the Corps

THE PEAK OF THE SCHNEEKOPPE towered over the Eleusinian symmetry of Buchwald, dominating the estate's carefully laid-out walkways and ornamental Roman ruins. Its owner's usual constitutional was cut short this September evening by the type of furious downpour one came to expect living next to the mountain. It was probably for the better, for Friedrich Wilhelm von Reden moved much more slowly and painfully than most well-to-do noblemen of fifty-five years. And then there were the migraines.

The Countess Friederike chose not to speak as Reden settled grimacing and angrily onto the divan. Although his beloved partner and equal, she knew what the events of the past year had done to his nerves and his moods—Jena, routed armies, fortresses surrendered without a shot, her husband compelled to pledge the allegiance of his beloved Mining Corps to the Corsican, and finally, dismissal for alleged disloyalty. Worse still, rumors were circulating that an angry king would disband the Corps itself. But there was good news to mix with the bad. Former colleagues Gerhard and Karsten had just written from Memel that Frederick William had seemed impressed with their impassioned defense of the bureau's financial and technological service to the kingdom. And now Baron Karl vom Stein, himself a former bearer of the dark, martial uniform of the Corps, would again assume the duties of chancellor.

"Our dear blackcoats are surely in good hands now," Frederike ventured when the moment seemed ripe. Reden's anger at the king gradually turned to wistfulness as he thought of his comrades. "They are doing so very much, the true ones," he said finally, eyes already welling with tears. "Perhaps I deserve their love, but they do more than I deserve. God bless them."[1]

Pride in technical excellence was central to Reden's deeply emotional bond with fellow members of the Mining Corps. Indeed, in its theoretical and practical knowledge of mineralogy, geology, surveying, tunnel-

[1] Countess von Reden relayed her words as well as her husband's remarks and emotions to Countess von Itzenplitz in a letter of 30 September 1807, cited in Wutke, *Aus der Vergangenheit*, 407. For the news from Memel, see Karsten (the elder) to Reden, 15 September 1807, and Gerhard (the elder) to Reden, 19 September 1807, cited on p. 409. For other evidence used in this scene, see pp. 141, 152–54, 162.

ing, mining, and metallurgy, the Prussian "*Oberberghauptmannschaft*" was unsurpassed in Europe. Out of the spirited camaraderie which reigned within this classless community of experts flowed an egalitarian critique of a classbound society. Bourgeois initiates were understandably excited about the equal opportunity which the Corps opened up to them in this waning aristocratic age. "Thank God we have no noblemen's pew in mining!"[2] exclaimed D.L.G. Karsten, one of the decidedly liberal "true ones," to Reden. But the chief captain was not offended. On the contrary, Silesian aristocrats found it "bizarre" that Reden was more relaxed in the company of middle-class professors, physicians, literati, and fellow corpsmen. With such an open-minded example to follow, other noblemen broke with occupational tradition, donned the cap of "mining cadet," and took their places next to the Karstens, Gerhards, and Rosenstiels, unconcerned with the *déclassé* status assigned to them by older elites.[3]

Experience, enthusiasm, and talent—not birth or class—explained the successes and accomplishments of this new technological aristocracy. "What ever would have become of mining if you had taken all of your officials from the upper castes?"[4] wrote Stein to Reden, his close friend, just weeks before the humiliation of Tilsit. There seems no reason to doubt that the Corps served as an important sociopolitical model for Stein as he began his famous reform administration that fall. It may also have added a liberal tint to the politics of other famous graduates like Johann August Sack and Alexander von Humboldt. It is the combined statism and egalitarianism of the Prussian Mining Corps which transcends two centuries to our present, bequeathing a definite modernity to this unusual institution.

And yet in another sense it is misleading to describe the early leaders of Prussian mining and metallurgy as modern, if by this we mean that they viewed their technological mission in a forward-looking, progressive sense. The Thirty Years' War (1618–1648), the Turkish invasion (1681–1697), and four European conflagrations between 1688 and 1763 brought terrible destruction and decline to once-flourishing mining centers in Hungary, Bohemia, Silesia, and Saxony. From the vantage point of the late 1700s, it was more reasonable to see the German past in brighter terms than its present or immediate future. The mission, in other words,

[2] Karsten to Reden, n.d. (1810), cited in Gustav Karsten, *Umrisse zu Carl Johann Bernhard Karstens Leben und Wirken* (Berlin: Druck von Georg Reimer, 1854), 90 n.1.

[3] For the "bizarre" quote, see Wutke, *Aus der Vergangenheit*, 163. See also p. 173. In general for the attitude of noblemen in the Corps, see Wolfhard Weber, *Innovationen im frühindustriellen Bergbau und Hüttenwesen: Friedrich Anton von Heynitz* (Göttingen: Vandenhoeck & Ruprecht, 1976), 199.

[4] Stein to Reden, 7 June 1807, cited in Wutke, *Aus der Vergangenheit*, 404.

was to restore Germany's mines to an *earlier* level of technical competence. Thus Abraham Gottlob Werner, an innovative geological specialist who from 1775 to 1817 taught at the prestigious mining academy in Freiberg, Saxony, based his mining lectures on Georg Bauer (Agricola), the Joachimsthal physician whose famous treatise on mining and metallurgy had appeared in 1556. The apprentice Prussian bureaucrats who sat at Werner's feet dominated the Corps until the mid nineteenth century. Friedrich Anton von Heynitz, chief captain of the Corps from 1779 to 1802, was even more preoccupied with the past. He believed, somewhat erroneously, that the sixteenth- and seventeenth-century mining technology which so impressed his contemporaries was itself behind the level of Roman antiquity. "Tunnel where the ancients did," the young men of his entourage were constantly reminded. From this circle came a line of four chief captains stretching into the 1850s.[5]

The pride which this later generation felt for its own accomplishments may have been measured, in fact, against the standard of former heroic eras. Thus Heinrich Steffens, a mining cadet in Freiberg in 1799, described his daydreams while following Werner through the five-hundred-year-old tunnel network of the Erzgebirge. Would there be anything in the future to compare, he wrote, "with the Cyclopean walls [of the Bronze Age], the remains of Susa and Palmyra, the Greek ruins or the Roman roads and aqueducts?" Would Europe's "lightly constructed cities, its luxurious palaces or its greatest factories" leave much trace? Steffens concluded that nothing much from his own time would survive which would truly impress future generations. Only the "mighty technology of the mines" would reach out to posterity. The imposing neoclassical gates which marked the openings to the Saar mines of Mining Captain Ernst von Beust, a close friend of Steffens in Freiberg, were monuments to this same mentality, persisting forty years later in the man who led the Corps into modern times.[6]

Fascination with the ancients and early moderns did not fade with ex-

[5] For the Heynitz quote, see Weber, *Innovationen*, 133; and for Roman mining in the area, C. N. Bromehead, "Mining and Quarrying to the Seventeenth Century," in Singer et al., *History of Technology*, 1:6. For Werner and Agricola, see ADB 42:35. Prosopographical information on chief captains Reden (1802–7), Ludewig Gerhard (1810–35), Franz von Veltheim (1835–40), and Ernst von Beust (1840–59) is found, respectively, in ADB 27:510–13; Rep.121, Abt.A, Tit X, Sect.7, Nr.2, and Sect.21, Nr.3, ZStA Merseburg; and Hans Arlt, *Ein Jahrhundert Preussischer Bergverwaltung in den Rheinlanden* (Berlin: Wilhelm Ernst und Sohn, 1921), 64. All studied with Werner and entered the Corps under Heynitz.

[6] For the quotes and Steffens' friendship with Beust, see Heinrich Steffens, *Was ich erlebte* (Breslau: Josef Max, 1840–44), 4:203, 205–6, 220–22, 229. For sketches of the neoclassical architecture of the Saar mines, see Max Noggerath, "Der Steinkohlenbergbau des Staates zu Saarbrücken," *Zeitschrift für das Berg-Hütten- und Salinenwesen im Preussischen Staate* 3 (1856/B): 159.

citing news from England of deeper mines, coke iron, and steam engines.[7] The protégés of Werner and Heynitz continued to see their challenge as one of equaling, perhaps in some cases surpassing, the accomplishments of past glorious ages. In the County of Mark, Stein completed a half-century of Prussian efforts in the 1780s by introducing shoring and tunneling methods developed centuries earlier in central Europe. Two of Reden's greatest successes in Silesia were to substitute the blast furnace—a sixteenth-century technology—for simpler hearth and hammer techniques (*Lupenfeuer*), and to reopen the lead and silver mines of Tarnowitz, abandoned since the 1500s.[8] Reden could write to the king in 1787 about the industrial "pearl" which Silesia would become "in the perhaps still very distant future,"[9] but the young Alexander von Humboldt was dismayed to learn a few years later that Reden's futuristic vision was essentially a mirror image of the past. Science had little to offer the Corps, Reden said, for "in technical matters we must adhere to the old ways" (*das Technische müsse beim Alten bleiben*).[10] In great contrast to Humboldt, the Count's ardent defenders were proud that Silesia's mining past had been "awakened and raised up."[11] Similarly, when Sack arrived in the Rhineland in 1814 he bemoaned the neglect which French administrators had shown toward "our venerable mining—that industry, rich in knowledge, whose technical language alone proves that it has been practiced by Germans with fondness and characteristic thoroughness since time immemorial."[12] After the wars one of Werner's last students, Franz von Veltheim, took great pride in reactivating coal mines near Wettin and Löbejün which had prospered in former times.[13] As we shall see, this "backward-looking progressivism," if you will, was characteristic of the Corps during the Era of Restoration.

One legacy of this worldview was a largely static concept of economic

[7] Reden was extremely interested in England and English developments. See Wutke, *Aus der Vergangenheit*, 97–98. See also Stein's letter to Reden about England, 4 May 1788, printed in Botzenhart, *Briefe*, 1:280.

[8] For these accomplishments, see Agnes M. Prym, *Staatswirtschaft und Privatunternehmung in der Geschichte des Ruhrkohlenbergbaus* (Essen: Verlag Glückauf, 1950), 16–17; M. Reuss, "Mitteilungen aus der Geschichte des Königlichen Oberbergamtes zu Dortmund und des Niederrheinisch-Westfälischen Bergbaues," *Zeitschrift für das Berg-Hütten- und Salinenwesen* 40 (1892): 322; and Henderson, *The State*, 13–14, 34–36. For Stein's early study of central European mining technology, see Heinrich Ritter von Srbik, "Die Bergmännischen Anfänge des Freiherrn von Stein 1779 und Ihr Nachklang 1811/12," *Historische Zeitschrift* 146 (1932): 480–88.

[9] Reden to Frederick William II, cited in Wutke, *Aus der Vergangenheit*, 93.

[10] Humboldt to Freiesleben, 7 March 1792, cited in ibid., 255.

[11] Wutke (ibid., 428) cites a speech by Friedrich Philipp Rosenstiel, a close associate of Reden, of March 1810.

[12] Cited in Arlt, *Ein Jahrhundert*, 16.

[13] ADB 39:586–87.

growth. Technological restoration of the economy to an earlier—and it was thought—higher level of activity meant only a temporary expansion of output before productivity leveled off. With the completion of Stein's improvements in the Ruhr around 1790, in fact, a ban was placed on new mine openings that was only lifted grudgingly in 1826. Empowered by the Prussian Legal Code of 1794, moreover, the Corps often refused to license new blast furnaces or finishing establishments.[14]

Two additional factors help to explain this reluctance to proceed quickly beyond a certain point. Eighteenth-century bureaucrats and political economists had little experience, and hence, little appreciation for the dynamism characteristic of modern economies. Accordingly, leaders of the Mining Corps expressed repeated concern that the ill-advised concessioning of ironworks would exhaust insufficient supplies of charcoal fuel, glut limited iron markets, send prices plummeting, and undermine tax revenues from state operations.[15] Reporting to Berlin about the intentions of private ironmakers to build new furnaces in 1832, for instance, a Corps metallurgist in Silesia wrote: "I fear there will be too many of these."[16] Mining experts expressed similar concerns about markets, prices, and revenues. In their case, caution was reinforced by the traditional technique of hillside tunnel-mining (see below), a practice which rarely exposed the vaster quantities of mineral deposits far below the earth's surface—or even knowledge of them. To Corps members unaware of these coal resources, a rapid pace of long-term growth seemed an inexcusable plundering of nature and an irresponsible mortgaging of future state income.[17] The captains of mining and metallurgy believed there was ample justification for strict supervision of "irresponsible" private producers and state ownership of mines and metalworks whenever possible.

As Reden paced the walkways of Buchwald in forced retirement, however, he had good reason to worry about the future of this mercantilistic approach. For the economic philosophy of his "black coats" was out of political favor. The laissez-faire advisers who now surrounded Frederick William referred derogatorily to Reden's contingent as "the mining

[14] Krampe, *Der Staatseinfluss*, 43. For the Corps and the Prussian legal code of 1794, see the *Motive zu dem Entwurf des gemeinen preussischen Bergrechts und der Instruktion zur Verwaltung des Bergregals*, 1841, Rep.84a, 11080, pp. 4–5, GStA Berlin (hereafter cited as *Motive zu dem Entwurf*-1841, GStA Berlin).

[15] Reil to Karsten, 15 May 1832, Rep.121, F.IX.3, Nr.103, Bd.1, Bl.55, ZStA Merseburg; Krampe, *Der Staatseinfluss*, 32, 34; Konrad Fuchs, *Vom Dirigismus zum Liberalismus: Die Entwicklung Oberschlesiens als preussisches Berg- und Hüttenrevier* (Wiesbaden: Franz Steiner, 1970), 102–5.

[16] Reil to Karsten, 15 May 1832, cited in n. 15.

[17] *Protokolle über die Revision des Berggesetzes in Folge der Gutachtlichen Bemerkungen der Provinzialstände Mai 1845 bis Dezember 1846*, 22 May 1845, Rep.84a, 11080, p. 14, GStA Berlin (hereafter cited as: Mining Corps Protocols 1845/46, GStA Berlin).

clergy," recommending severe curtailment of its activities. "There is absolutely no reason," wrote Karl von Altenstein in September 1807, "why all burdensome limitations to freedom should not also be removed from mining."[18] The Corps could maintain a few model operations and offer technical advice when required, but mining was no more complicated than farming and, like agriculture, required no complex regulatory apparatus to flourish. The outgoing chancellor, Karl August von Hardenberg, seconded Altenstein, and both reports fell on fertile ground at court. Disliking most state monopolies, the king considered selling the Corps' mining and metalworks and reducing the once extensive mercantile domain to an advisory service attached to each district government.[19]

It was only the forcefulness and persuasiveness of Stein that shielded his former comrades from these first assaults. And Reden was truly grateful. "I thank God with tears in my eyes," he wrote to Karsten in November 1808, "for the rehabilitation and maintenance of [the Mining Corps] and the good sense and unflagging zeal of its advocates." The corporation would "save itself and become important [once again], however slight the prospects, hcwever great the difficulties."[20]

The Count's words would prove prophetic indeed. The upcoming years saw both challenges and—what was always most important to members of the mining service—survival. Shortly after Stein's departure in November 1808, Frederick William decreed the sale of most state mines and ironworks at the earliest practical opportunity. Then in October 1810, after sacking most of Reden's remaining colleagues, Hardenberg demoted the Mining Corps to a subordinate section under the liberal Business Department.[21] Further dismantling of the once-autonomous Corps and the announced "privatizing" (*Veräusserung*) of state operations seemed likely to follow.

That they did not was the result of shrewd maneuvering and propaganda by the new chief captain, Johann Carl Ludewig Gerhard. The forty-two-year-old, second generation salt-mining expert had studied with Werner in Freiberg, then advanced through the organization during Heynitz's administration in the 1780s and 1790s. By 1806 he headed the mine office at Rothenburg, one of four in the Prussian service at that time. A loyal patriot and personal favorite of the king, Gerhard reached

[18] Altenstein to Frederick William, 11 September 1807, printed in Winter, *Reorganisation*, 453.

[19] Hardenberg to Frederick William, 12 September 1807, printed in ibid., 335; Frederick William to Massow, 3 September 1807, printed in Wutke, *Aus der Vergangenheit*, 408.

[20] Reden to Karsten, 23 November 1808, cited in Wutke, *Aus der Vergangenheit*, 411; for Stein's effectiveness, see Reden to Karsten, 20 November 1808, cited on p. 410.

[21] See Frederick William's decree of 15 April 1809, cited in ibid., 416; for the sackings, see pp. 417, 431–32.

the top by 1810. Once there he combined personal advantages with the rare opportunity provided by the wars of liberation to prove the worth of his organization. The seventy-nine iron cannons, thirty-eight bronze field pieces, and tens of thousands of cannon balls and other ordnance produced at Corps plants in Silesia operating day and night could not help but impress the generals and the king. Six weeks after the Battle of Nations at Leipzig, in fact, the mining "section" was upgraded to an autonomous bureau equal to the Business Department within the Ministry of Finance.[22]

The coming of peace found Gerhard and his fellow captains locked in struggle with economic liberals over the extent of state interference in mining and metallurgy (see chap. 1). The Corps relinguished most of its regulatory powers over metalworks and failed to extend Prussian mining law to the western provinces, but generally fared well in this internecine warfare. The years of "steady conflict,"[23] as Beuth put it, abated somewhat in 1817 when the Corps was transferred to the more hospitable environs of Schuckmann's Ministry of the Interior. Gerhard was sometimes treated roughly by the haughty aristocrat, but always knew how to present a good case for the captaincy and generally found a receptive ear in Schuckmann, the man Heynitz had first recommended to Hardenberg. Thus when Count von Wittgenstein tried to eliminate the Dortmund Mine Office, leaving Beust's officials in Bonn with the onerous burden of supervising the entire Rhineland and Westphalia, Gerhard and Schuckmann convinced Frederick William to reject the scheme. The Corps had not survived a challenge from economic liberalism only to fall victim to conservative hostility toward industrial technology.[24]

Just like Reden, Gerhard was surrounded by "true ones" who helped manage the Corps' complicated operations. In fact, the team which he assembled in Berlin headquarters after 1815 was run in a collegial fashion by experts whose functions were very specialized. Carl Wohlers managed Corps financial affairs, while Friedrich Skalley handled mining law. Carl Karsten and an associate, Heinrich Klügel, headed a section dealing with metallurgy and metallurgical machinery. Finally, Gerhard and his colleague, Karl von LaRoche, headed a division responsible for the vast area

[22] Ibid., 473–78; Georg Schmidt, *Die Familie von Dechen* (Rathenow: M. Babenzien, 1889), 97–99; Schulz-Briesen, *Der preussische Staatsbergbau*, 60–62; and Bülow to his bureau heads, 24 July 1814, Rep.120, A.I.1, Nr.23, Bd. 2, Bl.58–59, 105, ZStA Merseburg.

[23] Beuth to Bülow, 31 December 1817, Rep.120, A.I.1, Nr.2, Bd.1, Bl.64, ZStA Merseburg.

[24] Krampe, *Staatseinfluss*, 48–68; Reuss, "Mitteilungen aus der Geschichte," 352–53; Fuchs, *Vom Dirigismus*, 106–12. For Heynitz and Schuckmann, see Lotz, *Geschichte*, 345. For the unpleasantry between the Corps and Schuckmann, see Varnhagen Diary I, 21 February 1820, 1:85; and Karsten, *Umrisse*, 88–90.

of mining. These six persons were joined for the most critical decisions by the heads of the five regional mine offices: Hans Martins (Brandenburg), Franz von Veltheim (Lower Saxony), Ferdinand von Einsiedel (Upper Silesia), Hermann Bölling (Westphalia), and Ernst von Beust (Rhineland).[25] As chief mine captain, Gerhard was a *primus inter pares* whose main external function was to speak for the captaincy before Schuckmann and other outsiders in the bureaucracy.

On certain matters, collegiality led quickly to consensus within Prussia's "mining clergy." We have seen that a major item on the Corps' agenda was the task of "awakening and raising up" mining and metal centers to their former flourishing status. This challenge was now more daunting due to the acquisition of declining or neglected mining regions like Lower Saxony and the Rhineland. The wars had also delayed replacement of worn-out machinery in recently established centers like Silesia and the home province of Brandenburg. Topping the list for funding priority was Franz von Veltheim's Lower Saxon Office where 306,000 thaler were invested between 1819 and 1825, mainly into restoration work on the long dormant coal mines of Wettin and Löbejün. The prestigious Silesian Office was next with 247,000 thaler during this period, primarily for repair and expansion of metallurgical operations. Martins's Brandenburg Office was alotted 200,000 thaler for its iron furnaces, while the western offices of Beust and Bölling shared 271,000 thaler to expand mining and metal works (see below). Assuming that two-thirds of this amount was a *net* increase in capital stock, then Corps investments were responsible for 3.8 percent of net industrial investment in Prussia between 1819 and 1825. These figures help to demonstrate the by no means negligible role of the state in early Prussian industrialization. But these massive projects also paid large dividends for the state—an annual surplus from the Corps in 1827, for example, of 1.5 million thaler.[26] Roughly equivalent to the proceeds of the tax on business (*Gewerbesteuer*), these revenues enhanced the political security of Gerhard and his black coats. Indeed, the "rehabilitation and maintenance" of the Corps itself was another area where consensus was easily achieved.

As easy as it was to reach agreement on these matters, when other

[25] For Gerhard's specialty, see Neue Deutsche Biographie 6:274; for Karsten, see Karsten, *Umrisse*, 88–90; for La Roche, Skalley, and Wohlers, respectively, see their personnel files in: Rep.121. Abt.A, Tit.X, Sect.17, Nr.1000, Sect.18, Nr.101, and Sect.22, Nr.15, ZStA Merseburg. The reports of the chief captain to the ministry are located in Rep.121, A.XX.1, Nr.102, Bd.1ff, ZStA Merseburg.

[26] For the investment amounts, see Treue, *Wirtschaftszustände*, 193, 217–18, 232. For the method used in calculating percentages of net industrial investment in Prussia, see chap. 1, n. 63. Ratios of net to total industrial investment are found in Tilly, "Capital Formation in Germany," in Mathias and Postan, *Cambridge Economic History*, 7 (1): 426. For Corps surpluses in 1826–27, see Rep.121, A.XX.2, Nr.102, Bd.10, Bl.5, ZStA Merseburg.

issues were under discussion, the comaraderie of the Mining Corps could turn to ugly discord. Tariffs and railroads were two such topics which sparked internal controversy in the late 1830s and 1840s. The earliest and most persistent source of disharmony, however, came over the question of amending Prussian mining law to soften the *Direktionsprinzip*, or regulatory principle, which still guided a third generation of Corps leaders. Under Prussian law, mines were not considered private property, rather a temporary privilege to exploit treasures owned by the state. Corps officials were also empowered to mandate certain technologies, supervise the books of mine companies, and set prices for ore and coal. Opposition to these legal traditions centered around Carl Karsten, head of the metals division, and Friedrich Skalley, the legal expert. They circulated a handwritten draft of a new mining law in 1827 which proposed eliminating the Corps' managerial duties in private mines and reducing technological control to mere supervision.[27]

Karsten "the younger" had a solid scientific education and fourteen-year metallurgical practicum in Upper Silesia behind him when he took charge of the metals section in 1817. Initially an outspoken advocate of the *Direktionsprinzip*, his agenda turned liberal during the 1820s, including causes like low tariffs and liberalization of Prussian mining law. This conversion resulted in part from observing the impressive performance of private ironworks during years when his section—in contrast to Gerhard's mining division—had already lost most of its regulatory powers. Personal and political factors may also have affected a certain rebelliousness. Karsten came to Berlin under the impression that the metallurgical section would be an interim assignment before promotion to chief mining captain of the Upper Silesian office, probably the most prestigious post in the Corps. But his unpretentious nature, open dislike of aristocratic privileges, and friendship with political liberals and fellow members of the Spanish Society like Peter Beuth and Friedrich Link, Director of the Royal Botanical Gardens, lost him Schuckmann's favor. Karsten was also influenced by Friedrich Skalley whose early years at the University of Königsberg and later study of European mining law bred an appreciation for western practices. And both, finally, were swayed by the magnetic Beuth. Karsten joined Beuth's Association for the Promotion of Technical Activity in 1821 and soon befriended the liberal Privy Councillor. Skalley had moved in these circles even longer, doubling as a legal expert in the Business Department since 1818.[28]

The forces of change favored by Karsten and Skalley remained rela-

[27] Mining Corps Protocols 1845/46 (16 January 1846): 261, GStA Berlin.

[28] Karsten, *Umrisse*, 88–90, 96–99, 101; Wutke, *Aus der Vergangenheit*, 436; and Skalley's file in: Rep. 121, A.X.18, Nr.101, Bl.2, ZStA Merseburg. For Skalley and the Business Department, see *Handbuch über den Königlichen Preussischen Hof und Staat*, 1818.

tively weak within the command structure of the Corps during the late 1820s. Counterbalancing Karsten and Skalley within Berlin headquarters were traditionalists like Gerhard and LaRoche, while outside, a solid majority of the influential mine office heads opposed reform. Einsiedel and Veltheim had both come up through the old school,[29] and Beust, yet another protégé of Heynitz and Werner, made no effort to conceal his contempt for French mining principles which illustrated "how dangerous it was to abandon mining to the arbitrariness of the [otherwise] excellent concept of economic freedom."[30] The views of Beust's compatriot in the Westphalian Office, Bölling, can only be described as reactionary. In 1826 he recommended tightening state mining regulations to include stiff punishments for recalcitrant mine owners and readoption of controls for metalworks unwisely abandoned during the Hardenberg years. Only Martins, a metallurgist like Karstens, kept to a different course.[31] Not surprisingly, the 1827 draft law made no headway.

The climate for reform became more favorable in the early 1830s. Karl Albert von Kamptz, former inquisitor of the "demagogues," received the Ministry of Justice in 1831 and immediately set about the task of codifying Prussian laws—including the various provincial mining statutes. But the work of reform soon felt the influence of Kamptz's administrative second, Heinrich [von] Mühler, a former presiding judge in Breslau who, like Skalley, favored French legal practices. Supposedly promoted to free Kamptz of time-consuming official duties, Mühler quickly established himself as a cominister. By 1832 he was collaborating with Karsten and Skalley on a liberal revision of Prussian mining law.[32]

These developments coincided with significant changes within the Corps itself. Bölling retired in 1830 to Ludwig von Vincke's provincial government in Münster; replacing him was Toussaint von Charpentier, an advocate of deregulation from the Silesian mine office. In June 1833, Einsiedel, Charpentier's former superior, died and was replaced by Martins, another liberal who was transferred from the Brandenburg office. The vacated post was filled by none other than Karsten himself, who now wore two hats.[33] Two months later Karsten and Skalley published a com-

[29] For Veltheim, see his file in Rep. 121, A.X.21, Nr.3, ZStA Merseburg; and Weber, *Innovationen*, 198.

[30] Cited in Arlt, *Ein Jahrhundert*, 47.

[31] For Bölling, see his report to the Oberberghauptmannschaft, 29 March 1828, Rep.121, A.XX.1, Nr.102, Bd.2/3, Bl.350, ZStA Merseburg; and Friedrich Zunkel, "Die Rolle der Bergbaubürokratie beim industriellen Ausbau des Ruhrgebiets 1815–1848," in Wehler, *Sozialgeschichte Heute*, 133. Some indication of Martin's more liberal views is found in Mining Corps Protocols 1845/46 (24 May 1845): 25, 29.

[32] Lotz, *Geschichte*, 382–83; Mining Corps Protocols 1845/46 (22 May 1845): 1–2, GStA Berlin.

[33] For Bölling's retirement from the Corps in 1830, see Reuss, "Mitteilungen aus der

promise mining law advocating technological but not economic control of mines by the Corps. The two authors were assured of a more favorable reception than they had received six years earlier.

These moves, so potentially damaging to the forces of tradition within the Corps, are difficult to interpret. Schuckmann had to approve all assignments and promotions within his Ministry of Commerce and Business, yet there is no evidence telling us why he did so or who approached him. The assignments, however, bear all the characteristics of a palace revolution. After Schuckmann suffered a stroke in 1830, Beuth had functioned as a shadow minister. Preventing Gerhard from doing so was his reserved nature and the collegial system of his organization.[34] It is inconceivable that Beuth, the man who had struggled against Corps regulatory powers in the 1810s, was not intriguing against the black coats again—this time with his friend Karsten—against Prussian mining law, one of the strongest remaining pillars of mercantilism in Prussia. Behind Beuth, moreover, was Finance Minister Georg Maassen who believed that deregulation, like the low tariffs of his Zollverein, would stimulate the economy. Maassen acquired both the Mining Corps and the Business department for his ministry, in fact, shortly after Schuckmann's death in 1834.[35] It was the Indian summer of economic liberalism in Prussia. Survival of the Corps was no longer at stake, but continuation of its paternalistic role was now in jeopardy.

It is difficult to exaggerate the anger and resentment which the reform proposal of 1833 provoked among traditionalists. The technological reputation of the Prussian Mining Corps remained justifiably high in the 1830s. This was especially true among the many "shareholders" of older, smaller mining operations. In 1833, for example, there were 36 lignite and 528 iron ore mines in the Siegerland, east of the Rhine, each worked by four or five part-timers who extracted a few hundred tons a year. There were another 140 tiny shareholdings for anthracite in the Ruhr.[36]

Geschichte," 353; and for his later presence in Münster, Friedrich Harkort's polemic in *Hermann*, 2 May 1832 (clipping in Reg. Arnsberg B, Nr. 46, Bl.153–55, StA Münster). For Charpentier, see Franz Haniel, *Autobiographie*, n.d. (1858–1862), in Bodo Herzog and Klaus J. Mattheier, *Franz Haniel 1779–1868* (Bonn: Ludwig Röhrscheid, 1979), 80–81; and Hans Spethmann, *Franz Haniel: Sein Leben und Seine Werke* (Duisburg-Ruhrort: Lübecker Nachrichten GmbH, 1956), 200. For Martins's and Karstens's annual reports from Silesia and Brandenburg beginning in 1833/34, see Rep. 121, A.XX.1, Nr.102, Bd.7/8, ZStA Merseburg.

[34] Gerhard and La Roche died in 1835. Schuckmann's prominence diminished after suffering a stroke in 1830. ADB 32:647–49.

[35] See the decree of 28 April 1834, in Rep.90a, B.III.3, Nr.8, Bl.16, ZStA Merseburg. The Business Department was shifted to Rother's new ministry in January 1835 after Maassen's death, while the Mining Corps remained in Finance.

[36] Arlt, *Ein Jahrhundert*, 126–27, 130–31; Norman J. G. Pounds, *The Ruhr. A Study in*

"From the Ruhr to the Lahn," observed Thomas Banfield, an English traveler, "every male knows something of mining."[37] The same could be said of the Lower Saxon and Silesian districts. The typical miner pursued agriculture or another trade, wanted to preserve a secure, supplemental income from mining, and therefore regarded Corps regulations, management, and expertise as a welcome, costless service. Thus to captains like Ernst von Beust it was unthinkable that anyone would consider dismantling such a popular, valuable mechanism. In addition, he could point to the need to preserve exhaustible resources and maintain a steady revenue for the state. The chief mine captain of the Rhenish Office resisted the draft reform of 1833, forcing Karsten and Skalley to reinsert economic control of mines in exceptional cases where this appeared necessary.[38] But Beust was unappeased, emerging in the mid-1830s as a champion of the mining system which he had unquestioningly imbibed as a young man at Freiberg in the 1790s.

Over the course of forty years, however, industrialists had come on the scene who were less willing to rely on the technological expertise of the state. The new "captains" of industry showed even less respect for Prussia's black coats, in fact, than they did for Beuth. The difference is not explained by a greater degree of technological expertise in the Business Department—for the Mining Corps easily triumphs in any comparison of this sort. Rather in the Corps we have an elite body whose economic interests and ideological preferences clashed with those of the largest private companies. We are dealing here with the politics of technological choice—a theme whose importance is emphasized repeatedly in the recent historical literature on technological change.[39] In order to see how the politics of technological choice deepened animosity between liberals and conservatives within the Corps and increased friction between state

Historical and Economic Geography (Bloomington, Indiana: Indiana University Press, 1952), 65–66.

[37] Banfield, *Industry of the Rhine*, 2:56.

[38] For the remainder of this paragraph, see Mining Corps Protocols (16 January 1846): 262, GStA Berlin; the reports of the Upper Silesian, Saxon, Westphalian, and Rhenish mine offices, summarized at a meeting of Corps officials with the Justice Ministry, 8 April 1839, Rep.84a, Nr.11078, Bl.198ff, GStA Berlin; Krampe, *Staatseinfluss*, 197–200; Reuss, "Mitteilungen aus der Geschichte," 352–53; and Manfred Jankowski, "Law, Economic Policy, and Private Enterprise: the Case of the Early Ruhr Mining Region," *Journal of European Economic History* 2 (1973): 704, 715.

[39] See Bertrand Gille, *Histoire des Technique L'Encyclopèdie de la Plèiade* (Montreux: Editions Gallimard, 1978); David F. Noble, *Forces of Production: A Social History of Machine Tool Automation* (New York: Alfred A. Knopf, 1984); Charles Sabel and Jonathan Zeitlin, "Historical Alternatives to Mass Production: Politics, Markets and Technology in Nineteenth Century Industrialization," *Past & Present* 108 (August 1985); and Langdon Winner, *The Whale and the Reactor: A Search for Limits in an Age of High Technology* (Chicago: The University of Chicago Press, 1986).

mining officials and private industrialists, we need to examine more closely the mining and metallurgical technologies which the Corps claimed to master. We begin by looking at mining, the industry which gave the Corps its name.

. . . . .

The techniques which young mining cadets like Heinrich Steffens and Ernst von Beust learned and observed in the mountains around Freiberg, filling them with justifiable pride and admiration, can be described most easily as tunnel-mining (*Stollenbergbau*). This term refers to the practice of extracting minerals from a ridge or hillside by burrowing upward from the valley floor and utilizing gravity to facilitate water drainage and the removal of ore or coal. With few exceptions, the expertise of Prussia's top black coats was acquired at Freiberg and disseminated to Upper Silesia and the other mining districts of the kingdom.[40]

We possess the most technical detail from the Saar. Although the area's mines were in a neglected, disorganized state when obtained by Prussia in 1815, the Corps rebuilt and extended them in the 1820s, offering state employment to 1,925 hewers, haulers, sorters, and crushers by 1836. The seven valleys of the region were perforated with eleven major collieries, each possessing two or three tunnel systems. The adits were located hundreds of feet below the summits of the ridges and were marked, as noted above, by massive neoclassical gates. Six or seven feet high and four to seven feet wide, the tunnels angled upward at three to twelve degrees, shored every meter with wooden timbers. The coal was removed by the "underhand stoping" method—hacking or blasting in a steplike pattern for a few meters *beneath* the haulage tunnel, then shoveling the coal upward from platform to platform. The mineral was next lowered by rope or steel wire in railroad wagons to the mouth, where horses, and later, steam engines, powered them back to the top of the tunnel. Once sorted, the coal made its way along short railroads constructed by the Corps to the Saar and Mosel rivers, and from there to the markets of Rhenish Prussia and France.[41]

The English traveler, Thomas Banfield, left us another excellent description of Prussian tunnel-mining after touring the Siegerland in 1846. This region, like the Saar, had fallen into Prussian hands in a deteriorated state in 1814. The Corps set about its task of reconstruction with exhila-

[40] For tunnel mining in Silesia, see Hugo Solger, *Der Kreis Beuthen* (Breslau: Wilhelm Gottlieb Korn, 1860), 87.

[41] Noggerath, "Der Steinkohlenbergbau," 139–208; Henderson, *The State*, 64–69; J.A.S. Ritson, "Metal and Coal Mining 1750–1875," in Singer et al., *A History of Technology*, 4:69–76.

ration. Local records documented mining activity in the area as early as the fourteenth century, and legend spoke of Witikind, the Saxon chief, smelting weapons from the excellent iron ore of the area during his thirty-year struggle against Charlemagne. Yet little evidence remained of this former age—small groups of farmers worked hundreds of shallow hillside diggings (*Tagebau*) in the off-season.[42]

So the Mining Corps set to work. The 1820s witnessed the opening or reopening of numerous tunnel systems, including the Stahlberg, a showcase mine partially owned by the Corps, where a form of "overhand stoping"—working in *ascending* stages from the tunnel—was in use.[43]

> The entrance is by an adit cut from the lowest point in the valley, and carried 660 fathoms [3,960 feet] on end into the hill. . . . The large central mass formed one of those workings which the old miners were particularly fond of, and in which their ingenuity displayed itself by cutting out chambers of irregular dimensions supported by great pillars left standing at intervals, and communicating with each other by staircases that led from one story to the other. . . . The height from the adit by which we entered to the highest point excavated is 60 fathoms [360 feet], and is divided into 10 stories, the first and second of which are now worked out.[44]

Banfield discounted stories that the Stahlberg stood behind Witikind's eighth-century exploits, but there was no doubt that the Prussian Mining Corps took great pride in its work of restoration and identified with the warlike zeal which drove the workers of the "wild Saxon." For the miners of the Siegerland now wore the Prussian miners' uniform, "a black linen jacket, cut full in the round and compressed by a broad leather strap at the waist, to which is appended a short leather apron, which the miner wears behind, the band clasping before with brass studs, on which the pick and hammer, crossed in the German fashion, serve as a coat of arms or masonic symbol."[45]

The decentralized structure of the Corps made it an ideal mechanism for the diffusion of these time-honored techniques. Gerhard's division presided over a mining institute in Berlin which trained lower-level officials in Freibergian expertise. The semi-autonomous mine offices in Berlin (Brandenburg), Brieg (Upper Silesia), Halle (Lower Saxony), Dortmund (Westphalia), and Bonn (Rhineland) also ran institutes for young cadets. Under these provincial bureaus stood district offices which coordinated miners' schools and supervised showcase mines like the Heinitz-

[42] Banfield, *Industry of the Rhine*, 2:71–75.

[43] Schulz-Briesen, *Der preussische Staatsbergbau*, 82–83; Banfield, *Industry of the Rhine*, 2:71.

[44] Banfield, *Industry of the Rhine*, 2:72.

[45] Ibid., 2:75.

Stollen in the Saar, Saline-Schönebeck in Lower Saxony, the Königsgrube in Upper Silesia, and the Stahlberg described above.

Although difficult to quantify, there are nevertheless scores of examples of private operators following the state's lead by introducing steel cables, steam-powered wagons, and mine railroads.[46] But the Corps did not have to rely on instruction and example alone. The various ordinances and codes which made up Prussian mining law east of the Rhine empowered district officials and their subalterns, the mine field inspectors (*Revierbeamten*), to dictate proper tunnel construction and investment in precautionary devices like the Davy's safety lamp.[47]

During the 1820s and 1830s, however, tunnel-mining began to yield to deeper mining as hillside veins were worked out and owners turned farther downward in hope of exploiting larger deposits. The sole private operation in the Saar, the Hostenbach Mine Company, completed a sixty-foot shaft in 1825. The state followed suit, sinking its own "deep tunnels" (*tiefe Stollen*) at Schwalbach (1826) and St. Johann (1832) in the Saar, and at Zabrze (1837) in Silesia. As the name suggests, some of the first deep mines were bored below existing tunnel networks, descending no lower than fifty to one hundred feet before water overcame the weaker pumping engines.[48]

By 1830, more expensive and powerful steam engines permitted truly deep mines (*Tiefbau*). In the Ruhr by this date, the Sellerbeck, Kunstwerk, Gewalt, and Sälzer & Neuack companies worked deposits at depths of three hundred feet. The number of deep mines in the Ruhr quadrupled soon after this, with one entrepreneur, Franz Haniel, reaching 570 feet before flooding terminated his descent in 1840.[49] At these levels rich deposits of high quality coal were found, thus confirming the most optimistic geological reports of the early 1820s.

A new era in mining had opened in Prussia—and the reaction of the Corps was not at all positive. To some extent this was explained by the novelty of these developments, particularly to subordinate members whose training and experience with existing techniques did not permit a

[46] Oberbergamt Dortmund circular of 14 June 1815, BA Essen-Werden 160, Bergamt Saarbrücken to Oberbergamt Bonn, 15 February 1824, OBA Bonn 679a, and conference protocol of Oberbergamt Dortmund, n.d. (January 1819), BA Essen-Werden 130, HSA Düsseldorf. See also Krampe, *Staatseinfluss*, 59–68.

[47] Ministry of the Interior to Oberbergamt Bonn, 30 March 1826, Bergamt Essen-Werden 165, HSA Düsseldorf. See also Krampe, *Staatseinfluss*, 48–53.

[48] Kurt Wiedenfeld, *Ein Jahrhundert rheinischer Montan-Industrie* (Bonn: A. Marcus and E. Webers, 1916), 7–8; Schulz-Briesen, *Der preussische Staatsbergbau*, 88–89, 97–98; Noggerath, "Der Steinkohlenbergbau," 158; Hedwig Behrens, *Mechanikus Franz Dinnendahl (1775–1826)* (Cologne: Rheinisch-Westfälisches Wirtschaftsarchiv, 1970), 539–40.

[49] See the report of the head of the Essen-Werden office, n.d. (1840/41), printed in Behrens, *Mechanikus Franz Dinnendahl*, 538–43; and Hans-Josef Soest, *Pionier im Ruhrrevier* (Stuttgart: Seewald Verlag, 1982), 52.

more optimistic assessment. In 1832, for instance, an inspector from the Essen district office simply shook his head upon observing one of Haniel's early shafts. "It is hopeless, the effort will be in vain,"[50] he predicted as the boring machines passed one hundred feet. This particular shaft was indeed abandoned. And yet, a second effort soon approached six hundred feet. By the mid-1840s, depths of nearly one thousand feet were reached by a few of the largest Ruhr enterprises.[51]

Farther up in the administrative hierarchy officials possessed more familiarity with deep mining methods, but doubted their profitability. Thus in 1840 the longtime head of the Essen office submitted a highly detailed report on the capital costs of the deep mines, concluding that the comparative cheapness of tunnel-mining would give it the long-term competitive advantage. He based his findings on the alleged saturation of the market for coals. Admittedly, sales had doubled since 1830—but they would double again in the decade after his report was submitted.[52] One suspects in this case that the wish was father of the thought—that it was primarily the fear of rapid expansion which unnerved many mine captains. Banfield came across these attitudes when he visited the Stahlberg in 1846, finding them a curious throwback to another era:

> The ruling wish is to look upon the veins of metal in which nature has been liberal as a treasure belonging to the land which must be slowly and economically worked out [so] that the people [will] not be impoverished. On such a theory all the calculations of modern mining break down. It discourages all concentration of power and rapidity of work because the task set to the miner is one that must last for centuries if possible. Accordingly . . . it is gravely asserted that at the present rate of working the streak of ore in the [Stahlberg] mine will furnish employment to the miners for centuries.

"We were told," he added, "of a visit paid by some Cornish miners . . . [who] set about valuing the work in the English fashion and so frightened the captains and officials that they were requested to desist in order that the miners might not be set astray by their calculations."[53]

Phobias of this sort help to explain the lengthy, obstructionist campaign which the Corps waged against deep mining. Toward the end of his struggle with the Corps (1821), Hardenberg implemented legislation which permitted boring to unlimited depths (*in die ewige Teufe*) during the prospecting stage, then greatly increased the surface area of mining claims once minerals were found. Mine captains undermined this legislation, however, with a series of internal directives restricting the surface

[50] Cited in Soest, *Pionier*, 50.

[51] Ibid., 50–53; Pounds, *The Ruhr*, 65–66.

[52] The report, n.d. (1840/41), is printed in Behrens, *Mechanikus Franz Dinnendahl*, 538–48.

[53] Banfield, *Industry of the Rhine*, 2:73, 78.

area of prospecting fields to specific hills and insisting on visual confirmation of a mineral vein before granting claims for larger fields. These decrees amounted to a near prohibition of deep mining, for entrepreneurs had no assurance that a prospecting shaft lay directly above a vein nor that funds would last until coal was actually sighted. District officials added further disincentives by arbitrarily delaying or withholding final concessions or unpredictably interrupting costly boring operations by announcing bans on new mine openings for entire provinces. In cases where claims were approved, moreover, industrialists were usually subjected to condescending lectures on the unfeasibility of the endeavor. Despite capitalization difficulties, finally, Haniel and his fellow entrepreneurs were taxed more heavily and consistently than struggling small operators, who received a full tax exemption after 1832.[54] Small wonder that few companies ventured into the new era of mining, or that those which persevered complained of the "hostile, persecuting ways"[55] of the Prussian Mining Corps.

But there were countervailing forces at work which traditionalists in the Corps found increasingly difficult to resist. A decree from the king's privy cabinet of July 1826 lifted a prohibition on new mine openings in the Ruhr which had been enforced since Stein's day.[56] The transfer of Toussaint von Charpentier to the Dortmund mine office in 1830 was another storm warning signal. For the new mining captain alienated his entire staff by using Prussian mining law—which did not specifically mention deep mining—as justification for the concessioning of deep shafts.[57] As discussed later in this chapter, the mining law reform proposal of 1833 eliminated the ambiguities in an attempt to facilitate technological change in mining. The draft law only exacerbated the siege mentality of conservatives in the Corps.

Developments in ferrous metallurgical technology were similarly controversial and unsettling. It is to these that we now turn.

. . . . .

Iron was a rural industry in the early nineteenth century. The countryside of Brandenburg, Silesia, the Siegerland, the Lahn, the Hunsrück,

[54] For Silesia, see Schulz-Briesen, *Preussische Bergbau*, 99. For the Ruhr, see Krampe, *Staatseinfluss*, 32–48; Zunkel, "Die Rolle," 135–37; and Jankowski, "Law, Economic Policy, and Private Enterprise," 709–12. For tax exemptions after 1832, see Frederick William to Schuckmann, 31 August 1832, 2.2.1., Nr.28373, ZStA Merseburg.

[55] Haniel's remarks of 1837 are cited in Soest, *Pionier*, 48–49.

[56] Krampe, *Staatseinfluss*, 43–46.

[57] Franz Haniel, *Autobiographie*, n.d. (1858–62), in Herzog and Mattheier, *Franz Haniel*, 80–81. For a list of deep mines concessioned in the Ruhr under Charpentier, see Behrens, *Mechanikus Franz Dinnendahl*, 540–42.

and the Saar was dotted with hundreds of water-driven ovens and forges that turned out household goods, farm implements, and armaments for a growing population. Bedecked in traditional white leather, the kingdom's ironmasters supervised operations which took the iron ore through smelting and refining to its finished state. Along with water sources, charcoal was indispensable to the entire process, for this was the fuel which first melted the ore, allowing it to flow out of the furnace into molds, then later remelted the cast iron "pigs" and reduced them to low-carbon wrought iron bars.[58]

Each master therefore maintained a company of burners who stacked beech or oak cuttings in high, covered mounds where the wood slowly smoldered into charcoal. Once surrounding forests were thinned, it became necessary to acquire additional wooded areas, or purchase wood from neighboring lords, farmers, or villagers. Thus the Saynerhütte, a state furnace opposite Coblenz had contractional obligations with nine village collectives.[59] The Rheinböllerhütte, a private establishment in the Hunsrück, made "countless wood purchases" that extended "far from the Rheinböller forest to Argenthal, Mengerscheid, Ravengiersburg, and the Wildburg."[60] Because wood was used for many other essential purposes, it had become a controversial commodity by the 1820s and 1830s.

The Corps' earliest efforts to solve this growing energy shortage have now assumed legendary proportions. Somewhat familiar with mid-eighteenth-century English developments, Heynitz and Reden attempted in 1796 to substitute coke, coal which had been roasted to burn away impurities, into an experimental blast furnace at Gleiwitz, Upper Silesia. By 1806, three coke iron furnaces were in operation there. But these experiments, grandiose statements in the literature notwithstanding, were not a great success.[61] Quality problems were so great (see below) that only one of scores of private furnaces in the region emulated the technique. Nor did contemporary Corps leaders place great importance in what they had accomplished. During a highly political speech aimed

[58] For the charcoal iron industry, see Eric Dorn Brose, "Competitiveness and Obsolescence in the German Charcoal Iron Industry," *Technology and Culture*, 26:3 (July 1985), 532–59.

[59] See the furnace's purchase records for 1863 in WA IV-1902 (Sayn 137), HA Friedrich Krupp GmbH Essen.

[60] Robert Schmitt, *Geschichte der Rheinböllerhütte* (Cologne: Rheinisch-Westfälisches Wirtschaftsarchiv, 1961), 65.

[61] Ludwig Beck, *Die Geschichte des Eisens in technischer und kulturgeschichtlicher Beziehung* (Braunschweig: Frierich Vieweg und Sohn, 1884–1903), 4:933–34; Henderson, *The State*, 17; Fuchs, *Vom Dirigismus*, 67–68. Wolffsohn, *Wirtschaftliche und soziale Entwicklungen in Brandenburg, Preussen, Schlesien und Oberschlesien in den Jahren 1640–1853: Frühindustrialisierung in Oberschlesien* (Frankfurt am Main: Peter Lang, 1985), 88, goes too far, on the other hand, in labeling Reden's attempt "technologically absurd."

at liberal enemies of the Mining Corps in January 1810, one of Reden's "true ones" wisely omitted the subject, choosing to emphasize his colleagues' replacement of primitive ironworking methods with blast furnaces fueled by charcoal.[62]

The first of two more ambitious attempts to introduce coke pig iron began as Napoleon's shadow receded from Europe. The Corps' Silesian ironworks had gained new significance and prestige during the campaigning of 1813/14. With the acquisition of new territory in 1814/15, Gerhard began to envision a state mining and metallurgical complex in the Rhineland equal to that of Silesia. As we know, restoration of the western mine fields proceeded quickly. The metallurgical portion of the plan centered on three plants which had fallen into state hands: those at Loh near Siegen, Sayn across the Rhine from Coblenz, and Geislautern in the Saar. The first would serve as a model charcoal ironworks for the Siegerland; the latter two as *coke* iron centers that would "set an example for the entire province."[63]

Behind the strategy to introduce coke fuel was a self-serving motivation. Gerhard knew that advances with coke iron "would be of considerable benefit to the state's coal mines" and underscore "the importance of this branch of the state budget."[64] The chief mine captain obviously realized the contribution his section could make to the financial recovery of the kingdom. Criticized by liberals in the Business Department who disputed his organization's right to exist, Gerhard also knew that coke iron could become the political salvation of a hard-pressed Mining Corps.

The battle was lost by 1825. As explained earlier (chap. 1), the Corps preserved its authority to grant concessions for blast furnaces in the two western provinces. Hoping to exploit this right, the mine office in Bonn tried to deny licenses to new ironworks which refused to use coke as fuel. Details are sketchy, but Hardenberg's party apparently squelched the move in 1817, condemning it as a violation of the principle of freedom of enterprise established in 1810/11.[65] Unable to coerce private producers, mine officials tried mild incentives like discounted coal prices for entrepreneurs willing to abandon charcoal. The state furnaces at Sayn and Geislautern also conducted open experiments with coke pig and wrought

[62] See Friedrich Philipp Rosenstiel's speech of 24 January 1810, cited in Wutke, *Aus der Vergangenheit*, 427–29.

[63] Gerhard to Ingersleben, 18 February 1816, 402/827, LHA Koblenz. For these plans, see also Beust to Gerhard, 19 August 1818, Rep.74, K.XVII, Nr.17, Bd.2, Bl.31–47, ZStA Merseburg.

[64] Gerhard to Ingersleben, 18 February 1816, 402/827, LHA Koblenz.

[65] Oberbergamt Bonn to Regierung Arnsberg, 9 February 1818, Reg. Arnsberg I, 566, StA Münster.

iron between 1819 and 1824.[66] Insurmountable quality problems persisted, however, and by mid-decade there were still no private ironworks in Rhineland-Westphalia which made coke pig iron.

This is not to say that all state efforts were ignored by the private sector. The Corps kept abreast of new metallurgical breakthroughs through research and travel,[67] often modifying foreign methods after expensive experimentation in its model plants. Private ironmasters near Berlin, Gleiwitz, Loh, Sayn, and Geislautern closely scrutinized the results and sometimes adopted the proven techniques. This process of diffusion included the practice of roasting ore before smelting, the hot-air blast, the addition of calk in taller furnaces, and *charcoal* puddled and rolled iron. Closed coking ovens developed in the Saar (see below) were probably the most widely emulated Corps innovation. Thus Franz Haniel in faraway Duisburg procured technical information from the Corps before installing his own ovens in 1821.[68] But throughout the 1810s and 1820s most leading ironmaking firms continued to shun pig iron smelted with coke. Even Karsten, an early enthusiast of using the much more abundant mineral fuel, had become a skeptic. In the margin of a report from Gleiwitz praising the quality of Silesian coke pigs in 1828, the incredulous chief of the metals division placed a question mark.[69]

Clearly, the first coke iron was qualitatively inferior. The main culprits were substances such as sulfur, phosphorous, and silicon. All were present in the coke fuel used both to smelt ore and "puddle" pigs into wrought iron. The consequences for iron which absorbed even slight percentages of these elements were usually ruinous. Iron cast directly from the blast furnace was "soft" (less resistant to indentation) and "weak" (liable to break if stretched, placed under a heavy load, or subjected to repeated heating and cooling). Wrought iron was "red short" (crumbling during hammering or welding) or was itself soft or weak. These problems

[66] Bergamt Saarbrücken to Oberbergamt Bonn, 17 April 1838, OBA Bonn, Nr.698, HSA Düsseldorf; Oberberghauptmannschaft to Oberbergamt Bonn, 8 April 1819 and 14 January 1820, Rep.121, D.III.5, Nr.1, Bd.1, Bl.70, 99, ZStA Merseburg; Fritz W. Lürmann, "Ein Jahrhundert deutschen Kokshochofenbetriebs," *Stahl und Eisen* 16 (1896): 813.

[67] See the Corps' request for foreign travel, 10 April 1826, and the king's approval, 12 May 1826, in Rep.121, D.III.2, Nr.103, Bd.1, Bl.3–5, ZStA Merseburg.

[68] Beust to Gerhard, 11 July 1827, Bd.1, Bl.53, 55, 61, Beust to Gerhard, 26 March 1828, Bd. 2/3, Bl.22, Einsiedel to Gerhard, 2 April 1828, Bd.2/3, Bl.91–92, Annual Report of the Oberberghauptmannschaft, 27 April 1828, Bd.2/3, Bl.160, Einsiedel to Gerhard, 9 April 1832, Bd.5/6, Bl.210–11, Beust to Gerhard, 26 March 1834, Bd.7/8, Bl.109–10, Beust to Gerhard, 5 March 1835, Bd.8/9, Bl.9, 17, Beust to Veltheim, 4 March 1836, Bd.9/10, Bl.57–58, Rep.121, A.XX.1, Nr.102, ZStA Merseburg. And for Haniel, see Hans Spethmann, "Die Anfänge der ruhrländischen Koksindustrie," *Beiträge zur Geschichte der Stadt und Stift Essen* 22 (1947): 12–13, 22–23.

[69] See Karsten's marginalia to the report of the foundry at Gleiwitz, 6 March 1828, Rep.121, D.III.5, Nr.124, Bd.1, Bl.217–19, ZStA Merseburg.

were sometimes surmountable if castings were remelted and made thicker or if wrought iron was hammered longer and restricted to uses where such defects that remained were not disadvantageous. But the price increased with extra working, oftentimes beyond what the market would bear for an inferior product.[70] It is not surprising, therefore, that the Corps' first really "progressive" project failed so miserably to impress private ironmakers.

Indeed, there was a definite logic behind the older technology. Charcoal iron's greatest asset was a negligible sulfur content due to the sulfur-free nature of the fuel. The shorter charcoal iron furnace also avoided the intense heat of the taller coke-fired furnace where silicon and phosphorous combined more readily with iron. The typical charcoal iron was thus free of most impurities and therefore quite versatile, another asset in the rural, small-shop economy of the early 1800s. Castings were strong and durable; wrought iron bars maleable enough to permit easy conversion into differently shaped products used for different purposes.[71]

Wrought iron from the Lahn river valley and the foothills around Siegen, for instance, was sent north to the Sauerland where it was transformed into guns, locks, swords, pins, needles, nails, wire, sickles, and shovels. Bar iron from the Hunsrück was marketed in nearby Coblenz, then resold to metalworking shops along the Rhine, while the Eifel sent its product north for resale and similar finishing work in Cologne, Aachen, and Liège. Iron producers in the Saar, Saxony, and Silesia were similarly dependent on distant markets for general-purpose iron.

Coke iron was not in great demand, therefore, for it was usually too soft, weak, or red short to be a useful substitute for ironmakers who shaped their iron into many different products. The demand was so low in Germany, in fact, that producers and consumers were largely indifferent to a much cheaper and rapidly plummeting price. Despite the fact that imported coke wrought iron averaged 30–40 thaler less per ton *after* the Zollverein's 20 thaler tariff—60–70 versus 95–100 thaler—and toll-free, coke pig iron 16 thaler less—34 versus 50 thaler—coke iron imports

[70] The argument that charcoal iron was qualitatively inferior is presented in greater detail in Brose, "Competitiveness and Obsolescence," 535–45. For additional archival evidence—*not cited there*—to support this thesis, see Haniel to Regierung Düsseldorf, 22 July 1833, 20010/59, HADGHH Duisburg; Saynerhütte to Bergamt Düren, 28 August 1833, and Bergamt Düren to OBA Bonn, 19 November 1833, WA-IV 1818, HA Friedrich Krupp GmbH Essen; Saynerhütte to Oberbergamt Bonn, 6 November 1838, OBA Dortmund 1216, StA Münster; Bergamt Saarbrücken to Oberbergamt Bonn, 17 April 1838, Bl.30–33, and Stumm to Oberbergamt Bonn, 19 June 1839, Bl.60, OBA Bonn 698, HSA Düsseldorf; Martins to Gerhard, 22 March 1832, Rep.121, A.XX.1, Nr.102, Bd.5/6, Bl.9–12, ZStA Merseburg; Heusler to Vincke, 16 August 1834, cited in Hedwig Behrens, *Mechanikus Johann Dinnendahl (1780–1849)* (Neustadt: Ph.C.W. Schmidt, 1974), 81.

[71] See the sources cited in ibid.

represented only about 8 percent of the 121,000 tons of iron *consumed* in the Zollverein countries in 1834. Nor was a significant amount produced domestically. Altogether, coke iron constituted a mere 13–14 percent of German consumption.[72] It was the difference in quality between coke iron and charcoal iron which was reflected in this figure—for even at low prices, few were buying the readily available coke iron.

But beginning imperceptibly in the 1820s—and accelerating noticeably in the 1830s—different markets opened for coke iron. The new demand was industrial, not agricultural: cast iron for construction and machinery; and above all, wrought iron for rails. Economic and technical logic was on the side of coke iron in these markets, for it was inexpensive and avoided previous quality drawbacks by meeting specific, not general, needs. Thus a manufacturer of iron rails tolerated larger percentages of sulfur and phosphorous in order to maximize hardness, one advantageous property of these elements, not strength, their major disadvantage. Tests run on English puddled rails found percentages of sulfur (0.04–0.2%) much higher than permissible in good, general-purpose wrought iron (0.001–0.07%). Nor was coke iron ill-suited for other urban and industrial uses. If machine-tooling was required, for example, softer irons facilitated cutting operations. If the ironmaster made it thick enough, moreover, a softer, weaker cast iron sufficed for factory gates, ceiling beams, machine frames, and train cars because none was subjected to great shock or the constant expansion and contraction of heating and cooling. All of these factors help to explain why coke iron "took off" from 13–14 percent of the German market in 1836 to 47–48 percent of the Zollverein's total consumption of 285,000 tons of iron in 1842. Consumption stood at 414,000 tons by 1847 and coke iron's share had risen to 78–79 percent.[73]

Significantly, these developments occurred almost exclusively in the private sector. The Stumm complex in Neunkirchen, the refurbished Rheinböllerhütte in the Hunsrück, the Remy works in Rasselstein, the

[72] Brose, "Competitiveness and Obsolescence," 538, 542–43; Beck, *Geschichte des Eisens*, 4:690, 731, 995, 999; Max Sering, *Geschichte der preussisch-deutschen Eisenzölle von 1818 bis zur Gegenwart* (Leipzig, 1882), 290. Of Germany's 76,727 tons of wrought iron made in 1834, 12,500 (16.3%) was coke iron. Using the rate of pig iron consumed at this time to produce wrought iron (1.35/1), Germany required about 16,800 tons of coke pig iron. Of Germany's 134,538 tons of pig iron production, about 6,000 (4.5%) were made with coke. The Zollverein countries imported 10,360 tons of pig iron in 1834, most of which, presumably, was coke iron from England and Belgium. Figured in terms of pig iron, Germany consumed 121,039 tons of iron in 1834. Thus coke iron probably represented about 13–14 percent of consumption.

[73] Brose, "Competitiveness and Obsolescence," 546, 555; Beck, *Geschichte des Eisens*, 4:714–15, 730–32. The bulk of Germany's imports of 133,257 tons of iron in 1842 was coke iron from England and Belgium. Prussia also produced 15,571 tons.

great Hermannshütte near Dortmund, and the modern Laurahütte in Silesia were just a few of the scores of specialized firms established in the 1830s to supply Prussia's industrializing economy with coke wrought iron. There can be no doubt that the owners of these operations benefited from the Corps' early experiments in the Rhineland at Geislautern and Sayn, and Upper Silesia at Rybnick. Indeed, in the case of the Laurahütte, state engineers drew up the initial plans and were hired as the first furnace managers.[74] But knowledge of coke iron techniques did not come mainly from the state—which, after four decades, had not mastered this technology—but rather from Belgian and British technicians lured to Prussia at great risk and expense by the industrialists themselves. This they had to do to survive, for the bulk of Germany's coke iron—probably 88–89 percent in the early 1840s—came from innovative industrial firms in England and Belgium.[75]

In fact, it was the state which now played catch-up. When the Saynerhütte readied efforts to smelt with coke in 1839, its furnace manager first toured a few of the more advanced private works of the Rhineland. Similarly, when Gerhard's successor, Franz von Veltheim, finally approved Karsten's plan for Corps puddling operations in Silesia in 1835, Corps technicians mustered their humility and traveled to Rasselstein, Neunkirchen, and Liège to study the latest advances.[76] Only in 1837 did they proceed with construction of the imposing Alvenslebenhütte. There was no time to waste—by 1839 private works for the first time produced more coke pig iron than the state furnaces at Gleiwitz and Königshütte.[77]

The challenge from private industry gave liberals and traditionalists within the Mining Corps a rare chance to unite. Karsten's liberalism was usually a source of great discord, but he tended to be quite proprietary about the ironworks under his supervision and, like Beust, regarded the rapidly maturing private sector as "dangerous competition"[78] for state works. This explains the unmistakable urgency to Karsten's correspon-

[74] Gustav Felsch, *Die Wirtschaftspolitik des preussischen Staates bei der Gründung der Oberschlesischen Kohlen und Eisen-Industrie (1741 bis 1871)* (Berlin: Gebrüder Ernst, 1919), 57.

[75] Conrad Matschoss, "Die Einführung des englischen Flammofenfrischens in Deutschland durch Heinrich Wilhelm Remy & Co. auf dem Rasselstein bei Neuwied," *Beiträge zur Geschichte der Technik und Industrie* 3 (1911): 86–130; Ulrich Troitzsch, "Belgien als Vermittler technischer Neuerungen beim Aufbau der eisenschaffenden Industrie im Ruhrgebiet um 1850," *Technikgeschichte* 39 (1972): 142–58. Also see n. 73 above.

[76] Saynerhütte to Oberbergamt Bonn, 28 September 1838, Rep.121, D.III.5, Nr.2, Bd.1, Bl.8–27; Oberberghauptmannschaft to Oberbergamt Brieg, 28 September 1832 and 25 August 1837, Rep.121, F.IX.3, Nr.103, Bd.1, Bl.57, 108, ZStA Merseburg.

[77] See Karsten's report, n.d. (1842), printed in Karsten, *Umrisse*, 162–63.

[78] Beust to Veltheim, 5 March 1839, Rep.121, A.XX.1, Nr.102, Bd.12/13, Bl.10–12, ZStA Merseburg.

dence with regional subalterns after 1835. When the Saynerhütte hesitated to undertake coke pig iron experiments, he instructed it to forget immediate benefits and concentrate on the future.[79] The Laurahütte in Silesia was another source of anxiety, for its four coke iron blast furnaces and huge puddling complex were designed to commandeer railroad markets which the Corps could not afford to lose. Thus Karsten ordered Silesian officials to forward plans for puddling and rolling works to Berlin immediately—the matter was of "special importance."[80] Count Hugo Henckel von Donnersmark, the owner of the Laurahütte, lodged a protest with Berlin in 1837, complaining that Silesian industrialists no longer needed "model works."[81] But he must have realized that the Alvenslebenhütte was not designed with the commonweal in mind, rather for Corps profits and Corps survival. With fifteen separate puddling and finishing ovens upon completion in 1844, the great plant had cost the Corps about 200,000 thaler.[82] Considerably smaller than the huge private complexes arising in Silesia, the Rhineland, and Westphalia—the Hermannshütte in Hörde boasted sixty-three puddling and finishing ovens—it was nevertheless "one of the most beautiful installations on the continent."[83]

The mining section was also desperately guarding its own interests. The dawning railroad era meant new sales of coked coals to iron firms and railroad lines, but competition was intense. Thus the Corps' own coking plants in the Saar wrestled for western markets with private producers from Belgium, Aachen, the Ruhr, Alsace, Rive de Gier, and St. Etienne.[84] The outcome hinged on production costs and coking technology. Most desulfurization furnaces in the early 1800s burned away 60–65 percent of the coal by allowing too much air into the oven chamber. This was both wasteful of coal and injurious to quality, for, although much sulfur was eliminated, high *percentages* remained. In the 1830s the Saar mining office designed a more efficient oven by combining the best aspects of Silesian and French "closed" models. Less open to the air, the

[79] Karsten to Oberbergamt Bonn, 21 April 1840, Rep.121, D.III.5, Nr.2, Bd.1, Bl.196, ZStA Merseburg.

[80] Karsten to Oberbergamt Brieg, 18 March 1836, Rep.121, F.IX.3, Nr.103, Bd.1, ZStA Merseburg.

[81] See the petition of Count Hugo Henckel von Donnersmark, 18 October 1837, Rep.121, F.IX.3, Nr.103, Bd.1, Bl.121–22, ZStA Merseburg.

[82] The Corps requested 150,000 thaler for the Alvenslebenhütte in November, 1837 (Rep.121, F.IX.3, Nr.103, Bd.1, Bl.116–17, ZStA Merseburg). Inflation alone (for price indices of construction materials, see Jacobs and Richter, *Die Grosshandelspreise*, 78) raises the cost to the 200,000 thaler level.

[83] Beck, *Geschichte des Eisens*, 4:700. For the Hermannshütte, see Beck, 4:703.

[84] See the conference protocols of Saar mining officials, 26 September 1833 and 29 September 1843, OBA Bonn 681a, Bl.409–21 and OBA Bonn 681d, Bl.49–50, HSA Düsseldorf.

new furnace incinerated only 37–40 percent of the coal and consequently produced a superior coke. Between 1834 and 1847 about 60,000 thaler was invested in four hundred of these ovens. The sum was no longer a significant percentage of net industrial investment in a land whose industries added 70 million thaler of new capital stock during these years (1830–1847). The Corps' *total* net investment, for instance, was probably a mere 0.7 percent of the Prussian total at this time.[85] But until superior Belgian ovens surpassed the Corps' accomplishments in the mid-1840s, the Saar was probably the largest, most technically advanced, and most widely emulated coking center in Europe.[86]

Understandably concerned about the security of the investment which Berlin had made in his district, Beust—like Gerhard twenty years earlier—took steps to protect it. If favorable returns in the Saar depended on expanded use of coke in private blast furnaces, the state should use its power to accelerate adoption of this still-nascent technology. Therefore, in 1839 Beust returned to the idea of withholding licenses for new pig iron furnaces until owners adopted coke fuel, but was unable to make headway against the beleaguered forces of economic liberalism in the Ministry of Finance.[87] His annual reports of the late 1830s also advocated "vast and thorough promotional means"[88] to advance railroad construction between coal mines and iron-producing centers. The resultant drop in coal prices would facilitate the production of iron with coke and thereby insure the "long-term profitability"[89] of state mines.

Not surprisingly, Beust added iron tariffs to his list of demands.[90] For

[85] See Carl Jung's report over the Saar efforts, February 1839, OBA Bonn 681b, Bl.132–36, 149–54, and for oven prices, the *Ökonomiepläne* for Bergamt Saarbrücken, 1843 and 1846, OBA Bonn 681c, Bl.360, and OBA Bonn 681d, Bl.175–76, HSA Düsseldorf. For the total number of ovens, see Noggerath, "Steinkohlenbergbau," 202–3. For net industrial investment in Prussia, see Tilly, "Capital Formation in Germany," in Mathias and Postan, *Cambridge Economic History*, 7 (1): 426–27. For methods of calculating the Corps' share, see chap. 1, n. 63. I assume that the 260,000 thaler represented by the Alvenslebenhütte and the Saar ovens was 50 percent of the Corps' total net investment during this period.

[86] The new ovens were not too complicated for the Corps to master; rather they were tested and rejected on grounds of cost. See Oberbergamt Bonn to Oberberghauptmannshaft, 6 December 1841, OBA Bonn 681c, HSA Düsseldorf. For evidence of emulation of Saar ovens by the private sector, see Milecki to Veltheim, 4 March 1839, Rep.121, A.XX.1, Nr.102, Bd.12/13, Bl.328, ZStA Merseburg; and Spethmann, "Die Anfänge," 12–14, 22–23.

[87] Beust to Veltheim, 5 March 1839, Rep.121, A.XX.1, Nr.102, Bd.12/13, Bl.43, ZStA Merseburg.

[88] Beust to Veltheim, 4 March 1836, Rep.121, A.XX.1, Nr.102, Bd.9/10, Bl.91–92, ZStA Merseburg.

[89] Ibid., Bl.57.

[90] Beust to Gerhard, 27 March 1833, Rep.121, A.XX.1, Nr.102, Bd.6/7, Bl.93; and Dechen to Beust, 28 February 1842, Rep.121, A.XX.1, Nr.102, Bd.17/18, Bl.87–89, ZStA

state mines would not benefit from the importation of English and Belgian iron made with foreign coke. He received hearty backing from the heads of local mine offices in Saarbrücken, Sayn, and Siegen who already belonged to the lobbying committees of private pig iron manufacturers whose business interest in tariff protection was identical to that of the Mining Corps. Without tariffs, in fact, some of the most advanced Rhenish facilities were postponing plans to construct new coke iron blast furnaces, for the return on this investment was threatened by Belgian and British companies marketing their pigs toll-free in Cologne for 30–35 thaler per ton.

All of these demands offended the liberal sensitivities of those around Karsten. But there was more than ideological preference at stake here. Karsten could see that Beust—and his successor in Bonn in 1840, Karl von Dechen—were trying to promote the interests of the mining division by creating "dangerous" private competition for the metals section. Beust's was a game, of course, which any number could play. For parallel alignments existed between the metals section of the Corps and private puddling and finishing firms dependent on the free importation of cheap pig iron. So Karsten looked to his own interests and sided with industrial lobbyists opposed to higher tariffs.[91]

A commission was established during the spring of 1842 to investigate iron industry complaints about the deluge of Belgian and Scottish imports and prepare Prussia's position for an upcoming conference of Zollverein states in Stuttgart. As head of the metals division of the Mining Corps, Karsten, not Beust, the new chief captain of the Corps in Berlin, took the chair. Karsten's report concluded that charcoal ironmakers needed no protection from imported coke iron whose inferior quality made it useless for many traditional purposes. Karsten conceded that domestic coke iron manufacturers faced smothering overseas competition in the newer industrial markets, but he did not believe the nation could afford to tax foreign iron which Prussia could not yet produce for herself. Domestic producers would either have to learn to compete or yield to the greater good of the consumers—even if this meant bankruptcy. Karsten had neither compromised his liberalism nor jeopardized the Corps' metalworks, for he was confident that black coat metallurgical expertise and the best Silesian coals guaranteed survival. The members of the Business Department who represented Prussia in Stuttgart agreed with Karsten.[92]

Merseburg. Also see Hans Kruse, "Die Einfuhr ausländischen Eisens nach Rheinland und Westfalen 1820–1844," *Glückauf* 51 (6 February 1915): 146–47.

[91] See Dechen's report of 1 March 1843, Rep.121, A.XX.1, Nr.102, Bd.19/20, Bl.2, ZStA Merseburg; Brose, "Competitiveness and Obsolescence," 548; Kruse, "Die Einfuhr," 144, 146.

[92] See Karsten's report, n.d. (1842), printed in Karsten, *Umrisse*, 144–80; and Kruse, "Die Einfuhr," 147.

Consequently, pig iron continued to enter Germany tax-free and wrought iron at 20 thaler per ton. After more intense pressure from protectionist allies in and out of government, tariffs were raised slightly in 1844 to 7.5 thaler and 30 thaler per ton, respectively. But this afforded little relief from foreign competition—iron imports drifted downward from 57 percent of German consumption in 1843 to 49 percent in 1847.[93]

Karsten was never forgiven by conservative protectionists in the Corps for his part in the continuation of free trade in Prussia.[94] Indeed, he had prevented Beust's mining section from acquiring a great lever for boosting sales of Saar coke. The fresh memory of this betrayal greatly reduced any likelihood of restoring Corps solidarity before 1848. A source of even greater animosity, however, soon clouded the horizon. For in 1845 and 1846 the Corps debated final revisions of mining law reform proposals which the ministries had passed back and forth since 1833. This controversy is our final subject of discussion.

. . . . .

The Mining Corps exploited the patriotic euphoria of the early 1810s to shield itself from the onslaught of Hardenberg and his liberal aides. By 1827, however, intramural economic liberalism had surfaced with a draft reform of mining law by Karsten and Skalley. Peter Beuth and Heinrich Mühler helped give birth to a reworked proposal in 1833 which strove to bring Prussian mining law into conformity with the principle of freedom of enterprise which underpinned Hardenberg's earlier legislation. Indeed, every facet of the ancient trade was to be emancipated. Karsten and Skalley called for freer prospecting, elimination of burdensome taxes, abolition of price and bookkeeping controls in most instances, and introduction of limited liability for mine companies. The Corps would continue to supervise the technical aspects of mining, but owners would participate in the planning of these complex and dangerous underground operations. It was Karsten's intention that mine owners would *eventually* assume full responsibility in this area too. The liberals' first draft of 1827, the reader will recall, included neither economic nor technical controls. Karsten also saw the new law as a device to sweep away the hundreds of small-scale, collective mine companies which were so dependent on Corps expertise and make way for larger and more capable capitalist undertakings.[95]

[93] Beck, *Geschichte des Eisens*, 4:732; Kruse, "Die Einfuhr," 147.

[94] Karsten, *Umrisse*, 96.

[95] Mining Corps Protocols (1845/46; 22 May 1845; 4 July 1845; 16 January 1846; 22 May 1846): 1–2, 7–9, 96–100, 261–62, 451–52, GStA Berlin. See also H. Brassert, "Die Bergrechtsreform in Preussen," *Zeitschrift für Bergrecht* 3 (1862): 234–38.

The prospects for implementation appeared bright at first. With Schuckmann's retirement in April 1834, the Corps was transfered to Georg Maassen's Ministry of Finance. The ex-chief of the Business Department and cofounder of the Zollverein was not only a kindred spirit of the liberals under him, but he also possessed the integrity of character that assured the king's ear when controversial issues were under discussion.[96] After Maassen's death in November 1834, however, the great age of economic liberalism begun by Stein and Hardenberg quickly waned. J. G. Hoffmann's guild reform proposals were amended between 1835 and 1837 by a Ministerial Cabinet which included ascendant conservatives like Count Wittgenstein and influential mercantilists like Christian Rother. Simultaneously, statist forces were on the offensive.[97] Both the army (chap. 5) and the Seehandlung (chap. 6) undertook huge industrial investments and emerged alongside the Mining Corps as competitors of private industry. In keeping with all of these illiberal trends, Count Albrecht von Alvensleben was chosen as Minister of Finance. Karsten's plans for a liberalization of mining practices found no patron saint as capable as Motz or Maassen, for the new man possessed neither the ability nor the desire to sway the Ministerial Cabinet as his predecessors had done. In March 1836 that body expressed its doubts about reducing the Corps to a "mere supervisory"[98] economic role and instructed the Ministry of Justice to undertake another review of the law.

Heinrich Mühler, Justice's cominister, referred the matter back to the Corps. Impatient for progress in the rapidly unfolding railroad age, Karsten made a tactical error at this point by returning to the notion of technological independence for the mines. As the five mine offices responded to his proposal during 1837 and 1838, it became clear that the chief of the metals division had gone too far for his erstwhile supporters. Beust was the only captain to denounce the entire reform proposal, but Martins (now in Lower Saxony), Charpentier (now in Silesia), and Alexander von Milecki (Charpentier's replacement in Dortmund) deserted Karsten on the issue of entrepreneurial control of technological planning. All three believed that the notoriously inadequate technical knowledge of the nation's small operators justified maintenance of the Corps' paternalistic approach.[99]

[96] ADB 39:586–87; Walter Serlo, *Bergmannsfamilien in Rheinland und Westfalen* (Münster: Aschendorffsche Verlagsbuchhandlng, 1936), 16; Eylert, *Character-Züge*, 3:203–4.

[97] For the adverse reaction among economic liberals in the bureaucracy when Maassen died, see Lotz, *Geschichte*, 392. For the remainder of the paragraph, see ADB 1:376; and Treitschke, *Deutsche Geschichte*, 4:543–44.

[98] Quoted from the protocols of the Ministerial Cabinet, 26 March 1836, Rep.90a, B.III.2b, Nr.6, Bl.201, ZStA Merseburg.

[99] See Karsten's memorandum of 5 April 1839, the protocol of a conference between the

The stage was set for even more drastic modifications in 1840 when Beust succeeded Veltheim as chief captain and head of the mining division. With extra-ministerial support from Frederick William IV's reactionary Minister of the Interior, Gustav von Rochow, Beust drafted a new proposal in 1841 which embodied the spirit of the Corps as it had been in Reden's day.[100] The legal right of the state to control mining through state works or strict regulation of leaseholders was, according to this viewpoint, clear, unambiguous, and irrevocable. The Prussian legal code of 1794 guaranteed these controls, and five hundred years of legal practice throughout Germany strengthened the anticapitalistic case against unattenuated rights of private property in mining.[101]

More important than the legality of these statutes, however, was the wisdom that stood behind them. For they bestowed property rights upon "the supreme lord of mining" (*der oberste Bergherr*) in order "to preserve (*Nachhaltigkeit erhalten*) minerals hidden in the womb of the earth and in this way guarantee the welfare of the entire nation."[102] The "shortsighted profit interests of mine owners" had to yield to "the general interest of the state in the enduring use of minerals which cannot be reproduced."[103] Beust used these principles to justify continuation of grand Corps projects—which were compatible with the concept of responsibility to the general welfare—and hinder large capitalistic operations—which he claimed were not.[104] His proposal reintroduced unlimited liability, the tithe, accounting and price controls, and Corps planning and supervision of mining itself.[105]

Beust's draft also placed barriers before the Haniels and Krupps by insisting on certain "environmental" standards. Neither prospecting nor mining could damage gardens, farms, vineyards, meadows, or springs. Property owners could demand a security deposit before permitting digging, and in the event of damages, receive reparations of one sort or another determined by the Corps. "Economic considerations of state with

---

ministries of Justice and Finance, 8 April 1839, and the various mine office reports of 1837/38 (summarized in the protocol of 8 April 1839), in Rep.84a, 11078, GStA Berlin.

[100] See the position paper (*Votum*) of the Ministry of Justice, n.d. (February 1840), Rep.84a, 11078, GStA Berlin; and the protocol of the Ministerial Cabinet, 26 January 1841, Rep.90a, B.III.2b, Nr.6, ZStA Merseburg.

[101] *Motive zu dem Entwurf*-1841, pp. 4–5, 13–14, and *Entwurf der Instruktion zur Verwaltung des Berg-Regals nach den Bestimmungen des gemeinen Bergrechtes*, pp. 51–52, Rep. 84a, 11080, GStA Berlin (hereafter cited as Instructions-1841, GStA Berlin).

[102] *Motive zu dem Entwurf* -1841, pp. 13–14, GStA Berlin.

[103] Instructions-1841, p. 51, GStA Berlin.

[104] Mining Corps Protocols 1845/46 (16 January 1846): 262, GStA Berlin.

[105] *Motive zu dem Entwurf*-1841, p. 20, Instructions-1841, pp. 41–42, 52–53, and *Entwurf des gemeinen preussischen Bergrechts*, 1841, pp. 5–6, 14–15, Rep.84a, 11080, GStA Berlin (hereafter cited as *Entwurf*-1841, GStA Berlin).

regard to agriculture . . . are so manifold that they can only be left to the judgement of administrative officials."[106] As a convenient sop to the metals division, moreover, Beust advocated reinstitution of Corps licensing of all metalworks.[107]

Karsten was furious. Prussian mining law had long been antiquated, he charged, catering and appealing only to small collective mine companies which possessed neither the ability nor the desire to supervise their own affairs. But Beust's proposals would remold the law into something worse—an intentional barrier in the path of daring entrepreneurs who were ready for independence. "Had the state overseen all of the other branches of industry with equally fearful solicitude," he observed in 1846, "their rapid rise and strength would have been impossible." It was the duty of the state to remove legal inhibitions to growth, especially now "that railroads promised to speed economic development."[108]

Unlike so many of the early economic liberals, Carl Karsten possessed a fairly accurate vision of the coming world and was not afraid to peer into this industrial future. Beust was looking too, but saw something different—a haunting apparition of imprudent economic development which only rigid adherence to the old ways could exorcize. To argue that one was modern and one was not, however, is fruitless and misleading. The series of mining reforms undertaken between 1848 and 1865 embodied, for all of their caution and moderation, more of Karsten's liberalism than the old guard traditionalism of Beust. These reforms have even left their mark on contemporary German mining law.[109] And yet the concept of perpetual economic growth has found its critics today in developed countries which are rapidly depleting once-abundant supplies of coal, oil, and natural gas and finding no cheap or uncontroversial substitutes. The Club of Rome's message about the impossibility of exponential growth within a finite system—earth—would not have been lost on Beust. Perhaps he—like the "Greens" in Germany or environmentalists in the United States who have turned to the state for prohibitively restrictive solutions—was the modern one.

The conflict between Karsten and Beust was not resolved in the Pre-March. Beust's radical revisions of 1841 were debated by the provincial

[106] *Motive zu dem Entwurf*-1841, p. 25, GStA Berlin.

[107] *Entwurf*-1841, p. 5, GStA Berlin.

[108] Mining Corps Protocols 1845/46 (22 May 1845): 9, GStA Berlin.

[109] Wolfram Fischer, "Das wirtschafts- und sozialpolitische Ordnungsbild der preussischen Bergrechtsreform 1851–1865," and "Die Stellung der preussischen Bergrechtsreform im 19. Jahrhundert," in Wolfram Fischer (ed.), *Wirtschaft und Gesellschaft im Zeitalter der Industrialisierung: Aufsätze-Studien-Vorträge* (Göttingen: Vandenhoeck & Ruprecht, 1972), 139–47 and 148–59, respectively, emphasizes the moderate nature of the reforms as opposed to the exclusively liberal aspects highlighted by previous authors.

diets in 1841 and 1843 before the Corps began its own deliberations in May 1845. The version which went back to the ministries in early 1847 remained in all essential respects Beustian, but the reform was not implemented before the events of March 1848 brought the era to an end. For his service to the old cause, nevertheless, Ernst von Beust might have earned a tearful nod of approval from Friedrich Wilhelm von Reden, laid to rest some thirty years earlier amidst the decorative grottos, ruins, and natural harmony of Buchwald. But Reden would have wept uncontrollably over the shattered unity of his beloved Mining Corps.

V

Spartans of the North

FREDERICK WILLIAM was growing too portly and unsure of his equestrian skills to ride frequently in public by 1823. So it was that a stately coach bearing the Hohenzollern coat of arms delivered him and his trusted aide, Job von Witzleben, to Unter den Linden 74 on the morning of October 12. The day's preliminary round of political briefings behind them, the two had come to inspect the recently completed Artillery and Engineering School on its first day of instruction. Receiving them in front of the building's imposing neoclassical facade was a retinue of the appropriate personages—Prince August and Gustav von Rauch representing their respective corps; Friedrich Schinkel, who drew up the plans; First Lieutenant Schwink, who executed them; and a nervous-looking Headmaster.

One hundred cadets and eight instructors had moved into these exquisite surroundings only two days earlier. But, with a royal inspection looming, all had done their duty and done it quickly. Indeed, everything was in its place as the procession advanced through the edifice's five ornate lecture halls, a sampling of its thirty-three efficient living quarters, and finally, to the rooftop. Resting there before his descent, Frederick William took in the panoramic view of the royal palace, the famous tree-lined boulevard, and the Tiergarten. Soon they all turned to go.[1]

The grand structure at Number 74 was taking its place in a city that bore increasingly less resemblance to the one which Napoleon had taken in 1806. After Berlin became the sprawling capital of a German Empire, the Prussian War Academy occupied this—by then centrally located—building. Out of its seventy-year-old doors came the men who would put modern technology to deadly use in two world wars. At the time Schwink constructed it, however, the structure could not be seen as a harbinger of any such bold future for Prussia's technical corps. Nor did its erection signify a recent elevation of status for the artillery or engineers. The king, it is true, had commissioned his great architect to demolish the old pontoon building and design a new school on the site. But the 100,000 thaler

[1] For facts about the building and the details of the king's visit, see Bonin, *Geschichte des Ingenieurkorps*, 2:163–64; B. Poten, *Geschichte des Militär-Erziehungs- und Bildungswesens in den Landen deutscher Zunge* (Berlin: A. Hoffmann, 1896), 394; and Eve Haas and Herzeleide Henning, *Prinz August von Preussen* (Berlin: Stapp Verlag, 1988), 153.

allocated for the project in 1822 was a more accurate reflection of Frederick William's desire to beautify Berlin than any enthusiastic support for the technical branches of the Prussian Army.[2]

If the monarch's feelings toward cannon, shrapnel, and rockets were ambivalent, attitudes elsewhere in the officer corps ranged from total indifference to outright contempt. Biases ran strongest in conservative aristocratic circles surrounding Duke Carl of Mecklenburg and the elite Guard regiments of Berlin.[3] War was seen here as an affair of caste-honor best resolved by noblemen who understood points of honor—hence the rigid exclusion of lower and middle class officers from these units.[4] And while few Guardsmen wanted to dispense with cannon fire, there was something bourgeois and dishonorable about "the secret science of the black collar"[5] which seemed to lie behind artillery technology, something derogating about an officer who "reeked of axle grease."[6] The engineering and artillery corps were at best a necessary evil in the mentality of the Guard.

Circumstances allowed these elitist attitudes to prevail in the decade after Waterloo. The king might remark in private that he "hated"[7] the artillery, but he gave no thought to spiking the ten- and twelve-pounders. With peace seemingly assured and political exigencies dictating pared budgets, on the other hand, some cuts in the specialized branches appeared wise. By the early 1820s, artillery companies were receiving less than one hundred rounds of ammunition a year for target practice and horses were so scarce that only two guns could be moved at one time.[8] An even worse slight occurred in 1824 when Wittgenstein's "budget savings" commission—the same one which abolished the Ministry of Business and Commerce and attempted to downsize the Mining Corps—recommended a partial dismantling of the so-called "scientific corps":[9]

[2] Poten, *Geschichte*, 394.

[3] Clausewitz blamed the Guard, Duke Carl, and Prince William (the king's son) for their backwardness in a letter to Gneisenau, 1 October 1824, printed in Delbrück, *Gneisenaus Leben*, 5:507–8. In this letter he defended the "scientific corps"; on other occasions, however, he expressed his own bias against technology (see below, notes 69–72 and 74).

[4] See Varnhagen von Ense's history of the events leading to 1848, printed in Varnhagen Diary II, 4:201–2; Demeter, *The German Officer-Corps*, 17, 111–15; Gerhard Förster et al., *Der preussisch-deutsche Generalstab 1640–1965* (Berlin: Dietz Verlag, 1966), 20–21.

[5] Curt Jany, *Geschichte der Preussischen Armee vom 15. Jahrhundert bis 1914* (Osnabrück: Biblio Verlag, 1967), 4:161.

[6] J. Castner, Geschichtliche Studie WA Xa3, 105 (p. 3), HA Friedrich Krupp GmbH. Castner, a retired artillery captain writing in 1909, headed a team of historians at Krupp investigating the history of the firm.

[7] Varnhagen Diary I, 4 November 1820, 1:223.

[8] Dennis E. Showalter, *Railroads and Rifles: Soldiers, Technology, and the Unification of Germany* (Hamden, Connecticut: Archon Books, 1975), 147.

[9] Clausewitz to Gneisenau, 1 October 1824, printed in Delbrück, *Gneisenaus Leben*,

the General Staff would have to get by with half of its officers, the engineers two-thirds, and the artillery three-fourths.

The lowly status of the technicians was still more apparent during the fall maneuvers of mid-decade. Engineers were not permitted to participate except as workers to dig latrines and cooking pits or grade fields for cavalry drills. Gunners accustomed by now to the perennial shortage of horses and ammunition were subjected to the additional indignity of having to commandeer peasant carts and plow animals in order to keep up with the other units. Generally, maneuvers had deteriorated into parade ground exercises reminiscent of the pre-Jena era, replete with attack-in-line formations and the other large-unit tactics which had spelled disaster in 1806.[10] "It was as if 1813, 1814 and 1815 had never happened,"[11] writes one historian.

No wonder that Prussia funded few technological advances in the years immediately after Vienna. Cannon and howitzers of uniform design were hastily introduced in 1816 to retire the motley collection of foreign and domestic artillery pieces dragged victoriously to Paris and back, but improved axles, laying machines, and carriages did not compensate for the inaccuracy and cumbersomeness of the new weaponry.[12] The bronze guns of "System c/16" were expensive, but it would be the last significant allocation for the artillery. Major Dietrich's tinkering with rockets, fuses, and fireworks in Spandau Fortress after 1817 was conducted on a minuscule budget, and when the General Staff joined Lieutenant General Johann Braun's Experimental Artillery Department (*Artillerie-Prüfungs-Kommission*) in more extended testing eight years later, they were given only 800 thaler.[13] Braun's letter of appointment from Frederick William in 1824 illustrated the low priority placed on basic research in the early 1820s. The division's work, wrote the monarch, "was becoming somewhat necessary."[14] The king's words, usually so supportive in letters of this sort, could hardly have instilled confidence in Prince August's right-hand man.

More surprising is the caution—occasionally even outright hostility—

4:507–8. See also Bonin, *Geschichte des Ingenieurkorps*, 2:160–61 (originally published in 1877–78). Rauch was able to protect the Engineering Corps from some of the cuts.

[10] Bonin, *Geschichte des Ingenieurkorps*, 2:171–73; Showalter, *Railroads and Rifles*, 148; Colmar Freiherr von der Goltz, *Kriegsgeschichte Deutschlands im Neunzehnten Jahrhundert* (Berlin: Georg Bondi, 1914), 1:18–22.

[11] Goltz, *Kriegsgeschichte*, 2:20.

[12] Showalter, *Railroads and Rifles*, 146.

[13] August Genth, *Die preussischen Heereswerkstätten: Ihre Entwickling, allgemeine volkswirtschaftliche Bedeutung und ihr Übergang in privatwirtschaftliche Betriebe* (Berlin: Fr. W. Universität Berlin, 1926), 26–27; Priesdorff, *Soldatisches Führertum*, 4:58.

[14] Frederick William to Braun, 25 March 1824, cited in Priesdorff, *Soldatisches Führertum*, 4:130.

shown toward technology in the artillery itself. Extremely negative views were most evident in the horse artillery companies assigned to each of the army's nine artillery regiments. Riding artillerymen derived self-respect from the fact that they were mounted, not from the mathematical science and metallurgical art incorporated in their guns. Indeed, these aspiring cavaliers scorned the encumbering devices they were required to pull and dreaded the derogating love of work and machinery which, it was widely assumed, was the fate of every artilleryman.[15] Attitudes in the Guard Artillery stationed in Berlin were only slightly more positive. The regiment begrudgingly shared its firing range at Wedding when tests of new metals and designs were ordered, but refused to assign Guard officers to the Experimental Department after 1824 for what was seen as a boring and useless tour of duty.[16]

Enthusiasm for new technology was also restrained within the very unit responsible for it. The Experimental Department did not deem shrapnel worthy of testing in 1822, and rejected it again in 1824. In 1823, moreover, the artillerists cut short a research collaboration with the Mining Corps and General Carl von Helvig.[17] The former Swedish artilleryman had entered Prussian service in 1816 to conduct tests with the Corps on the relative strength and reliability of fortress cannon wrought from iron smelted with different fuels: charcoal, which was free of chemical impurities; coke, which was cheaper, but less pure; and mixtures of charcoal and coke. Joining this team in 1822, the Experimental Department soon grew weary of Helvig's painstaking experiments, opting quickly for iron from the Saynerhütte near Coblenz after two barrels produced there survived three thousand shots without exploding. Too much time had been spent on tests of "purely scientific interest"[18] which kept the department from more important work with practical application in war. The Pomeranian's decade-long efforts, widely regarded as an insignificant make-work assignment, finally came to an end when he fell victim to the budget cuts of 1825.

[15] Castner, Geschichtliche Studie WA Xa3, 105 (p.3), HA Friedrich Krupp GmbH; H. Müller, *Die Entwicklung der Feldartillerie* (Berlin: Ernst Siegfried Mittler & Sohn, 1893), 1:106; Showalter, *Railroads and Rifles*, 149.

[16] Hugo Denecke, *Geschichte der Königlichen Preussischen Artillerie-Prüfungskommission* (Berlin: Artillerie-Prüfungskommission, 1909), 4–5; Prinz Kraft zu Hohenlohe-Ingelfingen, *Aus meinem Leben* (Berlin: Ernst Siegfried Mittler & Sohn, 1897), 1:118.

[17] For shrapnel, see Müller, *Entwicklung der Feldartillerie*, 1:35–36. For Helvig's experiments, see Oberstlt. von Breithaupt, "Allgemeine Betrachtungen über die Anwendbarkeit des Eisens für Guss von Geschütz-Röhren jeden Calibers," November, 1824 (pamphlet in MgFa Freiburg); Varnhagen Diary I, 16 June 1822, 2:144; and Zeitschrift-Krieges 63 (1845): 189–93, MgFa Freiburg.

[18] Hake to Schuckmann, 13 October 1823, Rep.121, D.III.5, Nr.124, Bd.1, Bl.137–38, ZStA Merseburg.

It may be that shrapnel and Helvig's experimental irons were too tentative to justify expenditure of limited resources. But mixed here with an understandable financial caution one detects an underestimation of the benefits of technology in combat as well as a disregard for the importance of basic research. For shrapnel was an improvement over grapeshot, deadly in its potential. And only sixteen years had passed since so many iron cannon burst at Kolberg that French troops nearly broke Neithardt von Gneisenau's heroic defense of the fortress—it was Gneisenau, in fact, who had enticed Helvig to serve in Prussia.[19] Thus Frederick William was not alone in his ambivalence toward military technology. As we shall see, such attitudes were slow to expire in the Pre-March.

The king's letter to Braun was a reluctant admission, on the other hand, that Prussia could not afford to ignore new developments in the life-and-death business of military technology. Indeed, many novel devices had surfaced abroad since Napoleon's exile: steamships, hand grenades, and land mines in England; improved iron barrels in Holland; breach-loading cannon and fine-grained gunpowder in France; torpedoes, rockets without firing staffs, and firearms made from standardized parts in the United States.[20] The need to stay informed about such breakthroughs prompted the heads of the three "scientific corps" to request special military attachés in Prussia's European embassies in 1824.[21] "It is not just against England's warriors that we may one day have to take up arms," wrote Moritz Meyer, an artillery metallurgist, in 1826. "No, also against English machines which we cannot match—[machines] whose effect will be all the more devastating the more we consider them insignificant."[22] But peace in Europe and the likelihood of its continuation worked to banish anxious thoughts as the tenth anniversary of Waterloo approached and passed. The Great Aggressor was dead, the Bourbons were restored, and the Holy Alliance stood firm in the East.

One would expect to find Prussia's leading civilian reformers in a state of irritation over this neglect of military technology. Our historical re-

[19] See Gneisenau to Frederick William, 15 June 1807, printed in Fritz Lange (ed.), *Neithardt von Gneisenau: Schriften von und über Gneisenau* (Berlin: Rütten & Loening, 1954), 99; and Zeitschrift-Krieges 63 (1845): 189–90.

[20] Nohn, *Wehrwissenschaften*, 336–41; and Moritz Meyer, *Handbuch der Geschichte der Feuerwaffen-Technik* (Berlin: Ad. Mt. Schlesinger, 1835), 226, 230–31, 234–41, 243, 245. Meyer was a weapons and metallurgical specialist stationed with the Prussian artillery in Breslau.

[21] Müffling to Hake, 11 March 1825, Rep.92 Müffling, A.10, Bl.58, ZStA Merseburg. For the successful opposition of the Foreign Ministry, see Varnhagen Diary I, 6 March 1825, 3:245.

[22] "Ueber Anwendung des Dampfes auf Forttreibung von Projektilen," Zeitschrift-Krieges 7 (1826): 99, MgFa Freiburg. The article was signed "M": probably Moritz Meyer. For Meyer, see above, n. 20.

flexes are conditioned for this type of response when we read that German states turned their backs on the guilds "because of the economic and military advantages offered by the factory system."[23] Other historians have written that it was a "political imperative"[24] or a "power-political"[25] interest which drove continental states like Prussia to emulate the industrialization of England. In reality, the reformers were quite content to let the army lie dormant. Men like Scharnweber, Kunth, Vincke, Schön, and Stägemann praised peace as the fundament of prosperity and castigated standing armies as a drain on the economy.[26] For all of its boasting about patriotic services, moreover, the Mining Corps privately feared that military orders would undermine its ability to compete with private iron firms. Peter Beuth represented one of the major exceptions. In typically sarcastic fashion, he likened the impossibility of maintaining national wealth by means of outmoded machines to "a humanitarian field-marshall . . . [who] would achieve great feats in our time with bows and arrows."[27] The ex-free corps volunteer also devoted space in the Technical Institute in 1824 to work with Job von Witzleben on an early American design for a rapid-firing pistol. Characteristically, the army rejected the gun because of "a complicated construction which does not recommend itself for military use."[28]

As the decade drew to a close, however, the worsening Greek crisis made it harder for the army to maintain such complacency (see chap. 2). European relations deteriorated steadily between the naval clash at Navarino in October 1827 and the Russian crossing of the Pruth in April 1828. By the time peace returned to the Near East in September 1829, European attention had shifted to Paris where revolution and military violence drew menacingly near. Moral and public financial considerations made Frederick William III genuinely dread war. While he prayed for peace, however, the Prussian king made sure his army was ready.

[23] Hamerow, *Restoration*, 37.

[24] Landes, "Technological Change and Development," in *Cambridge Economic History of Europe*, 6:368.

[25] Helmuth Bleiber, "Staat und bürgerliche Umwälzung in Deutschland," in Gustav Seeber and Karl-Heinz Noack, *Preussen in der deutschen Geschichte nach 1789* (Berlin: Akademie-Verlag, 1983), 104.

[26] Scharnweber, "Vermehrung des Wohlstandes d. nied. Klassen," 1814, Rep.92 Hardenberg, H.11 3/4, ZStA Merseburg; Goldschmidt, *Das Leben*, 305; Stägemann to Olfers, 7 February 1828, and Schön to Stägemann, 27 June 1835, printed in Rühl, *Briefe*, 3:391–92, 560; Varnhagen Diary I, 4 and 24 July 1820, 1:161, 172; Niebuhr to Stein, 4 January 1808, in Gerhard and Norwin, *Die Briefe*, 1:431–34.

[27] Quoted in Matschoss, *Preussens Gewerbeförderung*, 63–64.

[28] Beuth recalled the military's reaction in a letter to Hake, 11 May 1831, Rep.120, D.XIII.2, Nr.22, Bl.9, ZStA Merseburg. See also the original correspondence of 1824 in Rep.120, D.I.1, Nr.13, Bl.55, 59, 76, ZStA Merseburg.

Under the pressure of these threatening developments, Frederick William's soldiers awoke from their lethargy. In fact, the years between 1827 and 1829 witnessed a veritable flurry of technological activity in Prussia.[29] "No time is richer than ours," wrote Major Louis Blesson, an instructor at the War School, "in innovations and modifications to the art of war." It was "the mission of the current generation," continued the young engineer and military journalist, "to examine the real value of these inventions" with calm, scientific detachment, adopting those with "practical value in war."[30] In keeping with Blesson's call for objectivity, the Artillery Department of the War Ministry sent its Head, Eduard Peucker, and Mortiz Meyer to Le Bouchet, France, after the outbreak of the Russo-Turkish War to evaluate the fine-grained black powder produced there by the Champy method. Orders were issued in July 1829, to plan for installation of the new water-powered manufacturing process in the citadel at Spandau.[31]

Research and experimentation were underway in a variety of other areas as well. Thus Frederick William approved a special commission to investigate optical telegraphs. The gun factory at Spandau also introduced a more "American" system of file jigs, gauges, and lathes while accelerating tests on a new firearm—the percussion musket.[32] The Experimental Department and the General Staff combined resources in November 1828, moreover, for expanded shrapnel tests. Johann Braun and Karl von Müffling were eager to combine the deadly new shell with another evolving technology—rockets—in support of attacking infantrymen. Rocket salvos launched every ten to fifteen seconds, reasoned Müffling, would rip apart defensive lines minutes before hand-to-hand

[29] Conflicting opinions in the literature about the technical competence of the Prussian Army result from a tendency to generalize about the entire era before 1848. Consequently, the importance of the late 1820s as a military-technological watershed is usually overlooked. Thus Jany (*Geschichte der Preussischen Armee*, 4:160–61) is very critical of the technical branches of the Prussian army, while equally harsh assessments are found in Goltz (*Kriegsgeschichte*, 2:29), Gerhard Ritter (*Staatskunst und Kriegshandwerk* [Munich: R. Oldenbourg, 1954], 1:142), and Manfred Messerschmidt ("Die politische Geschichte der preussisch-deutschen Armee," in Gerhard Papke, *Handbuch der deutschen Militärgeschichte 1648–1939* [Munich: Bernard & Graefe Verlag für Wehrwesen, 1975], 4 (1): 115). Gordon A. Craig (*The Politics of the Prussian Army 1640–1945* [New York: Oxford University Press, 1955], 77–78) offers a more positive judgement. The most balanced handling is Showalter, *Railroads and Rifles*, passim.

[30] Zeitschrift-Krieges 12 (1828): 272, MgFa Freiburg.

[31] Frederick William to Hake, 16 July 1829, 2.2.1., Nr. 32161, ZStA Merseburg; Priesdorff, *Soldatisches Führertum*, 4:148.

[32] Bertold Buxbaum, "Der deutsche Werkzeugmaschinen- und Werkzeugbau im 19. Jahrhundert," *Beiträge zur Geschichte der Technik und Industrie* 9 (1919): 100; Hugo Gothsche, *Die Königlichen Gewehrfabriken* (Berlin: Militärverlag der Liebelschen Buchhandlung, 1904), 25; Priesdorff, *Soldatisches Führertum*, 4:318.

fighting ensued, restoring the invincible reputation which Prussian attackers had enjoyed under Frederick the Great. Impressed, the king approved construction of an extensive rocket laboratory in the old watchtower on the Havel island of Eiswerder.[33]

The army also showed more interest in recent metallurgical developments. Technological missions visited Luxemburg, Holland, and Sweden in the winter of 1827/28 to gather more information on coke and charcoal irons and compare these with bronze cannon metal. To the dismay of Silesian mining officials who hoped to win contracts for their coke iron foundries, the artillery decided to retain the lighter, safer, more expensive bronze for field guns and continue to import high-quality Swedish charcoal iron, smelted and refined with the purer fuel, for most fortress cannon.[34] The Saynerhütte would still receive regular charcoal iron orders for the Rhenish fortress of Ehrenbreitstein, but closer attention was paid now to the "reliability and purity"[35] of the cast iron. Thus artillery examiners rejected Saynerhütte iron in 1829. By opting for bronze and charcoal iron over coke iron, the artillery reached the same conclusion that it had five years earlier. In 1829, however, a more open-minded scientific spirit governed the decision-making process.

This turnabout received official sanction in December 1828, with the establishment of the army's Commission on Science and Technology. The new body would investigate any area of scientific or technological knowledge with potential for military application. It was important, moreover, that these deliberations take place "at the highest level."[36] Representatives from the entire scientific, technological, and educational establishment of the military would participate and report directly to the king or Minister of War. Prince August took the chair, but there can be little doubt whose inspiration lay behind the Commission, as well as much of the new research undertaken in Prussia during the charged atmosphere of the late 1820s. In apparent recognition for his work, Lieutenant General Johann Braun received the Order of the Red Eagle, First Class with Oakleaves, in January 1829.[37]

[33] See Hake to Müffling, 6 March 1830, and Muffling's reports of 1 May 1829 and 18 March 1831, Rep.92 Müffling, Nr. "zu A.14," ZStA Merseburg. See also Priesdorff, *Soldatisches Führertum*, 4:131; and Genth, *Preussische Heereswerkstätten*, 27.

[34] Ministry of War to the Oberberghauptmannschaft, 24 January 1828, the Gleiwitz Foundry to the Oberberghauptmannschaft, 6 March 1828, and Koeppen to the Oberberghauptmannschaft, 22 September 1841, Rep.121, D.III.5, Nr.124, Bd.1, Bl.208–09, 217–19, 244–45, ZStA Merseburg; Priesdorff, *Soldatisches Führertum*, 4:148.

[35] Oberbergamt Bonn to Oberberghauptmannschaft, 28 April 1830, Rep.121, A.XX.1, Nr.102, Bd. 3/4, Bl.190–95, ZStA Merseburg.

[36] Frederick William to Hake, 28 December 1828, Rep. 90, Tit. xxxv, Nr. 195, ZStA Merseburg. See also Varnhagen Diary I, 5 December 1828, 5:145.

[37] Varnhagen Diary I, 20 January 1829, 5:161; Priesdorff, *Soldatisches Führertum*, 4:128.

The destruction of the Prussian Army Archive during World War II has hampered military scholars ever since. Along with so many other important documents, the fires of that conflagration consumed the records of the Commission on Science and Technology. Working through other sources, however, it is possible to identify some of the Commission's operations and internal workings. Thus Carl von Clausewitz, one of the early members, mentions a study done in 1830 on the crushing strength of cannon axles.[38] The revolutionary crisis which spread from Paris and Brussels that summer and gripped Europe until early 1833 immediately intensified labors and pushed the Commission to the center of a febrile national effort to gird for modern war. Prince August's men took over the shrapnel tests and advanced them quickly to the point of arming all howitzer batteries with the lethal new shell in February 1831. Moritz Meyer, the chief of Müffling's rocket team, complained about the priority afforded shrapnel that winter, worrying that "if war breaks out . . . we may not enjoy the fruits of Your Excellency's labors."[39] In March, however, the Commission assumed responsibility for all rocket experiments, including the troublesome business of fuses for rockets armed with shrapnel. Now it was Braun who complained, for August allowed Meyer, Dietrich, and Radowitz to control the experiments.[40] We also know that the Commission approved and undertook the first extensive testing of Nicolaus Dreyse's famous "needlegun" in May and June of 1832. Delegated for this assignment were Eduard Peucker of the War Ministry's Artillery Department and Georg Heinrich [von] Priem, a Jäger specialist.[41]

The army pushed forward with other new technologies during the war scares of the early 1830s without involving the Commission on Science and Technology. Presumably this was because all of these projects had passed the experimental stage. Thus Franz August O'etzel supervised installation of a semaphore telegraph line connecting Berlin, Cologne, and Coblenz in the winter of 1831/32. Although war-readiness demanded top priority for existing technologies, moreover, Prussia's gun factories

---

It seems probable, given Braun's area of expertise, that the decoration came in recognition of his technical work, not merely his fortieth anniversary with the army (April 1828). Varnhagen found it noteworthy that the bourgeois Braun had received the award. For the magnitude of Braun's role in the artillery, see Clausewitz to Gneisenau, 31 July 1830, printed in Delbrück, *Gneisenaus Leben*, 5:593–94.

[38] Paret, *Clausewitz and the State*, 326.

[39] Moritz Meyer to Müffling, 25 September 1830, and the king's decree of 28 February 1831, Rep.92 Müffling, Nr. "zu A.14," ZStA Merseburg.

[40] Braun to Müffling, 21 July 1831, Rep.92 Müffling, Nr. "zu A.14," Bl.119, ZStA Merseburg.

[41] Heinrich von Löbell, *Des Zündnadelgewehrs Geschichte und Konkurrenten* (Berlin: E. S. Mittler und Sohn, 1867), 13, 17–19. For the connection between the crises and the tests, see Brandt, *Aus dem Leben*, 2:43.

continued their conversion to the percussion-cap musket. The most impressive of these new undertakings, a huge gunpowder factory, was also begun in the midst of Europe's crisis. When construction finally started at Spandau in 1832, drawings called for a twenty-one-building complex replete with a saltpeter refinery, carbonization plant with hot-air-ventilated condensation apparati, roller presses and carding engines for finishing powder, and cheap, efficient water power throughout. The Champy method of fabrication at Spandau yielded 3,500 centner of fine-grained black powder during its first year of operation (1838).[42]

The crises of the early 1830s also triggered a reorganization of the Artillery Experimental Department. In the autumn of 1830 Prince August freed Braun's division from many of the nonessential tasks which had hampered practical research during the 1820s. The Department, for instance, had been responsible for the curriculum of the Artillery and Engineering School. These matters were now transferred to the school's Studies Commission. The mundane task of bookkeeping went to artillery headquarters in Berlin, while investigation of promising scientific developments was left completely to the Commission on Science and Technology. The division now had more time for tests on existing techniques. Consequently, Braun and his colleagues gave top priority to improving shrapnel and redesigning bronze cannon for enhanced mobility.[43]

Another important reorganization occurred in May 1832. Previously, Prussia maintained two separate chains of command for armaments production. Reporting to Eduard Peucker in the War Ministry were the directors of the (1) powder works (Berlin [later Spandau] and Neisse), (2) gun and cartridge factories (Spandau [later Potsdam], Danzig, Neisse, Suhl, Saarn/Hattingen, and, after 1840, Sömmerda), and (3) cannon foundries (Berlin and Breslau), while Braun supervised the (4) artillery workshops (Berlin, Danzig, Neisse, and Deutz) which assembled the cannon, carriages, laying machinery, and other artillery accessories.[44] Kindred spirits, the two men shared a love of antiquity, early friendships with Gneisenau, enthusiasm for the latest military technology, and a dedication to the artillery corps.[45] But dangerous times mandated one person in command of weapons' procurement. The king appointed Braun, who

[42] Priesdorff, *Soldatisches Führertum*, 6:81–82; Müller, *Die Entwicklung*, 36; Denecke, *Geschichte*, 6; M. Thierbach, *Die geschichtliche Entwicklung der Handfeuerwaffen* (Graz: Akademische Druck- und Verlagsanstalt, 1965), 314–16 (first published in 1886). For the Spandau powder works, see Nieden to Beuth, 14 January 1836, Rep. 120, D.XVII.1, Nr. 14, Bd. 1, ZStA Merseburg; and Genth, *Preussische Heereswerkstätten*, 15.

[43] Denecke, *Geschichte*, 5. Again, for the connection between the changes and the crises, see *Militär-Wochenblatt*, Nr.37, 16 September 1843, MgFa Freiburg.

[44] The chains of command can be discerned from Priesdorff, *Soldatisches Führertum*, 4:131, 148; and Gothsche, *Königlichen Gewehrfabriken*, 66–68.

[45] For Peucker, see Priesdorff, *Soldatisches Führertum*, 4:147–48.

chose his liberal-minded son-in-law, Eduard [von] Kunowski, as his first adjutant. Nepotism had little to do with the selection, for Kunowski had studied under Rühle von Lilienstern and Clausewitz at the War School, served four years in the Artillery Experimental Department, and conducted valuable research on shrapnel, rockets, cannon metals, and fine-grained gunpowder.[46]

Until 1835, father, son, and an omnipresent Prince August presided over the thirteen manufacturing complexes which comprised the army's industrial establishment. With the nation's security at risk, the army believed that it had to take command of the production process as closely as Prussia's Spartan budgets would allow. Most of the gunpowder works and artillery workshops were owned directly by the state, managed by line artillery officers, and staffed by special artisan companies of worker-soldiers who were not liable for call-up to the militia. The army was unable to banish unpredictability, however, for Prussia depended on foreign suppliers for sulfuric acid (England), raw nitrates (Chile), and the majority of its cast-iron cannon barrels (Sweden).[47]

Elsewhere in the establishment, command of production was even less complete. The cannon foundries were owned by the state and leased on a long-term basis to private entrepreneurs who purchased their own raw materials and hired their own laborers. Quality was preserved by refusing to pay for barrels which exploded during extensive test-firing, and occasionally, by placing an officer in charge of production. The gun and cartridge factories were privately "owned," but subjected to an equally rigid system of army supervision. Officers usually managed production and closely examined muskets and cartridges before purchase, while sometimes artisan companies manned the files, jigs, and lathes.[48]

The army's preference for an autarkical, command economy was neither illogical nor new.[49] Like other nations on a continent torn by war in the previous two centuries, Prussia had attempted to concentrate weapons production in the reliable hands of the state. Europe's successive crises after 1828 only strengthened this predilection. None of the Great European Powers, however, had mastery over, or could dispense with, a private industrial sector which was in the throes of "take-off." And Prus-

[46] For Kunowski, see ibid., 4:421–22; Denecke, *Geschichte*, 34; and for his politics, see Varnhagen Diary I, 26 November 1823, 2:441.

[47] Genth, *Preussische Heereswerkstätten*, 15, 16–17, 23–24, 29; Hans-Dieter Götz, *Militärgewehre und Pistolen der deutschen Staaten 1800–1870* (Stuttgart: Motorbuch Verlag, 1978), 57–59; Wilhelm Hassenstein, "Die Gewehrfabrik Spandau im Übergang aus der privaten in die staatliche Leitung," *Beiträge zur Geschichte der Technik und Industrie*, 27 (1938): 61–68.

[48] See the sources cited in n. 47.

[49] In general, see William H. McNeill, *The Pursuit of Power* (Chicago: University of Chicago Press, 1982), passim, but especially 144–222.

sia was no exception. As the Russo-Turkish War entered its climactic stage in the early summer of 1829, the War Ministry undertook an initial survey of the location, availability of transportation, means of production, and total output of all private powder mills in Rhineland-Westphalia.[50]

Prince William (the king's brother) instructed his staff in early 1831 to widen this search, especially to iron foundries. "For a situation could easily arise where the War Ministry requires a supply of iron munitions which cannot be manufactured in the state's Rhenish foundries as quickly as perhaps appears necessary for special purposes."[51] After Braun assumed central command of armaments production, he required annual reviews of private gunpowder output and added tanneries to the growing list of private works surveyed. Army pressure, moreover, explains the sudden experimentation with coke iron at the Saynerhütte during the spring of 1832, while Braun—or perhaps the anxious adjutants under Prince William in Cologne—may have been behind a simultaneous policy of enlivening western iron-making through tax exemption.[52]

The war scares of the early 1830s had clearly underscored the military potential and necessity of private industry. This was a troubling development, however, for leaders of an institution who were not yet convinced that industrialism was a boon to Prussia. This was especially true among the older generation of officers. Opinions ranged from conservatives like Ludwig Gustav von Thile, who castigated industry as "a cancerous growth on the land"; to moderates like Witzleben, who recognized new industries and factories as sources of employment, public revenue, and prosperity, but never fully accepted the passing of guild life and handicraft methods of production as a sacrifice worthy of industry's benefits; to more progressive spirits like Gneisenau, who accepted freedom of enterprise and fought the reactionaries who wanted to undermine industry and technology by bringing back the guilds. Even Gnei-

[50] See Oberpräsidium Coblence to Regierung Trier, 17 July 1829, 442/3917, LHA Koblenz; and 3rd Artillery Inspection Coblence to the Landräte of Berleburg, Hagen, and Altena, 4 August 1829, Reg. Arnsberg I Nr. 612, StA Münster.

[51] 3rd Artillery Inspection Coblence to Regierung Arnsberg, 13 February 1831, Reg. Arnsberg I Nr. 612, StA Münster. The pressure, writes the artillerist, stemmed from Prince William. See also Prince William Papers, D-22, Nrs.29–6 and 29–10 (Bl.135), HeStA Darmstadt, for a list of private iron output in the Rhine Province in 1830 and a list of all private manufacturing firms producing for the army in early 1831. Both lists were part of the mobilization material used by William's staff.

[52] For powder works and tanneries, see General Inspection Berlin to Oberpräsident Westphalia, 18 January 1833, Oberpräsidium 5899, StA Münster. For the Saynerhütte, see the annual report of the Oberbergamt Bonn for 1832 (Rep.121, A.XX.1, Nr.102, Bd.6/7, Bl.66, 69, ZStA Merseburg) which places Sayn's coke iron tests—the first since the early 1820s—squarely in the context of military orders. For tax exemptions, see Frederick William to Schuckmann, 31 August 1832, 2.2.1., Nr. 28373, ZStA Merseburg. For the anxiety of William's staff, see the materials in n. 51 above.

senau was concerned, however, about the effects of affluence on the physical condition of the middle classes.[53]

Worries about the lower classes were much greater. During the campaigning of 1814, Prussia's generals were shocked at factory conditions in the western provinces. And in 1828, the commander of the Westphalian Corps, General von Horn, doubted whether factory districts could produce good recruits. It is not surprising, therefore, that the first piece of social legislation in Prussia—a law for the protection of working children in the 1830s—resulted from pressure by the military, not the civilian, authorities. Leading army personalities like Gneisenau and Boyen were fearful, moreover, that Prussia's growing industrial towns would witness ugly social revolutions as unemployed, uneducated factory operatives took to the streets. Such premonitions led Boyen to propose in 1823 that new factories be restricted to rural districts in eastern Prussia. He resubmitted the proposal in 1832 after the revolutionary disturbances of the preceding years, adding the idea that entrepreneurs also grant workers a plot of land. The related notion of basing the nation's defense on industrial centers dispersed throughout the countryside was one that Boyen shared with Grolman, Krauseneck, and Clausewitz. But the precaution of rural industry offered no prophylaxis against the problem, mentioned above, of physical deterioration. Would factory workers who monotonously tended textile machinery in the countryside or supplied human power in rural step-mills be able to overcome French peasants in hand-to-hand combat? The unreassuring answers to these questions prevented many in the army command structure from fully accepting modern industrialism.[54]

[53] For this paragraph and the next, see Alphons Thun, *Die Industrie am Niederrhein und ihre Arbeiter* (Leipzig: Duncker & Humblot, 1879), 190 (Thile quote); Witzleben to an unnamed person, 29 June 1821, cited in Dorow, *Job von Witzleben*, 4 (see also pp. 73–74); Gneisenau to Schön, 24 February 1827, cited in Treitschke, *Deutsche Geschichte*, 3:377; Petersdorff, *Friedrich von Motz*, 2:237; Meinecke, *Das Leben*, 1:27, 2:153, 309, 334, 345–46, 424–25, 430–32; Hansjoachim Henning, "Preussische Sozialpolitik im Vormärz? Ein Beitrag zu den arbeiterfreundlichen Bestrebungen der Preussischen Seehandlung unter Christian von Rother," *Vierteljahrsschrift für Sozial- und Wirtschaftsgeschichte* 52 (1965), 524; Lange, *Neithardt von Gneisenau*, 288; Clausewitz to Gneisenau, 28 April 1817 and 5 October 1822, printed in Delbrück, *Gneisenaus Leben*, 5:214–15, 475–76. Also in Delbrück, see Gneisenau to Clausewitz, 6 July 1817 and 1 November 1822 (5:228, 479–80); Gneisenau to Gibsone, 3 July 1829 (5:561); Clausewitz to Gneisenau, 28 October 1817; Gneisenau to Clausewitz, 23 December 1817, 2 November 1820, and 17 November 1830; and Gneisenau to Gibsone, 31 December 1823 (5:265, 280, 447, 492–93, 623); see also 5:35.

[54] For Grolman, Krauseneck, and Clausewitz, see Meinecke, *Das Leben*, 2:156; Bonin, *Geschichte des Ingenieurkorps*, 2:138–42; Karl von Clausewitz, *Vom Kriege: Hinterlassenes Werk* (Frankfurt: Verlag Ullstein GmbH, 1980), 408, 410, 416, 420; and E. Folgermann, *Der General der Infanterie von Krauseneck* (Berlin: E. S. Mittler und Sohn, 1852), 76. Gneisenau, Scharnhorst, and O'etzel belonged to Beuth's Association for the Promotion of

The performance of private armaments producers during the Belgian crises reinforced these prejudices and strengthened the army's preference for command over weapons production. This represented an anticapitalism of sorts in the sense that the army was inclined to challenge and abuse the private property rights of military suppliers. Indeed, from the army's perspective, private weapons makers had shown themselves to be highly unreliable. Some of the more craft-oriented gun factories resisted pressure to implement mechanization, arguing that the investments required were risky, unprofitable, and provocative to gunsmiths proud of their older traditions. Private iron works were very reluctant to fill state orders, insisting that a state of emergency be proven before agreeing to interrupt service to regular customers. Even the Mining Corps fell short of military expectations. Coke iron cannon balls and grenades from the Silesian Mine Office were declared brittle and useless, for instance, even after remelting.[55] Experiences like these explain the desire expressed by artillery specialist Moritz Meyer for army control of Mining Corps foundry operations.[56]

This longing for technological command created friction between the military and all outsiders. Proud black coats in the Mining Corps reacted irritably and anxiously to pressures placed upon them by pushy artillery officers completely indifferent to the fact that the Corps' competitors were attempting to lure away its customers. Private entrepreneurs were no less incensed at the treatment they usually received. One inventor of a new wood-cutting lathe regretted the day that he approached the artillery workshops. After preventing him from selling the device to private industry, the artillery "tested" it in Berlin for over a year without paying the man a pfennig. Another owner of a powder mill in Hagen, having dutifully informed the army for years of every technological detail of his own operations, was denied access to the new Spandau works in 1840. The army, he was told, had taken years to develop its own process, invested a considerable sum in new buildings, water wheels, and machinery, and therefore reserved the right to keep the benefits to itself. No wonder that Alfred Krupp turned a deaf ear to Eduard Kunowski's pleas to view the family's foundry works near Essen. The clever Cannon King

Technical Knowledge in Prussia and had probably taken part in his Sunday discussions. The three, like Boyen and the others, may therefore have favored rural industrialization. See Matschoss, *Preussens Gewerbeförderung*, 37; and Priessdorff, *Soldatisches Führertum*, 6:82. In general for this paragraph, see the sources in n. 53.

[55] Genth, *Preussische Heereswerkstätten*, 24–25; Gothsche, *Königlichen Gewehrfabriken*, 22, 31–32; Saynerhütte to Oberbergamt Bonn, 9 May 1831, Rep.120, D.XIV.1, Nr.21, Bd.1, Bl.62–64, and Oberbergamt Berlin to Oberberghauptmannschaft, 22 March 1832, Rep.121, A.XX.1, Nr.102, Bd.5/6, Bl.9–11, ZStA Merseburg.

[56] Nohn, *Wehrwissenschaften*, 320, 326.

suspected—with good reason—that Kunowski merely wanted to memorize the layout and techniques for immediate implementation in army foundries.[57] Private industry was truly important and army technocrats like Kunowski did not doubt its potential in war. As will be demonstrated, however, they took every available opportunity in the waning Pre-March to limit this importance for future wars.

The military's fixation with command over the means of production found clear expression in the drive to perfect and produce the needlegun. Dreyse constructed, tested, and repeatedly redesigned the new firearm in the 1830s with the unstinting backing of prominent officers like Peucker, Priem, and Witzleben, who was himself a former Jäger officer. When the prototype was finished in 1840, the master machinist built a new gun and cartridge factory astride the Unstrut River in Sömmerda to complete the army's order of 60,000 guns and 30 million shells. Fully capitalized with interest-free government loans and manned by 423 artisan-soldiers, Dreyse's factory turned out the first needleguns in 1842, completing 39,200 by 1847. These efforts were carried out behind a tight veil of secrecy: workers had to take oaths of silence about all operations, including the machine tool breakthroughs which were making the weapon possible. Even other quasi-private gun factories in the army's establishment like the Schickler Brothers in Potsdam, Prussia's longest-standing gun production firm, were denied information and access.[58]

Meanwhile, rumors spread among private gunmakers and metal firms about novel technical developments in Sömmerda. The needlegun was a lightweight, breech-loading, bolt-action weapon with a rapid firing pin—hence the "needle" in its name. One unique—and largely overlooked—aspect of the firearm was the near interchangeability of its metal parts. In order to mass-produce thousands of pins, bolts, springs, screws, gun-

[57] For the Mining Corps, see Saynerhütte to Oberbergamt Bonn, 9 May 1831, cited in n. 55; Oberbergamt Bonn to Oberberghauptmannschaft, 31 March 1831, 25 March 1832, and 27 March 1833, Rep.121, A.XX.1, Nr.102, respectively in Bd.4/5, 5/6, and 6/7, ZStA Merseburg. For the wood lathe, see Sachse to the Ministry of Finance, 12 May 1841, Rep.120, D.XIV.1, Nr.63, Bd.1, Bl.85–86, ZStA Merseburg. For the Spandau works, see the War Ministry to Oberpräsident Westphalia, 6 March 1840, Oberpräsidium 5899, StA Münster. And for Krupp, see J. Castner, Geschichtliche Studie WA X a 3, 71 (p. 13), HA Friedrich Krupp GmbH.

[58] Löbell, *Des Zündnadelgewehrs*, 5–34; M. Thierbach, *Die geschichtliche Entwicklung der Handfeuerwaffen* (Graz: Akademische Druck- und Verlagsanstalt, 1965), 311–26; Jaroslav Jugs, *Handfeuerwaffen: Systematischer Überblick über die Handfeuerwaffen und ihre Geschichte* (Berlin: Deutscher Militärverlag, 1968), 1:532; Werner Eckardt and Otto Morawietz, *Die Handwaffen des brandenburgisch-preussisch-deutschen Heeres 1640–1945* (Hamburg: Helmut Gerhard Schulz Verlag, 1973), 115–17; Günter Thiede, "Zur Geschichte des Zündnadelgewehrs," *Militär-Geschichte* 12/4 (1973): 444–45; Showalter, *Railroads and Rifles*, 76–139; Hans-Dieter Götz, *Militärgewehre und Pistolen der deutschen Staaten 1800–1870* (Stuttgart: Motorbuch Verlag, 1978), 299.

locks, and plates with the requisite precision, Dreyse had been compelled to advance beyond the technological level of contemporary German machinists. His experimentation resulted in more accurate milling machines, jigs, and gauges, as well as rotating "cylinder" lathes capable of repeating the same series of cuts on thousands of workpieces. Dreyse also devised an innovative process for boring low-carbon, cast steel barrels at a time when most gunmakers were still using welded wrought iron plates. Craft traditions and microeconomic realities explain why these techniques did not evolve in the private sector—for machinists producing for smaller, less predictable markets had no incentive to invest heavily in such technological innovation themselves. Once these useful tools were developed, however, it was army secrecy which blocked their diffusion among entrepreneurs bold enough to employ them. This restrictive situation stands in stark contrast to the rapid, salutary emission of similar techniques from government arsenals in the United States. It was unfortunate for a state like Prussia, where relations with private businessmen were rapidly deteriorating, that army operations in Sömmerda were not secret *enough* to prevent industrialists from wondering enviously and angrily about the pathbreaking advances made there.[59]

[59] For American developments, see Nathan Rosenberg, *Technology and American Economic Growth* (New York: Harper & Row, 1972), 87–116.

Volker Mollin, *Auf dem Wege zur "Materialschlacht": Vorgeschichte und Funktionieren des Artillerie-Industrie-Komplexes im Deutschen Kaiserreich* (Pfaffenweiler: Centaurus-Verlagsgesellschaft, 1986), 58–60, argues that the eventual spread of machine tool technology from state gun factories around 1870 was as important to the growth of the German machine tool industry as the railroads. We can assume that the lack of such a stimulus in the quarter-century prior to 1870—because of secrecy—must have inhibited the growth of this industry to a significant extent. For the generally low state of German machine tools before 1850, see Alfred Schröter and Walter Becker, *Die deutsche Maschinenbauindustrie in der industriellen Revolution* (Berlin: Akademie-Verlag, 1962), 89–92, 109–10, 116–17; and Buxbaum, "Deutsche Werkzeugmaschinen," 103–18. The sample of firms studied by Schröter possessed no cylinder lathes and only one milling machine.

For the cast steel smooth-bore barrels, see J. Castner, Geschichtliche Studie WA VIIf, 694 (pp. 24–25), HA Friedrich Krupp GmbH Essen. Dreyse purchased the cast steel from Krupp, but used his own boring techniques, not Krupp's hollow-forging process (for hollow-forging, see Showalter, *Railroads and Rifles*, 156). Information about Dreyse's machine tools has been pieced together from the following sources: Löbell, *Des Zündnadelgewehrs*, 23; Siguard Rabe, *Das Zündnadelgewehr greift ein* (Leipzig: Lühe & Co., 1938), 47; Fritz Pachtner, *Waffen: Ein Buch vom Schaffen und Kämpfen im Waffenbau* (Leipzig: Wilhelm Golsmann Verlag, 1943), 211–12; Jugs,*Handfeuerwaffen*, 1:519; and especially Rolf Wirtgen, *Handfeuerwaffen: Die historisch-technische Entwicklung* (Freiburg: Militärgeschichtliches Forschungsamt, 1980), 63.

For partial knowledge of Sömmerda in the private sector, see Castner (cited above); Thiede, "Zur Geschichte," 444; and Showalter, *Railroads and Rifles*, 156–57. For a general contemporary critique of government-owned factories, see *Hamburgische Unparteiische Correspondenz*, 24 December 1841 (newspaper clipping in Rep.121, E.VII.1, Nr.102, Bd.1, Bl.3–4, ZStA Merseburg) .

The needlegun expansion in Sömmerda was also significant in economic terms. Hermann von Boyen's War Ministry poured loans totaling 90,000 thaler into the facility before it was "tooled up" and ready to produce serviceable guns. This was no mean amount—roughly equivalent to the annual budget for all of the Business Department's operations. But Dreyse's factories on the Unstruth were only one expanding part of a burgeoning military-industrial network in Prussia. The massive new gunpowder works in Spandau began secretive operations in 1838 after investments of at least 500,000 thaler. The gun factory at Saarn/Hattingen added new metal workshops and boring mills during the late 1830s to complete the army's conversion to percussion-cap muskets. In order to enhance the quality of iron barrels, the manager's lease was canceled in 1840 and the entire complex repossessed by the army. Similar considerations led the military to commandeer the Neisse gunpowder works in 1844. Moreover, steam engines were installed in the gun factories at Potsdam and Danzig, and elaborate steam hammers in the artillery workshops at Cologne-Deutz and Berlin.[60] While only partial statistics are available, net investment in the army's industrial establishment since 1832 probably exceeded one million thaler—or about 1.4 percent of net industrial investment in Prussia between 1830 and 1847.[61] Only the Seehandlung invested greater amounts in this timespan (see chap. 6).

The lower class revolution of 1848 did not alter army attitudes toward reliance on suspicious private suppliers. In fact, the trend toward autarky strengthened. Eduard Kunowski, a determined opponent of outside arms contracts, played a central role in this ongoing campaign for self-sufficiency. In May of that critical year he took over the Artillery Department of the War Ministry. With the gun factories reporting again to this office, Dreyse's patent permitting exclusive production in Sömmerda was renegotiated to allow expanded output of the needlegun in Spandau, Danzig, and Saarn. To preserve secrecy, enhance control, and facilitate mechanization dreaded by craftsmen, the Spandau and Danzig works

[60] For the loans to Dreyse, see Thierbach, *Geschichtliche Entwicklung*, 324; for Saarn, Neisse, Potsdam, and Danzig, see Gothsche, *Königlichen Gewehrfabriken*, 25–27, 29–30, and Genth, *Preussische Heereswerkstätten*, 15, 25; and for the artillery workshops in Deutz and Berlin, see the correspondence and drawings from 1847 in 2103/10 and 20330/61, HADGHH. The conservative estimate of 500,000 thaler for the powder works at Spandau is based on the number of buildings (21) and the capital intensity (see Genth, *Preussische Heereswerkstätten*, 15) in comparison to investment figures on eight Seehandlung factory buildings (machinery included) given in Radtke, *Preussische Seehandlung*, 108, 142, 146, 157. The average amount for a factory building with machinery was 36,250 thaler. Thus the powder works may have cost upward of 761,250 thaler.

[61] For the methods used in estimating these percentages, see chap. 1, n. 63. Again, for net investment figures in the Pre-March, see Tilly, "Capital Formation in Germany," in Mathias and Postan, *Cambridge Economic History*, 7 (1): 427.

were converted into full army operations—as Saarn had been in 1840. Not only were these three complexes nationalized, they were also expanded and completely retooled with American machines even more impressive than those developed by Dreyse. To bring munitions manufacture into line with gun production, moreover, special gunpowder plants for the needlegun were erected by the army in Coblenz, Spandau, and Danzig. Representing a net investment of about one million thaler, Kunowski's buildup by itself comprised around 1.8 percent of net industrial investment in Prussia during the boom years between 1847 and 1854. The army's total must have been in the 2–3 percent range.[62]

While such trends strengthened the intolerance of merchants and industrialists for state encroachment in the economy, animosity generated by economic and technological issues was not limited to the captains of industry. Indeed, as we have seen, the army had mixed feelings of its own about the emergence of modern capitalistic industry. The ambivalence and occasional hostility of soldiers and businessmen toward their respective institutions was perpetuating a general political crisis among the upper orders of Prussian society which had worsened in the last decades of the Pre-March. Such exclusion of private arms producers could not long endure, however, in Europe's hostile atmosphere. As argued in the Conclusion to this study, Kunowski's autarkical preferences eventually had to yield to the altered military, economic, and political imperatives of the unification era.

. . . . .

The gunsmoke and smell of powder hung thick in the autumn air of the Hasenheide outside Berlin. Fifty shots had cracked out from the marksman's stand before Rudolf Dreyse turned to his brother, a look of anxiety beginning to crowd confidence from his face. The weapon was so hot from firing that he could no longer hold it. Nicolaus Dreyse, the proud master machinist and brother of the shooter, glanced awkwardly toward the small throng of examiners who were still keeping their distance for fear a round would explode within the overheated barrel. The leader of

[62] J. Castner, Geschichtliche Studie WA X a 3, 71 (p.13), HA Friedrich Krupp GmbH; Priesdorff, *Soldatisches Führertum*, 4:421; Götz, *Militärgewehre*, 299–300; Gothsche, *Königlichen Gewehrfabriken*, 34. For net industrial investment, see Tilly, "Capital Formation in Germany," in Mathias and Postan, Cambridge Economic History, 7 (1): 426. I assume that the "list of twelve" industries given here by Tilly represented 80 percent of total net industrial investment in Prussia, which corresponds to the ratio of "list of twelve"/total net industrial investment for the 1840s. Investment in the Spandau works totaled 358,000 thaler (Gothsche), while investment in the Danzig and Saarn factories probably totaled another 350–400,000 thaler, for both factories had half the productive capacity of Spandau. The new powder works would easily raise the total to 1 million thaler.

the commission, Prince August of Hohenzollern, rode forward, angry at the irregularity of an interruption.

August's countenance had changed too. Ten minutes earlier he had displayed a look of pity for the two Saxons who made incredible claims. Now the prince's face had the flushed appearance of one who was losing heavily at cards. Nearby, the other commissioners exchanged looks of concern and incredulity. Only Job von Witzleben, the old Jäger Corps officer, smiled telepathic encouragement to the two brothers. Eventually, August broke the silence and ordered Dreyse to resume firing.

There were still fifty cartridges on the table. Quickly but calmly repeating the manual exercise he had perfected in Sömmerda, Rudolf Dreyse sent one bullet after another speeding toward the center of the target. Unable to keep still, August trotted back and forth between his colleagues and the shooter's stand, counting and recounting with disbelief and disapproval the rapidly shrinking pile of shells. The others, seeming to lose their fear of explosion, nudged closer themselves to the mechanical marksman. With seven shots left, one of them scooped up a handful of sandy Brandenburg dirt and threw it at the open chamber of the weapon to see if it would malfunction under simulated field conditions. But, like time itself, the needlegun would not be stopped.[63]

. . . . .

The tests on the Hasenheide in October 1836 were one of a series of contemporary events which were advancing warfare to its twentieth-century form. The progress, however, was largely imperceptible to contemporaries. As the critics of the needlegun rightly pointed out, the weapon was technically imperfect, inaccurate at long range, and unproven in the field. Logical worries were also voiced about the difficulty of supplying ammunition to undisciplined soldiers who would fire it too quickly. Such doubts understandably outweighed the potential military advantage of tripling firepower. It was not the skeptical commissioners, but rather Witzleben and his avant garde of Eduard Peucker and Georg Heinrich Priem who seemed rash and immoderate in the mid-1830s.[64]

One has to wonder, however, whether the officer who hurled dirt at the needlegun hoped it would malfunction. For how many officers could sympathize with Witzleben's wish to have had such a gun to use against

[63] All of the details of this scene, including the facial expressions and emotions of the participants, are found in Nicolaus Dreyse's recollection of the test, printed in Rabe, *Zündnadelgewehr*, 25–26. See also Löbell, *Des Zündnadelgewehrs*, 28. Rabe notes erroneously that Dreyse's short memoir describes a test in 1839, but Witzleben's presence clearly identifies the test as that of October 1836—for Witzleben was dead by 1839.

[64] This is the thesis of Showalter, *Railroads and Rifles*, 76–80.

Napoleon at Dresden? How many actually wanted the needlegun to determine the extent of Prussian territory, as Priem brashly asserted it could to the Crown Prince?[65] How many believed that mortal men, not unfeeling machines, should determine the outcome of battles? Or that excessive reliance and emphasis on engines of war could undermine the manly virtues that were really required to win wars?

Such questions, of course, have no quantitative answers. But it is safe to conclude that the advent of modern weaponry gave rise to a variety of views in the Prussian officer corps. As we saw earlier, there were soldiers in the horse artillery, cavalry, and Guard Corps who regarded the mechanization of war as unmanly. On the opposite side were officers like Witzleben and Priem who were genuinely excited about technological advances in weaponry; or generals like Ernst von Pfuel who can reasonably be described as a "futurist." The liberal nobleman liked to predict what the upcoming decades held in store—occasionally with uncanny insight, as shown by his fascination with the development of railroads, submarines, and aircraft.[66] The emotions and feelings of the vast majority of officers were probably mixed and ambivalent or complex to the point of sublimation. For many, the unwillingness to fully embrace modern warfare found expression in praise for ancient fighters whose exploits were unaided by machines.

Eduard Peucker provides a fascinating example. He was one of the leading weapons specialists in the army, usually central to research efforts in new areas like the rocket, fine-grained gunpowder, the needlegun, and railroads. Like so many bourgeois officers educated in the nation's gymnasiums, however, Peucker's mental being was inseparable from the ancient past. He believed that there was substance to legends which placed the ancestors of the early Germanic tribes on the plains of Hellas. The warlike spirit and fighting practices which built the Greece of Agamemnon, Leonidas, and Alexander lived on in the German warriors who united for centuries to beat back the mighty Romans.[67] But two millennia of "intellectual progress" had been accompanied by a certain "loss of character and resolve." Although "advancements in the various areas of technology" and the scientific treatment of warfare had contributed greatly to success in military endeavors, "the warlike virtues of a people have always outweighed this contribution." The latter depended upon "simple mores, unattenuated bodily strength and participation in

[65] For the statements of Witzleben (1836) and Priem (1838), see Löbell, *Des Zündnadelgewehrs*, 26, 30.

[66] Bernhard von Gersdorff, *Ernst von Pfuel* (Berlin: Stopp Verlag, 1981), 36, 135–37; Rochow, *Vom Leben*, 121–24.

[67] Eduard [von] Peucker, *Das Deutsche Kriegswesen der Urzeiten in seinen Verbindungen und Wechselwirkungen mit dem gleichzeitigen Staats- und Volksleben* (Berlin: R. Decker, 1860), 4–5.

public affairs."[68] Peucker wrote two tomes on the early Germans, dedicating each to the proposition that knowledge of history would serve the cause of German unification as no weapon could.

The great theorist of war, Carl von Clausewitz, was another soldier who harbored lingering doubts about the application of modern technology to warfare. He appreciated the distinct weaponry and unique mode of fighting which had dominated each separate historical epoch. War in the early nineteenth century, he realized, was conducted on a much larger scale and with more lethal tools than had been the case in either early modern, medieval, or ancient times. The concentration of artillery fire, musket volleys, and cavalry attack at the weakest point in an enemy line could often rout a foe superior in overall numbers. The "art" of modern warfare therefore broke down into categories of mass arms-production, clever campaign strategy, and instinctive maneuvering to affect superior numbers and firepower at a particular point on the battlefield.[69] Yet Clausewitz was never overly impressed with a modern military technology which he felt could never replace human heroics. He wrote to Johann Gottlieb Fichte in 1809 that "as important as . . . firearms are in modern times for preparing success in battle, it is still the enthusiastic rush forward with glistening cold steel which captures victory."[70] In lectures to young Crown Prince Frederick William around 1811, Clausewitz criticized muskets and pistols with the sarcastic observation that Prussia's soldiers could not dispense with firearms, for "if we could, why should we carry them at all?"[71] And later, around 1815, he wrote in metaphoric derision of modern artillery that close combat with "the dagger and the battleaxe" aroused the passions of a fistfight, unlike long-range weapons such as "the lance, the javelin or the sling" which allowed "the instinct for fighting . . . to remain almost at rest."[72] Clausewitz was no stranger to artillery, having served as Prince August's adjutant and instructor from 1803 until 1808—and as his quick-studying apprentice in 1829/30. He was angered, moreover, at conservative enemies for attempting to undermine the engineering corps and artillery in 1824.[73] When war threat-

[68] Ibid., 3, 4.

[69] Clausewitz, *Vom Kriege*, 82–90, 651–61; Clausewitz to Hardenberg, 12 February 1811, and Clausewitz to Gneisenau, 2 September 1811, printed in Werner Hahlweg (ed.), *Carl von Clausewitz: Schriften, Aufsätze, Studien, Briefe* (Göttingen: Vandenhoeck & Ruprecht, 1966), 147–48, 657; and Clausewitz to Fichte, 11 January 1809, printed in Hans Schulz (ed.), *J. G. Fichte: Briefwechsel* (Leipzig: H. Haessel, 1925), 2:520–25.

[70] Clausewitz to Fichte, 11 January 1809, printed in Schulz, *J. G. Fichte*, 2:524.

[71] See Clausewitz's instructions for the Crown Prince, n.d. (1810–12), printed in an English edition of *On War* (London: Kegan, Paul, Trench, Trübner & Co., 1911), 3:195.

[72] Cited in an appendix to ibid., 3:250.

[73] Clausewitz to Gneisenau, 1 October 1824, printed in Delbrück, *Gneisenaus Leben*, 5:507–8.

ened in the summer of 1830, however, Clausewitz wrote Gneisenau that he would rather "hang himself"[74] than serve in the artillery during wartime.

Clausewitz's friend harbored similar views about the glories of hand-to-hand combat. "When in history," he asked Boyen about the Spartan stand at Thermopylae, "was more achieved?"[75] The recipient of the letter sympathized with Gneisenau. Hermann von Boyen, who returned to head the Ministry of War in 1841, did not ignore the fiscal importance of the industrial bourgeoisie or the vital contribution of machine tools in constructing a revolutionary weapon like the needlegun. But his ideal remained the yeoman farmer-soldier who would rally to the nation's defense and dominate the invader with superior "warlike spirit." All of his proposals and legislative initiatives were consistent with this central concept. Boyen opposed primogeniture, energetically supported peasant emancipation, and advocated a more extensive system of rural schools because these policies would allow the class of small, independent landowners to grow in numbers, moral strength, and loyalty to the state. His national militia would draw its recruits from the small towns and villages of these ready patriots—and an active program of domestic resettlement would swell their ranks. Boyen believed that the western provinces were overfilled with factories and urban dwellers unfit for combat. Migration to farms or rural manufacturing centers in eastern Prussia would alleviate this problem and reduce the threat of revolution. Fortresses should be constructed in the provincial heartland, not on the frontier, to defend farms and rural industries and provide a staging ground for a popular war against the next Napoleon.[76]

Each province or department would also possess political representation in a national senate. Extra seats would be awarded to provinces which had distinguished themselves in war. Moreover, each social estate would hang special medallions around the necks of its own representatives. The ideas came from the early Romans—as did his concept of a national militia, a nineteenth-century version of the *secunda acies* of republican Rome.[77] Boyen did not overestimate the significance of his own era. Although he realized the extent to which gunpowder had changed warfare and admitted that the printing press had enabled modern warri-

[74] Clausewitz to Gneisenau, 20 August 1830, printed in ibid., 5:607.

[75] Gneisenau to Boyen, 9 September 1826, I HA Rep.92 Boyen, GStA Berlin.

[76] For Boyen's views, see his unpublished essays, "Über kriegischen Geist," n.d. (1820s), and "Die geistigen und körperlichen Eigenschaften des Kriegers," n.d. (1820s), in I HA Rep.92 Boyen, Nrs. 439 and 442, GStA Berlin. Also see Meinecke, *Das Leben*, 2:152–53, 156, 331–32, 336, 339–41, 344–46, 349, 366–67, 424–25, 432, 526–29.

[77] Ritter, *Staatskunst*, 1:135.

ors to improve upon military science,[78] he still maintained that the Romans had much to teach—above all in the area of moral instruction. The officer's challenge was to develop a character which enabled him to balance strength and wisdom. "To do this a leader must carry great ideals in his breast and familiarize himself with the heroes of remote antiquity."[79]

Clearly the unfolding marvels of modern industrial and military technology did not diminish the impression, widely held in the army's reformist fraternity, that "modern times" were overshadowed in some important ways by the greatness of former eras. In order to acquaint themselves with the heroic ages, for instance, Boyen encouraged students of the War School to read all twenty-four volumes of universal history by Johannes von Müller, a famous professor from the University of Göttingen.[80] Curricular emphasis on the classics strengthened under Georg von Valentini and his successor as Inspector of Military Schools, August von Luck. Under the latter's tutelage in 1842, the army required classical gymnasial training for all officer candidates. Luck's old friend from the wars, Johann Krauseneck, Chief of the General Staff until 1848, applauded from the wings, for since his early days in the gymnasium, "Greek ideals and Roman deeds, the grand forms of the ancient world, had fueled a fire in his soul."[81] In Rühle von Lilienstern, Inspector after 1844, the office received a man whose gaze was fixed even more keenly on remote antiquity. History, Latin, and Greek were important, but Prussia's top soldier-scholar believed that the most "imaginative and poetically inclined" cadets would benefit exceedingly more from studying Homer and other ancient sagas and myths with their "mysterious mixture of poetry and [historical] truth."[82] Peucker advanced to this post in 1854, fighting a holding action against modern critics of ancient language and literature. This true Spartan of the North found many reasons to study the classics, above all to appreciate "the connection between modern European culture and antiquity."[83]

Karl Wilhelm von Grolman, commander of the Fifth Army Corps in Breslau until 1843, often scoffed at such notions. Ignorance of Tacitus, he

[78] See his "Die geistigen und körperlichen Eigenschaften des Kriegers," cited above in n. 76.

[79] Cited in Scharfenort, *Königlich Preussische Kriegsakademie*, 40.

[80] Ibid., 40.

[81] Cited in Erich Weniger, *Goethe und die Generale* (Leipzig: Im Insel Verlag, 1942), 184–85. For Valentini and Luck, see ADB 19:355; Poten, *Geschichte*, 220–22; and Eduard [von] Peucker, *Denkschrift über den geschichtlichen Verlauf welchen die Vorschriften über das von den Offizier-Aspiranten darzulegende Mass an formaler Bildung und die Erfolge dieser Vorschriften seit der Reorganisation des Heeres im 1808 genommen haben* (Berlin: R. Decker, 1861), 42–43.

[82] Cited in "Rühle von Lilienstern," 180, MgFa Freiburg.

[83] Peucker, *Denkschrift*, 92.

boasted, would not have diminished his military exploits in the least. Clausewitz also liked to ridicule the learned philologists and historians. In both cases, however, appearances were deceiving. Like Johann Braun, Grolman loved classical literature, an affair which began in the gymnasiums of Berlin. And Clausewitz's mockery of the academicians stemmed from his belief that they were too dilettantish, idealistic, and impractical to possess a true appreciation of great civilizations like those of Greece and Rome.[84] Beneath the caustic comments, in fact, lay the same unmistakable reverence for the distant past. In *On War* he writes that the "excellent" army of Alexander the Great achieved "intrinsic perfection," that Rome "stands alone," and elsewhere, that ancient soldiers were unsurpassed in the development of "individual warlike spirit."[85] Nor was this surprising, for his mentor, the elder Scharnhorst, had also praised the "cruel austerity" and "virtuous self-sacrifice"[86] of the Romans.

The proud, yet backward-looking mentality of Prussia's military reformers was most evident in the allusions and metaphors they used to describe themselves and their own times. Oldwig von Natzmer, Witzleben's friend since childhood, knew no greater complement when Prince William (Frederick William III's brother) retired to a Silesian estate in 1822 than to compare his friend with Cincinnatus, the consul who returned to farming after defeating Rome's enemies. Similarly, Valentini's protégé, Heinrich von Brandt, flattered Gneisenau with the reminder that "we cannot all be Scipios [like you]."[87] The "petty kings" of modern times were Bronze Age "Agamemnons"[88] to Valentini, a man whose favorite pastime was to read and reread the allegedly more impressive history of Rome. To Clausewitz, conquerors like Gustavus Adolphus, Charles XII, and Frederick the Great were "three new Alexanders."[89] During the heyday of Napoleonic victories, Scharnhorst and his disciples wondered if the French were the new Romans and the Prussians a Sparta

[84] For Grolman, see Zeitschrift-Krieges 68 (1846): 34–35, MgFa Freiburg; and Weniger, *Goethe und die Generale*, 183–84. For Clausewitz, see Clausewitz, *Vom Kriege*, 146; id., "Umtriebe," n.d. (1822/23), printed in id., *Politische Schriften*, 166.

[85] Clausewitz, *Vom Kriege*, 651–52; Clausewitz to Fichte, 11 January 1809, printed in Schulz, *J. G. Fichte*, 2:524.

[86] Cited in Reinhard Höhn, *Scharnhorsts Vermächtnis* (Frankfurt: Bernard & Graefe Verlag, 1972), 96.

[87] Brandt, *Aus dem Leben*, 2:11. For Natzmer's remark, see Natzmer to Prince William, n.d. (1822), printed in Natzmer, *Unter den Hohenzollern*, 1:107.

[88] Cited in Weniger, *Goethe und die Generale*, 77. For his long-standing awe of the Romans, see Berenhorst to Valentini, 30 November 1809, printed in Eduard von Bülow (ed.), *Aus dem Nachlasse von Georg Heinrich von Berenhorst* (Dessau: Verlag von Karl Aue, 1847), 2:306–7; and Brandt, *Aus dem Leben*, 2:17.

[89] Clausewitz, *Vom Kriege*, 654.

whose zenith had passed.[90] A similar Greco/Roman dichotomy with its anti-French symbolism adorned Peucker's historical works. But for some the imagery shifted after Waterloo. Thus Gneisenau and Clausewitz waited for the inevitable "Third Punic War"[91] which must destroy a New Carthage in the West. This was the most flattering metaphor of all, for a half century would elapse before Prussia was mighty enough to sow salt in France's soil.

Those who accomplished the feat, moreover, were too busy toasting their own achievements to continue to pay homage to the ancients. Like the sculptor in Friedrich Spielhagen's *Sturmflut* who turns a sedate statue of Homer into a proud Germania by switching heads, the new generation was preoccupied with the present and future, not the past.[92] This typically modern orientation had first become evident decades earlier. Arriving at the War School in 1828, Heinrich von Brandt was dismayed to find the young cadets "completely indifferent" to the history of ancient Greece and Rome. Only the military history of recent times interested them "somewhat more." A new, pragmatic generation of officers was emerging which found more solace in the teaching of instructors like Carl Ritter, the School's geographer. "Ideals arise from the weaknesses of people," he noted in 1830. "We want to raise life—reality—to an ideal, then reality becomes richer than any ideal."[93] Students of his like Albrecht von Roon, Edwin von Manteuffel, and Helmut von Moltke were as committed in their opposition to parliamentarism as they were dedicated to the systematic exploitation of technology for military purposes. In stark contrast to the action of an older generation of conservatives in 1824, the new breed would never have considered a cutback in the army's three scientific corps.[94]

Prince William (Frederick William III's son), commander of the elite Guard Regiment in Berlin, and his adjutant, Wilhelm von Reyher, emerged as the champions of this early variant of "reactionary modernism," Jeffrey Herf's term for those who welcomed the technological fruits

[90] Höhn, *Scharnhorsts Vermächtnis*, 96; Clausewitz to his wife, Marie, 27 June 1807 and 5 October 1807, printed in Linnebach, *Karl und Marie von Clausewitz*, 126, 142, 144.

[91] Treitschke, *Deutsche Geschichte*, 4:45.

[92] See the discussion in Katherine Roper, *German Encounters with Modernity: Novels of Imperial Berlin* (New Jersey and London: Humanities Press International, Inc., 1991), 65–66.

[93] For Brandt, see *Aus dem Leben*, 2:18; for Ritter, see Michael Karl, "Fabrikinspectoren in Preussen," unpublished dissertation, University of Konstanz, 1988, chap. 3, p. 9.

[94] For Ritter, Roon, and Moltke, see Albrecht von Roon, *Denkwürdigkeiten aus dem Leben des General-Feldmarschalls Kriegsministers Grafen von Roon* (Breslau: Verlag von Eduard Trewendt, 1892), 1:46, 55–57. For the well-known political conservatism of Roon and Moltke, see Craig, *Politics of the Prussian Army*, 138–60.

of modernity but rejected political offshoots like parliamentarism.[95] Gneisenau's son-in-law ridiculed Reyher's politics as "antediluvian,"[96] but the antiliberal cavalry officer was thoroughly in tune with the defensive implications and offensive potential of the new technologies. Prince William himself was a die-hard opponent of his brother's reformist politics after 1840, but again, he combined these views with a mind which was open to rapid-firing muskets, rifled steel cannon barrels, and railroads. By 1848, in fact, the "leathery practitioner"[97] had developed close ties to many industrialists and possessed a reputation as an advocate of industrial, technological, and scientific development.

It was under the tuteledge of these reactionary modernists that the liberals' old dream of a united Fatherland came to fruition. The War School, renamed the War Academy, eventually shed all pretense to humanistic, universal education and concentrated on an expanded range of the narrow, technical subjects which, it was felt, were all that a soldier needed in modern combat.[98] The day when German officers would say that "our weapon is a science—our science is a weapon" (*die Waffe eine Wissenschaft, die Wissenschaft eine Waffe*) was drawing near. The hour of the Gneisenaus, Rühle von Liliensterns, Boyens, and Peuckers had passed.

[95] Jeffrey Herf, *Reactionary Modernism: Technology, Culture and Politics in Weimar and the Third Reich* (New York: Cambridge University Press, 1985). And for Reyher, see Meinecke, *Das Leben*, 2:490, 510–11; General von Ollech, *Carl Friedrich Wilhelm von Reyher* (Berlin: Ernst Siegfried Mittler und Sohn, 1879), 86, 95–97; Bernhard Meinke, "Die ältesten Stimmen über die militärische Bedeutung der Eisenbahnen, 1833–42," *Archiv für Eisenbahnwesen* 42 (1919): 65; and Showalter, *Railroads and Rifles*, 37. Standing in for a sick Prince William, Reyher transported the entire Guard Corps from Potsdam to Berlin during the 1839 maneuvers. For William, the needlegun, and rifled steel cannon, see, respectively, Thierbach, *Geschichtliche Entwicklung*, 314; McNeill, *Pursuit of Power*, 247; and William Manchester, *The Arms of Krupp* (Boston: Bantam, 1970), 87, 95–101, 121. And for William, industry, and science, see Varnhagen Diary II, 13 November 1842; 22 May 1844; 19–20 July 1844; 4 January 1845; and 24 July 1845, 2:118, 299, 328–33, 3:3, 134.

[96] Brühl to Gneisenau, 21 April 1830, printed in Heinrich von Sybel, "Gneisenau und sein Schwiegersohn, Graf Friedrich Wilhelm von Brühl," *Historische Zeitschrift* 69 (1892): 257.

[97] This was Prince William's own description, noted in the Diary of Josua Hasenclever, 20 July 1844, printed in Hasenclever, *Josua Hasenclever*, 98.

[98] Manfred Messerschmidt, "Die preussische Armee," in Papke, *Handbuch*, 4 (2): 106–9, 117–21.

VI

The Masonic Vision

CHRISTIAN ROTHER squinted through the window of his writing room onto the snow-covered grounds of Rogau, his estate in Lower Silesia. A slightly wrinkled face and gray hair were the unmistakable signs of a man who had already passed middle age. The unfinished business of state which usually lured him away from the stressful distractions of Berlin did not fully explain his presence at home on this particular occasion. The Chief of the Seehandlung, the state's merchant-banking empire, expected weekend guests. No longer able to concentrate, Rother had put the desktop in order and begun his vigil.

Finally, they were coming. A mud-spattered carriage pulled past the icy gatehouse and opened its doors for two warmly dressed gentlemen who waved to their host above. By the time he descended Rogau's ornate staircase, the two businessmen, Wilhelm Oelsner and Heinrich Ruffer, were already seated in a front parlor continuing their discussion. The war scare in Greece was on everyone's mind in early 1828.

As Rother strode into the room he motioned facetiously for the conversation to end. It had been years since the three brothers were together at Rogau and all knew that greetings took precedence. Rother moved toward the elder Oelsner and welcomed him in the masonic way. Oelsner and Ruffer also extended hugs and salutations from their lodge brethren in Breslau and Liegnitz.

With formalities out of the way, the three settled into armchairs next to a well-stoked fire. Before long, they were caught up in personal stories, gossip, local politics, and more talk of war. Rother waited until these topics were exhausted before adroitly steering the discussion to the business which lay behind the visit—and the matter which stood closest to their hearts. Somehow the means had to be found to enliven Silesia's dying textile industries.[1]

. . . . .

Rother, Oelsner, and Ruffer were freemasons. The Seehandlung Chief belonged to the Grand National Lodge of the Freemasons of Germany,

[1] For Rother's discussions with Ruffer and other Silesian businessmen at Rogau in the late 1820s, see Radtke, *Seehandlung*, 104–5; and Gerhard Webersinn, "Gustav Heinrich Ruffer," *Jahrbuch der Schlesischen Friedrich-Wilhelms-Universität zu Breslau* 11 (1966): 171, 177.

while the two entrepreneurs were members of the Grand National Mother Lodge of the Three Globes. Together with a third brotherhood in Prussia, the Grand Lodge of the Royal York Friendship, these societies believed in a supernaturalistic, humanitarian strain of freemasonry which had originated in England a century earlier.[2]

Like other deists, the English freemasons saw God as a glorious architect whose ordered, mechanical universe held the clues for human progress on earth.[3] Constituted in 1723, the Grand Lodge of London promoted scientific advance and taught the value of social equality, Christian charity, and brotherly love. An official brochure of 1739 claimed that "the brothers are not comfortable with pedants and crepe-hangers in their midst."

> Preferring serene and cheerful faces, [it continued], they seek to determine the causes of things with grace and without arrogance, affectation or prejudice. They are all equal and call themselves brothers, comrades and friends. Quarrels, obstinacy, envy and jealousy are banned from their assemblies. One is accustomed there only to science, a thirst for knowledge, politeness and warmth—what I call true Christian love. Truth is their ultimate goal and reason is the leader they follow.

Members also learned, the author went on, "how one should use the things which are created."[4] Indeed, as Margaret Jacob has shown in two recent monographs, the freemasons played a major role in propagating Newtonian physics and in attempting to find industrial applications for these mechanical principles.[5]

In English Freemasonry, however, we possess another instructive example of progressives who were really looking backward. "One should not be led to the false assumption," the pamphlet boasted, "that freemasonry has come into the [modern] world as something new." On the contrary, it was "patterned after the model of those societies which existed

[2] For the Masonic affiliations and close friendship of Rother, Oelsner, and Ruffer, see Trende, *Im Schatten*, 21, 28; Friedrich Kneisner, *Geschichte der deutschen Freimaurerei* (Berlin: Alfred Unger, 1912), 147; and Ulrich von Merhart, *Weltfreimaurerei: Ein Überblick von ihrem Beginn bis zur Gegenwart* (Hamburg: Bauhütten Verlag, 1969), 223–24. Trende identifies the local lodges correctly, but mistakenly implies that all of these were affiliated with the Royal York Friendship.

[3] For English Freemasonry, see Eugen Lenhoff, *Die Freimaurer* (Zürich: Amalthea-Verlag, 1929), 11–21, 66–70; Epstein, *The Genesis of German Conservatism*, 84–87; but especially, (a) Margaret C. Jacob, *The Radical Enlightenment: Pantheists, Freemasons and Republicans* (London: George Allen & Unwin, 1981), 109–37, and (b) Ludwig Keller, *Die Freimaurerei: Eine Einführung in ihre Anschauungswelt und ihre Geschichte* (Leipzig: B. G. Teubner, 1918), 9–16.

[4] Quoted in Keller, *Die Freimaurerei*, 13.

[5] Jacob, *Radical Enlightenment*, 109–37; id., *The Cultural Meaning of the Scientific Revolution* (New York: Alfred Knopf, 1988), 126–28, 186–87.

in Rome, Athens, Lacedaemon [Sparta] and other cities where the arts and sciences were in a flowering state."[6] The leaders of the Grand Lodge wrote—and were genuinely convinced—that Greek and Roman brotherhoods had once possessed oriental secrets which predated the classical era itself by more than a millennium. Brought to ancient Italy by Pythagoras, a sixth-century Greek mathematician who had studied for two decades with Egyptian priests, these geometric formulas and symbols allegedly expressed numerical relationships which unlocked the mysteries of nature and pointed the way toward harmony among men. To English masons, the study of science in modern times was a quasi-religious quest for long forgotten knowledge. As one historian of freemasonry observes, the brothers "sought valiantly, in the ruin of time, for the lost keys to the sacred sciences."[7]

The English cult spread to Germany in 1737 with the establishment of a lodge in Hamburg. From there the zealous colonizers disembarked to Göttingen, Königsberg, Berlin, Frankfurt am Main, and other German towns.[8] The claim to represent an ancient heritage remained an important element of English freemasonry as it spread through Germany. Thus lodge masters issued the brochure quoted above in eight German language editions between 1739 and 1792.[9] Moreover, in 1812 we find Maximilian von Castillon, grand master of Rother's Grand National Lodge, looking into the future through the rearview mirror of the past.

> The fashionable expression—"one must progress with the times"—carries no weight. In the true sense of the word, we want—we must—progress with the times. But that means to strive to become complete according to the spirit of our teachings and laws. [For] true freemasonry is a truth which is eternal and immutable, but, like a mathematical truth, is also capable of multiple applications. We can and want to adapt this application to the spirit of the times—in other words, "[to] progress with the times."[10]

Lodge names provide another clue to the continuing allure of antiquity to the masonic elite. In the 1820s, for example, the Royal York Friend-

[6] Quoted in Keller, *Die Freimaurerei*, 12.

[7] Quoted in Manly P. Hall, *Freemasonry of the Ancient Egyptians* (Los Angeles: The Philosophical Research Society, Inc., 1937), 6. Also see ibid., 14–15, 47; Lenhoff, *Die Freimaurer*, 38–42; Merhart, *Weltfreimaurerei*, 22.

[8] For the spread of English Masonry to Germany, see Jacob, *Radical Enlightenment*, 135–36, 194; Manfred Steffens, *Freimaurer in Deutschland. Bilanz eines Vierteljahrtausends* (Flensburg: Christian Wolff, 1964), 119–215; and especially Gerhard Krüger, *Gründeten auch unsere Freiheit: Spätaufklärung, Freimaurerei, preussisch-deutsche Reform, der Kampf Theodor von Schöns gegen die Reaktion* (Hamburg: Bauhütten, 1978), 11–33, 42–46, 138.

[9] Keller, *Die Freimaurerei*, 11.

[10] Cited in *Die ersten 150 Jahre des Grossen Ordens-Kapitels der Grossen Landes-Loge der Freimaurer von Deutschland* (Berlin, 1926) (#3083), 69, A-BDDFM, Bayreuth.

ship (RY) had active lodges named after Hippocrates (Magdeburg); Pythagoras (Berlin); Socrates (Kalisch); Isis, the Egyptian goddess of fertility (Lauban); and Urania, the Greek Muse of astronomy (Berlin). Rother's Grand National Lodge (GLL) dedicated chapters to Urania (Bützow); Pegasus (Berlin); Hercules (Schweidnitz); Hippocrates (Schwerin); Prometheus (Rostock); the Phoenix (Königsberg); and Minerva, the Roman goddess of wisdom (Potsdam). Similarly, Ruffer's chapter in the Mother Lodge of the Three Globes (3WK) was named after Pythagoras (Liegnitz), while other lodges paid homage to Urania (Stargard), Minerva (Cologne), Memphis (Memel), and Apollo (Güstrow).[11] It was common by this time for leading freemasons to make bold public claims about the advantage which brothers possessed over secular scientists as a result of "the secret teachings" and "the deep and fundamental knowledge"[12] which the lodges had preserved from ancient times. Internally, however, a debate was raging among masonic officers over the exact source and meaning of masonic traditions. Thus Johann Gottlieb Fichte believed that modern masonic ceremonies had ascended from the murky depths of remote antiquity, but sought proof for his beliefs. "We want, on the one hand, to convince ourselves of the genuineness of these orally transmitted usages and rituals; [but want to] inform ourselves, on the other, about their origin and true significance."[13]

When we shift our focus from the German leadership to the rank and file, it is extremely difficult to determine with any certainty why English freemasonry appealed. Presumably the English emphasis on social equality was attractive to the "subordinates"[14] of German society, as one observer of the German Freemasons described typical members. Of the 1,592 freemasons in Berlin's fifteen lodges in the late 1830s, for instance, 1,513—over 95 percent—were bourgeois. The need for surrogate religious experiences in a secular, more skeptical age; the desire for cama-

[11] A. Flohr, *Geschichte des Grossen Loge von Preussen, genannt Royal York zur Freundschaft im Orient von Berlin* (Berlin: P. Stankiewicz Buchdruckerei, 1898), Part II, 136–37; *Nachweiss der Grossen Landes-Loge der Freimaurer von Deutschland zu Berlin für das Jahr 1843 und 1844* (Berlin: E. S. Mittler, 1844), 36–44, 70–73, 74–75, MLM-Penn, Philadelphia.

[12] See K. A. Varnhagen von Ense, "Franz von Baader," n.d. (1841), printed in *Ausgewählte Schriften von K. A. Varnhagen von Ense* (Leipzig: F. A. Brockhaus, 1875), 18:290.

[13] Fichte's remarks of 1800 are cited in Ferdinand Runkel, *Geschichte der Freimaurerei in Deutschland* (Berlin: Reimar Hobbing, 1932), 2:262–64. For the internal debate, see the following sources from A-BDDFM, Bayreuth: Johannes August Freiherr von Starck, *Ueber die alten und neuen Mysterien* (Berlin, 1817) (#7388), first published in 1782, *passim*; *Versuch eines alphabetischen Verzeichnisses der wichtigen Nachrichten zur Kenntnis und Geschichte der Freimaurerei* (Jena, 1817) (#658), 131–32; and *Encyclopädie der Freimaurerei* (Leipzig, 1828) (#902), 3:175–90.

[14] Leo to Hengstenberg, 3 June 1852, printed in Johannes Bachmann, *Ernst Wilhelm Hengstenberg: Sein Leben und Wirken nach gedruckten und ungedruckten Quellen* (Gütersloh: C. Bertelsmann, 1876–1892), 3:262–63.

raderie with other men; the gratification that comes from being a part of something secretive; and the opportunity to make useful contacts in one's profession were also probable motivations. With regard to the latter point, there appears to have been a great deal of recruiting and networking within certain vocational groups. Thus the Pilgrim Lodge (GLL) was known as the "doctor's lodge" because of the thirty-five physicians—22 percent of the membership—who belonged. The Aries Lodge (GLL) had a large concentration of thirty factory owners (24%), while the Golden Ship (GLL) was a "soldier's lodge" with forty-three army officials (52%). Similarly, Rother's Three Golden Keys (GLL) was a haven for thirty-one civilian bureaucrats who made up 46 percent of the lodge membership.[15]

The emphasis in masonic literature on the mechanistic aspect of nature and the claims made about the antiquity of the brothers' quest for scientific knowledge may also have impressed entrepreneurs, civil servants, and army officers whose work was scientific or technical in nature. Thus Friedrich Dannenberger, Berlin's leading cotton manufacturer, and Johann Hempel, owner of the capital's biggest chemical works, were freemasons, as were Silesian industrialists like Oelsner, Ruffer, and Henckel von Donnersmark. The first head of the Business Department (Theodor von Schön); the chemical expert on the Technical Deputation (Sigismund Hermbstädt); the chiefs of the mining and metals divisions of the Mining Corps (Ludewig Gerhard and Carl Karsten); the Director of the Royal Botanical Gardens (Friedrich Link); the Chief of the Seehandlung (Christian Rother); the Chief of the Army Medical Corps (Johann Nepomuk Rust); and six of the army's leading technologists—Johannes Braun and Eduard Peucker (weapons testing), Moritz Meyer (metallurgy), Louis Blesson (army engineering), Franz August O'etzel (telegraphy), and Karl Turte (explosives)—were all masonic brothers. Of Berlin's 1,592 freemasons in the late 1830s, in fact, 537—or 34 percent of the membership—were either factory owners (274), medical doctors and pharmacists (182), or specialized technocrats, technicians, and engineers (81). The

[15] In general for the appeal of masonry by class in Germany, see Krüger, *Unsere Freiheit*, 26, 180; Steffens, *Freimaurer*, 198, 571; Ernst Mannheim, *Aufklärung und öffentliche Meinung*, first published in 1933 (Stuttgart, 1979), 93; and Jacob, *Radical Enlightenment*, 208. For the statistics in this paragraph, see *Mitgliederverzeichniss der unter Sr. Majestät des Königs von Preussen allerhöchstem Schutze dirigirenden Grossen Loge der Freimaurer Royale York zur Freundschaft im Orient von Berlin . . . 1833–1834*, and *Haupt-Uebersicht der Grossen National-Mutter-Loge der Preussischen Staaten, genannt zu den drei Weltkugeln . . . Für das Jahr 1839–1840*, A-BDDFM, Bayreuth; and *Nachweiss der Grossen Landes-Loge* (GLL membership list, 1843/44, cited above in n. 11), MLM-Penn, Philadelphia. For recruitment of doctors into the Pilgrim Lodge, see Otto Rosenthal, *Beiträge zur Geschichte der Loge zum Pilgrim in Berlin 1776–1901: Bei der Feier des 125 jährigen Stiftungsfestes am 24. Februar 1901 den Brüdern gewidmet* (Berlin: Johannis-Loge zum Pilgrim, 1901), 16–17, MLM-Penn, Philadelphia.

"preindustrial," generally antitechnological *Mittelstand* of artisans, handicraftsmen, and shopkeepers was less significantly represented with 207 members (13%).[16]

The urge to serve humanity with one's scientific and technical expertise was another motivating force within the Prussian lodges. In fact, as we narrow our focus to the present subject of analysis—Christian Rother and the Seehandlung—the humanitarian factor takes on particular significance. Rother's Grand National Lodge placed increasing emphasis on the social responsibility of the individual as the late 1700s gave way to the Era of Restoration.[17] Accordingly, we find Samuel Rösel, a professor of architecture and speech-giver at the cross-town Pilgrim Lodge during the 1830s, encouraging brothers to let masonic teachings "bear fruit in deeds." Two of these "truly heavenly fruits" were "love for the truth" and a striving for "fairness in all things" (*Billigkeit*). Members who failed to pass along benefits received from freemasonry to others were "neither brotherly, nor Christian, nor humane." It was quite characteristic, therefore, when the Grand National Lodge collected donations in 1826 for the widows and orphans "of the brave, courageous and steadfast defenders of Hellas" against the Turks.[18] A dedicated mason for over two decades, Rother established his own charities for the upbringing (*Erziehung*) of "morally depraved children," and later, after he dropped off the active membership rolls, for orphaned and impoverished "daughters of faithful state servants."[19] It is significant for our purposes, moreover, that a friend

[16] Kaelble, *Berliner Unternehmer*, 131, finds that 68.8 percent of Berlin businessmen in the 1830s and 1840s who described themselves as "Kaufmänner" were actually engaged in industry, not trade. I have used this percentage in determining the number of freemasonic "Kaufmänner" who were industrialists.

In addition, see from A-BDDFM, Bayreuth: *Mitgliederverzeichniss* and the *Haupt-Uebersicht* (RY and 3WK membership lists, 1833/34, 1839/40, cited above in n. 15), plus: Heinrich Brüggemann, *Bausteine zur Geschichte der Johannis-Loge Zu den drei goldenen Schlüsseln Berlin* (Berlin, 1919) (#3891), 112; *Verzeichniss der Brüder der gesetzmässigen, verbesserten und vollkommenen St. Johannis-Loge zum Goldenen Schiff in Berlin (März, 1825)*; *Verzeichniss sämmtlicher Mitglieder der unter Constitution der Hochwürdigen grossen Loge der Freimaurer in Berlin Royale York zur Freundschaft arbeitenden St. Johannis Loge Horus im Orient zu Breslau beim Anfange des Maurerjahrs 5816.*

From MLM-Penn, Philadelphia, see also: *Nachweis der Grossen Landes-Loge* (GLL membership list, 1843/44, cited above in n. 11), 45–69; *Namentliches Verzeichniss sämtlicher zu dem Bunde der grossen National-Mutter-Loge zu den drei Weltkugeln gehörigen Brüder Freimaurer* (Berlin, 1813), 13–15, 53; and Rosenthal, *Beiträge*, 11, 14–15, 17. Also see Krüger, *Unsere Freiheit*, 38, 81; Trende, *Im Schatten*, 21, 42; Kneisner, *Geschichte*, 178, 180; and Radtke, *Seehandlung*, 19, 22–23.

[17] Merhart, *Weltfreimaurerei*, 117–18; Kneisner, *Geschichte*, 173–74.

[18] The Rösel and Hellas quotes, respectively, are from Rosenthal, *Beiträge*, 22, 16, MLM-Penn, Philadelphia.

[19] ADB 29:360–61. For Rother's dedication to the practice of Freemasonic principles, see

could describe Rother's "higher purpose" with the Seehandlung as "lifting Germans out of their misery."[20] To this task Rother brought a masonic enthusiasm for technical and mechanical progress, a love for his Silesian homeland, and a peculiarly masonic willingness to aid disadvantaged people.

. . . . .

The Seehandlung was in no position to help during most of its first decade under Rother. The corporation received new life in January 1820 as "an independent financial and commercial institute of the state."[21] It was given four specific charges from the king, namely: supervision of the salt monopoly; purchase of indispensable foreign products; provision of financial support for the ministries; and management of the state's foreign debts. Between 1820 and 1822, Rother was preoccupied with two of these regular assignments which fell in the delicate area of public finance. He negotiated two huge loans through the Rothschilds of London, but the state's budgets were still precariously out of balance. Accordingly, profits had to be transfered in trouble-shooting fashion from the corporation's salt monopoly to the ministries and district governments most in need.[22]

It is important, however, that Rother and Frederick William did not intend to restrict the Seehandlung to the four charges specified in 1820, especially if the institute could serve in "extraordinary" ways to bolster state finances and improve the kingdom's economy. Indeed, both men placed special importance on the "independent" (*selbständigen*) status of the Seehandlung. Before submission to Frederick William, Rother's proposals for extraordinary projects were to be examined by a supervisory committee consisting of the king's liaison with the ministers, Count Lottum; the director of the Statistical Bureau, Johann Gottfried Hoffmann; and the current President of the Council of State—Duke Carl after 1825; General Müffling after 1837.[23]

The Seehandlung's first special ventures began soon after the second London loan.[24] Frederick William was personally involved with many of

Radtke, *Seehandlung*, 19, 22–23; and Henning, "Preussische Sozialpolitik," 519–23, 534–35.

[20] Vincke to Rother, 28 June 1824, Rep.92 Rother, Nr.18, ZStA Merseburg.

[21] Radtke, *Seehandlung*, 54.

[22] Rother to Hardenberg, 14 April 1821, printed in Trende, *Im Schatten*, 102; Radtke, *Seehandlung*, 58–77.

[23] Radtke, *Seehandlung*, 50–52.

[24] For this paragraph, see Gerhard Webersinn, "Christian von Rother: Ein Leben für Preussen und Schlesien," *Jahrbuch der schlesischen Friedrich-Wilhelms Universität zu*

these initial projects, affording him an opportunity to implement personal economic notions about the importance of agriculture and rural by-industries. In 1824 and 1825, for instance, the Seehandlung found itself under the king's orders to revitalize agriculture in the northeastern provinces. It expended the regal sum of 3 million thaler to convert estates there from grain to sheep farming, which was considered more lucrative. Although there were initial successes—prompting Pomeranian farmers to speak of "the time of the golden fleece"[25]—wool prices plummeted in 1826. This setback induced Rother's corporation to deepen its involvement by purchasing large amounts of Prussian wool for resale abroad. But exploding Australian exports to Britain soon undercut this early experiment in market manipulation. The Seehandlung also chartered three sailing vessels to expand foreign markets for Prussian linen. This was "one of the country's most excellent manufactured products,"[26] as Frederick William's decree worded it, and probably represented the economic cause closest to the king's heart. During the 1820s, moreover, the Seehandlung expanded its interests to road construction, angering private contractors by outbidding them for completion of one thousand kilometers of highways linking Berlin to Hamburg, Stettin, and East Prussia. This was an early sign of the tension which the zero-sum game of state enterprise could generate.

Rother was not unhappy with these operations. He knew the importance of a modern transportation system and was eager to participate in its completion. Rother had also spent his first ten years in government under the administration of Hardenberg, a man with many Physiocratic beliefs.[27] This experience undoubtedly strengthened the peasant son's appreciation for economic progress based on agriculture. But Rother never drank deeply of the chancellor's economic liberalism, nor did he feel that the initial degree of Seehandlung intervention in the Prussian economy was sufficient. He hoped to convince the king of the need to do much more for the Prussian state, economy, and people.

Thus in November 1823 Rother encouraged Frederick William to permit the Seehandlung to found modern, mechanized industrial ventures. It was the "higher purpose" of the institute "to find and show the way to private industry by starting up branch after branch of industry." He rejected "the unfortunate idea of political economists"[28] who believed that

---

Breslau 10 (1965): 163–65, 173–74; Radtke, *Seehandlung*, 85–90, 112; and Pruns, *Staat und Agrarwirtschaft*, 1:91–96, 291–96, 2:39–40, 76–77.

[25] Walter Görlitz, *Die Junker* (Limburg: C. A. Starke Verlag, 1964), 209.

[26] Cited in Radtke, *Seehandlung*, 253.

[27] Recruited by Stägemann, Rother joined Hardenberg's staff in 1810. See ibid., 22–23.

[28] Rother to Frederick William, 15 November 1823, cited in Rolf Keller, *Christian von*

entrepreneurs had to rely mainly on themselves to survive in the business world. Most Prussian manufacturers lacked the capital, entrepreneurial spirit, and courage to risk competition with foreign counterparts receiving various forms of state support for the latest technology. Rother's low estimation of the private sector, however, was based on more than economics. Unlike his friend Peter Beuth, who believed that the masses, like businessmen, had to help themselves through education, hard work, and savings, the Chief of the Seehandlung expected more social responsibility from capitalists. But he doubted whether there were enough Oelsners and Ruffers with the humanitarianism required to guard the welfare of their male, female, and child employees. In more than one sense, therefore, the Seehandlung should "find and show the way to private industry." The institute would undoubtedly have to make material sacrifices, but the state would benefit directly from the "expansion of trade" and indirectly "from the welfare of its subjects."[29]

The king shared many of Rother's doubts about businessmen and empathized with the "beneficent purpose"[30] of his adviser's masonic movement. Indeed a few years earlier Frederick William had rejected conservative entreaties that he ban the freemasons, citing the fact that he knew many masons "who I have to count among my best and truest subjects"[31]—a clear reference not only to Rother, but Hardenberg and Witzleben as well.[32] Rother's memorandum failed to convince Frederick William, however, because the monarch still doubted that machines were the answer to Prussia's economic needs. In September 1826, for instance, he denied a weavers' petition seeking a ban on the importation of automatic looms, but ordered the Business Department to be ready with corrective measures should "the dubious phenomena which have already accompanied the use of power looms in England"[33] threaten to reappear in Prussia.

Rother als Organisator der Finanzen, des Geldwesens und der Wirtschaft in Preussen nach dem Befreiungskriege (Rostock,1930), 14.

[29] Ibid., 15. For Beuth's views on social welfare in the 1820s, see *Verhandlungen* 3 (1824): 165. For Rother's opinions at that time, see Webersinn, "Christian von Rother," 175; and Henning, "Preussische Sozialpolitik," 520–21. For later statements and actions, see ADB 29:361; Rother to Alvensleben, 4 March 1839, and Rother's half of a joint report with Alvensleben of 3 December 1839, Rep.120, D.V.2c, Nr.9, Bd.1, Bl.155–56, 200–207, ZStA Merseburg.

[30] For a good and balanced discussion of Frederick William III's attitudes toward the freemasons, see Hans Riegelmann, *Die europaeischen Dynastien in Ihrem Verhaeltnis zur Freimaurerei* (Berlin: Nordland Verlag, 1943), 171–83. The quote (p. 183) is from a letter of the king, n.d. (1838), to a lodge in Düsseldorf reminding the freemasons there of their real purpose.

[31] Riegelmann, ibid., 180, cites a letter of the king, n.d. (1821–22).

[32] Trende, *Im Schatten*, 19, 48.

[33] Frederick William to Schuckmann, 19 September 1826, Rep.120, D.V.1, Nr.11, Bd.1, Bl.14, ZStA Merseburg.

Nor did he react enthusiastically a year later when Rother proposed greater Seehandlung involvement in the promotion of textiles. Declining sales of Prussian wool abroad had motivated Rother to suggest state-aided expansion of domestic wool fabrication. In particular, he envisioned the formation of joint-stock companies with major Seehandlung participation. The ambivalent monarch replied that he "did not wish to let practical proposals go unconsidered."[34] But for one as eager as Rother to serve the good cause, this response could only be interpreted as an invitation to come forward with more specific suggestions.

Thus in January 1828, Rother turned to Peter Beuth for counsel. What was the best means for the Seehandlung to stimulate domestic production of woolen cloth? Should the Seehandlung merge its capital with that of "the biggest firms which already exist?" Or "establish in the various provinces one or two great enterprises which combine all stages of manufacture from the first purchase of material to the finished product?" The original capital for these new joint-stock factories would come from the Seehandlung as well as "merchants, capitalists, land owners and other private persons."[35] Rother left open the question of management.

The new Head of the Business Department was clearly flattered that his expertise was valued so highly by an enemy of economic liberalism. He tactfully refused to expound at great length upon "the great number of well-known arguments against such state involvement."[36] Instead Beuth mentioned a few of his concerns about the ruinous effect which state-owned or state-controlled factories might have on private competitors, then placed these doubts aside. For Prussia, he wrote, possessed only one or two existing factories worthy of Rother's plans. In choosing the second option of founding a number of great *new* cloth-producing enterprises, Beuth cast his glance at Silesia. The Privy Councillor trusted that Cockerill, Busse, and other entrepreneurs in Berlin and the Mark Brandenburg would respond to his own feelers about private joint-stock projects. Hopes for the Eichsfeld and Rhineland were also high. But in Beuth's opinion, Silesia lacked the existing private factories to exploit its natural advantages in wool, water power, and river transport. He offered his services to Rother in evaluating those manufacturers who might come forward and promised to help him select the best machines for the new ventures.

A storm of protest against the threatening idea of state enterprise arose during the spring of 1828 after the scheme became known publicly. The outcries reinforced the contempt which Rother felt for selfish business-

[34] Rother quoted the king indirectly in a letter to Beuth, 14 January 1828, Rep.120, D.IV.1, Nr.26, Bd.1, Bl.1–10, ZStA Merseburg.

[35] Ibid.

[36] Beuth to Rother, 19 January 1828, ibid., Bl.11–14.

men and strengthened Beuth in his own growing conviction that only a few Prussian manufacturers were worthy of the state's efforts. To Beuth fell the depressing task of responding to a full portfolio of indignant petitions from factory owners and chambers of commerce. Each received the brief, by now standard reminder that Prussia could not allow other nations to pull farther and farther ahead technologically.[37]

But there was one positive development which survived that winter. Our opening scene suggests that it came as a result of private dealings between Rother and his Silesian brothers at Rogau. We know that Rother, Oelsner, and Ruffer met there on a number of occasions in the 1820s to discuss the revitalization of Silesia. Because Rother's proposals centered around the idea of promoting Prussian woolen manufacture directly through textile factories, it seems that it was the businessmen who countered with the idea of *indirect* promotion by means of a joint-stock company to manufacture steam engines and all types of textile machinery. We know that Ruffer attempted to raise capital for such a plant in 1828 and 1829. Based in Breslau, the heart of Lower Silesia, Ruffer's plant would undercut the local market for British machines exported over vast distances and a 25 percent Prussian tariff on imported machinery. Breslau's cheaper machines would spread throughout the province, modernizing its hardpressed linen and woolen industries. If this interpretation is correct, Rother saw the wisdom of Ruffer's modifications and—what is historically certain—pledged Seehandlung support.[38]

It took five years before the plan came to fruition. Ruffer's original effort collapsed after a year of distressing Near Eastern news contributed to the unwillingness of Silesian investors to participate in his ambitious undertaking. Then in February 1830, Beuth revived the idea. Significantly, his proposal came only two weeks after explaining to Cockerill that the Business Department's Dobo wool-spinning machines were too expensive to entice Prussian wool manufacturers (see chap. 3). Reverting to contractual arrangements more in keeping with his liberal ideological convictions, the rising star of the Business Department wanted a private joint-stock company to own and operate the new factory. Ruffer, Oelsner, and three additional businessmen responded enthusiastically to the plan—made bold, no doubt, by a gift from Beuth of machine tools worth 22,000 thaler. But diplomatic uncertainty again interfered with a daring business venture—this time caused by the persistent Belgian crisis. With the company still badly undersubscribed in early 1832, Beuth's ministry

[37] The letters are in ibid., Bl.17ff.

[38] It seems unlikely that the idea originated with Rother or Beuth, for neither mentioned it in their exchange of letters (in ibid., Bl.1–16) . For Rother's amenability to Ruffer's suggestions, and Ruffer's efforts of 1828/29, see Webersinn, "Gustav Heinrich Ruffer," 177–78, 190; and Radtke, *Seehandlung*, 105.

turned to the Seehandlung. Eager for nearly a decade to initiate extraordinary business projects and pursue a "higher purpose," Rother provided the firm's entire capital of 70,000 thaler. Far more than Beuth could afford to donate out of his total annual budget of 100,000 thaler, the amount was easily found in Rother's massive coffers. In November 1833, Silesia's first machine-building factory could finally begin production.[39]

The Breslau venture opened a period of closer cooperation between Beuth and Rother. Most of Schuckmann's departments were placed under Maassen in Finance in April 1834, then, after his death, moved to a new ministry under Rother responsible for commerce, transportation, and manufacturing. During their years together (1835–1837), each man pursued his own dream. For Beuth this meant rural, aesthetic industrialization spawned by a vanguard of helpful, yet not overweening technocrats. For Rother, on the other hand, the 1830s provided an opportunity to blaze a trail which, he was sure, would have strained the economic and moral capabilities of most private industrialists. As discussed earlier (chap. 2), the turning point came in the summer of 1832 with a Seehandlung lottery that channeled 8 million thaler to a nearly bankrupt state and established a fund of 4.6 million for Rother's mighty institution to invest.[40]

The Seehandlung's business empire was already beginning to expand when the Breslau machine-tool plant opened its doors in 1833. Earlier that year an impressive chemical works in Oranienburg and a steamship line between Berlin and Hamburg were acquired, while 1834 saw the completion of an American-style "automated" grain mill in Ohlau (Silesia) and the purchase of property in Hohenofen (District of Potsdam) for construction of a paper factory. When the latter plant began production in 1838, the Seehandlung had added a second new machine works in Berlin-Moabit and a new zink rolling mill near the grain mill in Ohlau. In total, Rother had invested 2.8 million thaler in seven operations by the end of the decade.[41] If we use the average Prussian ratio of net to total industrial investment for the 1830s (62.8%), the Seehandlung was responsible for 5.4 percent of net industrial investment during this decade. This was a truly significant industrial expansion.[42]

[39] See Beuth's correspondence with Merckel, February–June 1830; Beuth (for Schuckmann) to Rother, 30 January 1832; and an inventory of machinery for the Breslau plant, in Rep.120, D.XIV.1, Nr.35, Bd.1, Bl.1–3, 4–5, 7–10, 133–34, and Bd.2, Bl.75, ZStA Merseburg; Webersinn, "Gustav Heinrich Ruffer," 177–78; and Radtke, *Seehandlung*, 102–11. Radtke largely overlooks the fact that Beuth was writing for Schuckmann and generally underestimates Beuth's role in the whole episode.

[40] Schrader, *Geschichte*, 10–12.

[41] For the total, see ibid., 17; for a full description of the Seehandlung's ventures, see Radtke, *Seehandlung*, 135–244.

[42] See chap. 1, n. 63; and Tilly, "Capital Formation in Germany," in Mathias and Postan,

While presiding over this period of exceptional growth for his organization, Silesia's loyal son had not forgotten about the problems of his native province. He knew that one machine plant, one zink rolling mill, and one grain mill would not reverse the decline of Silesia's unfortunate linen industry. The District President in Liegnitz underscored this point in February 1838, by reporting to Frederick William that Silesia's linen spinners and weavers required nothing short of emergency measures. The aged king transferred 13,000 thaler to Liegnitz for relief of the poor, then ordered Rother to come forward with long-term ameliorative proposals.[43]

In a series of reports and memoranda over the subsequent year, Rother returned to his initial proposal of 1823—and to the teachings and beliefs of freemasonry. He explained that the steady deterioration of Silesia's linen industry was due to the rise of cotton fabrics and the coming of machine-made linen products which were cheaper and more uniform. The province's small shop and cottage operations were doomed by these trends. Only "the creation of an enduring and adequate living wage" (*nachhaltigen und auskömmlichen Arbeits-Verdienstes*),[44] he stated, could bring about a "lasting improvement" in the lot of the spinners and weavers. And only the Seehandlung could create jobs quickly in Silesia, for provincial capitalists were too preoccupied with profits and losses to put machinery into the service of the poor. Eventually, he predicted, private industrialists would follow his beneficent example.

Rother had maneuvered himself onto a collision course with the Head of the Business Department. The division had moved again in March 1837—this time to the Ministry of Finance. From that vantage point, Beuth could observe what seemed to him an improvement in Silesia's industrial fortunes. The Breslau machine factory was prospering and accomplishing its broader mission. The linen firms of Kramsta and Kopisch were installing new machines and joining the older company of Alberti as promising operations. Beuth was extremely proud of these businessmen "who moved triumphantly forward out of the old rubble, had no use for charity and showed themselves the way."[45] It was understandable,

Cambridge Economic History, 7 (1): 426. On the surface, it would seem conservative to use the average ratio of net to total industrial investment because the bulk of the Seehandlung's ventures were *new* factories. Reducing the percentage to the 62.8 percent range, however, was the fact that the Seehandlung included nonrelated real estate purchases in the 2.8 million total (see Schrader, *Geschichte*, 12, 17).

[43] For the background, see Rother and Alvensleben to Frederick William, 3 December 1839, Rep.120, D.V.2c, Nr.9, Bd.1, Bl.200–201, ZStA Merseburg.

[44] Ibid. Rother's reports of August and November 1838 are also paraphrased in this report. See also Rother to Alvensleben, 4 March 1839, Bl.155–56.

[45] Beuth to Delius, 14 November 1836, Rep.120, D.V.2d, Nr.12, Bd.1, Bl.221, ZStA Merseburg.

therefore, that this entrepreneurial troika would protest what struck them as unnecessary, intolerable competition from the Seehandlung—and that Beuth championed their cause. In a terse note to Rother which exposed the strains in their friendship, he reminded his comrade that the Breslau machine works had been a necessary exception and urged him to abandon "the promotion of technology according to Egyptian principles," a derogatory reference to Pasha Mohammed Ali's state-sponsored industrialization of the 1820s and 1830s.[46] But Rother would not listen. By mid-1839, Beuth and his new ministerial boss in Finance, Albrecht von Alvensleben, were waging a bureaucratic skirmish with the Seehandlung.[47]

Frederick William sided with Rother at the end of the year. The needs of the Silesians were so great that quick intervention was mandatory. The government could not leave any means untried, especially "after that which the state has done so far, at considerable sacrifice to the treasury, has resulted in no success at all." Alluding to expensive programs dating back to the time of Baron Kottwitz and before, the sixty-nine-year-old monarch was finally giving up his last reservations about mechanization. Rother received permission to construct "mechanized flax spinneries in combination with improved bleaching, weaving, and printing works." It was perhaps a minor concession to Beuth that the Seehandlung Chief was ordered to sell the plants to private industrialists "as soon as this can take place without risk and sacrifice."[48]

By 1845, Seehandlung investments in Silesia alone stood at 3.6 million thaler. Most of this was tied up in eight new textile factories in the countryside and small towns of that province. Seven of these works manufactured linen or woolen goods, while only one produced cotton fabrics.[49] The new factories were either owned outright by the Seehandlung or run in partnership with local freemasons whom Rother trusted—Heinrich Ruffer, for instance, participated in three of these ventures. Indeed, their operations strove to realize the humanitarian ideals of freemasonry. Over fourteen thousand employees found gainful employment, providing more hope for tens of thousands of their dependents. Benefits included free meals, factory schools, subsidized savings accounts, medical care, health and funeral insurance, and payment of Prussia's class tax. Here were good examples for all manufacturers to follow.[50]

[46] Beuth to Rother, 24 April 1839, Rep.120, D.V.2c, Nr.9, Bd.1, Bl.188, ZStA Merseburg. See also the protests of Kampsa (13 February 1839), Alberti (23 February 1839), and Kopisch (20 March 1839), Bl.135–36, 138, 166–67.

[47] See the correspondence of April–November 1839 in ibid., Bl.175–99.

[48] Frederick William to Rother and Alvensleben, 5 January 1840, in ibid., Bl.213.

[49] A list of Seehandlung investments in Silesia on 3 October 1846 is located in Rep.109, Nr.4126, Bl.7, GStA Berlin.

[50] Henning, "Preussische Sozialpolitik," passim; and Radtke, *Seehandlung*, 135–73.

The Seehandlung Chief's sincerity about creating "an enduring and adequate living wage" was most evident in his benevolent style of managing production. One of his largest operations was located in Wüstegiersdorf on the Silesian side of the Sudeten Mountains. The first steam-powered weaving, dyeing, and finishing complex for worsted woolen cloth in continental Europe, the plant was equipped with an interesting assortment of machines. One floor housed some two hundred automatic looms of Jacquard and Schoenherr design. Half had been built cheaply in the Breslau machine factory and the rest imported at greater expense from England. A second floor of the factory employed over twelve hundred hand-operated machines, while one thousand households in the surrounding hills wove cloth for bleaching or dyeing at Wüstegiersdorf.[51] Not only were overhead, wages, and operating costs minimized in this way, but the putting-out system offered the possibility of cutting back here—not at the factory—during recessions. To its credit, however, the Seehandlung opted to ignore the ebb and flow of the business cycle, produce at a more constant level, and provide steadier employment to a maximum number of local inhabitants. The Wüstegiersdorf factory was an excellent example of masonic principles "bearing fruit in deeds." It was some measure of this commitment that the plant had accumulated an inventory worth well over 200,000 thaler by the late 1840s and could not point to a single profitable year.[52]

It was consistent with Rother's love of Silesia, however, that he placed greater emphasis on linen—the province's staple industry of old—than wool. Indeed, by 1848 the Seehandlung had erected complicated flax-heckling installations at Patschkey and Suchau, and modern linen-spinning factories at Patschkey, Landeshut, and Erdmannsdorf. Rother wanted these operations to provide an enlightened social and technological example for all of Silesia.[53] Despite Beuth's bruised ego and disagreement over policy, he would have applauded if his friend had succeeded. Regardless of whether it came under liberal or "Egyptian" auspices, however, rural industrialization centered around linen was slipping farther and farther beyond the state's reach. Consumption patterns in 1800 had still favored linen over wool and cotton, the former commanding 90 percent of the textile market in Germany. By 1843, however, linen retained only 24 percent of the market. In stark contrast, cotton consumption had increased twenty-sixfold, ascending rapidly from 5 percent to 69 percent

[51] See the correspondence of 1840–43 over purchase of machinery in Rep.109, Nr.191, Bl.26ff, Nr.206, Bl.179–82, and Nr.208, Bl.163–68, 314–15, GStA Berlin; and Radtke, *Seehandlung*, 147.

[52] Radtke, *Seehandlung*, 350. It should be noted here that Radtke does not draw this conclusion from his evidence.

[53] Radtke, *Seehandlung*, 148–49.

of the total.[54] Foreign penetration of the home market for linen further compounded matters. British linen goods were mule-spun and power-woven by the mid-1840s, expanding output, driving down per unit costs, and sending prices to one-third of their 1815 level. Just as important, Prussian producers faced higher construction and operating costs—40 percent and 20 percent, respectively.[55] Yet the bureaucracy refused to follow the French example of high tariffs. Imported linen cloth was taxed at 11 thaler per centner and yarn at a mere one-sixth of a thaler. Accordingly, while overall linen consumption in Prussia declined from 237,000 centner in 1831 to 141,000 centner in 1845, imports nearly doubled and the foreign share of the market rose from 16 percent to 42 percent.[56] Given the declining domestic share of sales which were rapidly shrinking in absolute terms, new Seehandlung factories were destined to prevent marginal private capital from entering this industry. Rother's emergence as a major linen producer also presented existing linen manufacturers like Alberti, Kopisch, and Krampsta with one more source of anxiety.

The Seehandlung, however, was not "in business" in a strict sense. Like the worsted woolen operation at Wüstegiersdorf, losses mounted continuously for the corporation's five flax and linen factories, exceeding 400,000 thaler by the end of the decade. Freemasonic beliefs stood behind much of this red ink. Thus the spinning mill at Erdmannsdorf built up huge stockpiles for humanitarian reasons rather than curtail production during slumps and sell in more "rational" fashion from inventory. The masonic mindset, with its mystical attachment to science and mechanization, also helps to explain Rother's consistently exaggerated emphasis on the use of sophisticated techniques. The flax-heckling works at Patschkey and Suchau represented daring but ill-advised attempts to introduce Belgian cultivation methods and fabrication processes in a sagging market which could not sustain the higher investments and operating costs associated with these intensive technologies. Except for Landeshut, moreover, all of the Seehandlung's flax and linen factories were high-overhead, steam-powered operations which suffered from the dwindling market for linen fabrics.[57]

[54] Banfield, *Industry of the Rhine*, 2:22; Landes, *Unbound Prometheus*, 171–73.

[55] Conrad Gill, *The Rise of the Irish Linen Industry* (Oxford: Oxford University Press, 1925), 326; W. G. Rimmer, *Marshalls of Leeds, Flax Spinners 1788–1886* (Cambridge: Cambridge University Press, 1960), 208, 309; Herbert Kisch, "The Textile Industries in Silesia and the Rhineland: A Comparative Study in Industrialization," in Kriedte et al., *Industrialization before Industrialization*, 186.

[56] Dieterici, *Statistische Übersicht*, 1838, 1844, 1848, for imports of linen and population. I have figured consumption using Dieterici's population figures with the per capita data cited above in n. 54.

[57] See the contract between the Seehandlung and B. R. Scheibler, a Belgian flax expert, 24 November 1845, Rep.109, Nr.196, Bl.1; Kaselowski's estimate of construction costs in

The Seehandlung's second woolen operation was another combination of awesome technology and business myopia. Constructed in Breslau to spin worsted yarn, the plant's problems began when it produced more fine yarn, which appealed to a limited number of wealthier customers, than coarse yarn, which Silesians demanded in greater quantity. This marketing error was compounded by plain microeconomic incompetence—the plant ignored the higher production costs and less elastic demand for fine cloth by pricing each type of yarn simply according to weight. But high overhead was the basic problem. Rother had spared no expense in constructing a multistoried spinning factory which was equipped with the latest British designs and powered by steam engines. The costly, ill-managed plant operated for six years deeply in the red.[58]

The question of substituting steam for water power demonstrates how faraway technocrats with different ideological agendas could easily disagree with cost-conscious industrialists about what constituted a good investment in technology. One of the Seehandlung's only industrial ventures in the western provinces provides a well-documented example. In 1841 Rother began to construct an ironmaking factory near Remscheid in partnership with the firm of Josua Hasenclever. The merchant-industrialist had agreed to join hands with the Seehandlung as long as the iron works "promises us gain," was conducted "in a solid fashion," and—probably most important of all—"does not risk too much."[59] Accordingly, his original plans called for water as the sole source of power.[60] This struck Rother as unwise. "It would be [more] practical,"[61] he wrote, to have a small auxiliary steam engine on hand to run the factory during droughts. But when the Chief of the Seehandlung pressed for full steam power in order to provide a technological showcase for factory-owners throughout Rhineland-Westphalia, Hasenclever rebelled. One could reduce the initial investment from 57,300 to 35,300 thaler, he protested, by eliminating the steam engine and costly building additions required to house it. Annual operating budgets would fall by 1,250 thaler, Hasenclever added, without the highly paid mechanics and coal costs of steam power. All of the disadvantages "fall completely by the wayside," he concluded, "if we

---

Patschkey, 20 January 1846; Haenel to Rother, 18 May 1846; Scheibler to Rother, 21 October 1846, Nr.485, Bl.115, 157–58, 208, GStA Berlin; and Radtke, *Seehandlung*, 348, 351–52. It should be noted again that Radtke draws none of these conclusions.

[58] Kaselowsky to Seehandlung Zentrale, 6 December 1848, Rep.109, Nr.92, Bl.4–9, GStA Berlin.

[59] See Hasenclever's "Memoirs," written in 1841, printed in Hasenclever, *Josua Hasenclever*, 12.

[60] Hasenclever to Rother, 2 January 1841, Rep.109, Nr.55, GStA Berlin.

[61] Rother to Hasenclever, 11 January 1841, in ibid.

use water power."[62] The two eventually compromised: Rother agreed to the installation of an impressive twenty-eight-foot, overshod, iron waterwheel; Hasenclever to the purchase of a steam engine at Seehandlung expense "if it were deemed necessary."[63] But it never was.

As the Remscheid example demonstrates, the Seehandlung's industrial expansion of the early 1840s was not limited to Silesian textiles. Joining Hasenclever's firm on the corporation's books, in fact, were six renovated grain mills in Bromburg and eight retooled grain mills in Potsdam. Between 1840 and 1848, in fact, Rother invested a total of 7.7 million thaler in all of his ventures, almost triple the Seehandlung's initial investment wave of the 1830s.[64] It is important to appreciate how gargantuan a sum this was at the dawn of the railroad era. Using the Prussian average ratio of net/total industrial investment for the 1840s (64.8%), the Seehandlung now accounted for 9.6 percent of net industrial investment in Prussia for the entire period since 1830—and 13.2 percent during the 1840s.[65] Together with the Mining Corps' massive new puddling works and coking ovens, and the army's new powder mills and gun factories, Rother's spreading complex of enterprises was helping to push the state's share of annual net industrial investment from the 4–5 percent level of the early 1820s to 11–12 percent between 1830 and 1848. In the immediate years leading up to midcentury this figure probably stood at 17 percent.[66]

These figures illustrate the lurch toward statism in Prussia. It was no exaggeration, therefore, when businessmen complained that they were being pressed on the one side by state enterprises and squeezed on the other by foreign imports which laissez-faire liberals around Beuth and Karsten refused to tax. As we shall see in chapter 7, the coming of railways created additional ill-feeling in the business world when the somewhat contradictory pleas of many entrepreneurs for government subsidies were ignored by technocrats like Beuth and Rother whose love of technology did not extend to railroads. The technical expertise of the Business Department, Seehandlung, and Mining Corps, moreover, was not always above the reproach of businessmen who believed they knew better. The Seehandlung's industrial expansion of the 1830s and 1840s

[62] Hasenclever to Rother, 31 March 1841, in ibid. See also Hasenclever to Rother, 14 February 1843, in the same portfolio.

[63] See the Seehandlung contract with Hasenclever of 3 August 1843, in Rep.109, Nr.60, GStA Berlin.

[64] Schrader, *Geschichte*, 17.

[65] See chap. 1, n. 63, and Tilly, "Capital Formation in Germany," Mathias and Postan, *Cambridge Economic History*, 7 (1): 426–27. See also above, n. 42.

[66] The reader will recall from chaps. 4 and 5 that the Mining Corps' share for the period 1830–48 was 0.7 percent and the army's 1.4 percent, rising to 2–3 percent between 1847 and 1854.

was contributing, in other words, to a worsening *general* crisis in Prussia over the economic and technological role of the state.

The magnitude of Seehandlung investments and the corporation's often dubious technological wisdom accelerated demands from the private sector that Rother scale back his operations. Not surprisingly, public attacks against him as an unnecessary competitor were far more frequent in the 1840s than praise for his enlightened example. Increasingly beleaguered, Rother felt compelled to publish a defense of his good deeds and investments in 1845. For two decades the Seehandlung had advanced "technology and fabrication" in order to protect domestic interests from "foreign undertakings which outstrip and injure the industrial activity of the Fatherland." In the process, "permanent employment was created for thousands of hard-working subjects and much of that was realized which associations dedicated to the betterment of the conditions of the working class are now trying to accomplish."[67] The aging freemason was saying that he had done his brotherly duty. For Prince William, the "Protector" of the Prussian lodges and himself a member of the Mother Lodge of the Three Globes, had issued a personal appeal to all Prussian freemasons to join "the associations which are forming in all towns [to enhance] the welfare of the working classes."[68]

The storm of protest did not abate. With no end to the controversy in sight, Frederick William IV banned the establishment of new Seehandlung enterprises. Its chief published another defensive brochure in 1847, but his day was nearly over. Disillusioned and bitter, Christian Rother finally resigned under pressure after the revolutionary disturbances of 1848.[69] The bulk of his industrial empire was sold into private hands during the 1850s.

[67] Rother's pamphlet of 18 February 1845 is located in Rep.109, Nr.3260, Bl.1ff., GStA Berlin. For the most detailed discussion, see Radtke, *Seehandlung*, 303–40.

[68] See Prince William's letter of 1845 to the Prussian Freemasonic orders, printed in *Politische Correspondenz Kaiser Wilhelm's*, 72–73.

[69] For Rother's last years, see Radtke, *Seehandlung*, 303–41.

VII

No Man's Hand Could Halt the Cars

THE LATE AFTERNOON SUNLIGHT created an impression of heat as it reflected off the glossy finish of a coach suitable only for someone of high rank or social position. The driver of this stately vehicle slowed it nearly to a halt, pulled off the road, and headed for a nearby hill. The horses, already underway for many hours on this unseasonably warm day in May, strained noticeably up a steep path to the summit.

The two men conversing inside were inspecting the lay of the land along a projected railroad route.[1] Once on top of the hill they could peer southeast to Elberfeld, its church spires barely discernible in the rural distance. To the northwest lay the small town of Vohwinkel, one potential station on the way to Düsseldorf over the horizon. The shorter man stepping out first was Ludwig von Vincke, Provincial Governor of Westphalia for the past fourteen years. Following was his guest and friend of nearly three decades, Peter Beuth. Soon the friends fell silent and walked back to the coach. Vincke ordered the driver to head for Elberfeld.

Beuth was excited about the possibility of connecting the booming industrial town of Elberfeld and the marketing and administrative center of Düsseldorf with Ruhr coal, hence his presence away from Berlin on this spring day in 1829. In general, however, he was very skeptical about the new means of transportation. Few railroad lines in Prussia would place enough freight in their cars to pay for all of the viaducts, bridges, embankments, and tunnels required to flatten out hilly terrain like that of the Sauerland visible from the window. Vincke saw greater possibilities. He had spoken with the great industrial pioneer from Wetter, Friedrich Harkort, and now believed that short lines like the Düsseldorf-Elberfeld could be linked to others in the area. He was also enticed by the prospect of connecting the salt mines of Reyme, near Minden, with the provincial town of Lippstadt. But Vincke, like Beuth, doubted the feasibility of longer lines. In Berlin, Prussia's dynamic Minister of Finance, Friedrich von Motz, was attracted by the bolder idea of running a line from Minden across Westphalia to the banks of the Rhine.[2] But none of

[1] Vincke and Beuth set out together on the morning of May 30, 1829. See the Ludwig von Vincke Diary, May 30, 1829, Ludwig von Vincke Papers, A-I-20, StA Münster. See also Matschoss, *Grosse Männer*, 51.

[2] Schuckmann (Beuth) to Motz, 9 April 1829, Rep.121, D.VI.2, Nr.108, Bd.1, Bl.40–41, ZStA Merseburg; Motz to Schuckmann and Vincke, 28 February 1829, Oberpräsidium B-

the bureaucrats had a real sense of what the future held in store for railroads.

As their coach rolled into the streets of Elberfeld, Vincke and Beuth readied their papers and maps and turned to small talk. Soon the driver pulled into the courtyard of an inn where they would spend the night. A prescient Harkort stood ready to greet them.

. . . . .

It took a lively mind to imagine railroads of even moderate length pulled by steam locomotives in May 1829. Dozens of companies were chartered in England by this time, but only one, the Stockton-Darlington, was in operation. Forty-one kilometers long, its cars were pulled variously by horses and experimental locomotives along flat portions of the route, and by stationary steam engines up two hills. The directors of the Manchester-Liverpool Railroad, whose fifty-four kilometers were nearing completion, favored steam-driven rope haulage of cars on steep as well as level grades.[3] The twenty-eight kilometers of terrain between Düsseldorf and Elberfeld which Vincke and Beuth visited were to be part of a line three times as long connecting the Ruhr with Holland. The stretch across sparsely populated eastern Westphalia from Minden—or nearby Reyme—to Lippstadt was roughly eighty-five kilometers. From Lippstadt through the hills of the Ruhr and Sauerland to Cologne-Deutz was another 160. Friedrich Harkort could write futuristically about English lines of nearly seven hundred kilometers and herald the day in Germany when "the triumphant car of technological progress is harnessed to smoking colossuses."[4] But continental observers could be forgiven for refusing to take him seriously.

Indeed the Prussian experience provided little hint that railroads were the wave of the future. In 1820, after four years of planning and experimentation, the Mining Corps had opened a colliery line running two kilometers from its Grosswald mine to the Saar. Black coats pulled the heavy cars on wooden rails until 1823 when the local mine office substituted horses and iron rails. High hopes of installing a steam locomotive

1136, StA Münster; G. Fleck, "Studien zur Geschichte des preussischen Eisenbahnwesens," *Archiv für Eisenbahnwesen* 19 (1896): 32; and Vincke to Harkort, February 1829, cited in Petersdorff, *Friedrich von Motz*, 2:314.

[3] Eugene S. Ferguson, "Steam Transportation," in Melvin Kranzberg and Carroll Pursell (eds.), *Technology in Western Civilization* (New York: Oxford University Press, 1967), 1:296–97.

[4] Cited in Arthur Fürst, *Die Hundert-Jährige Eisenbahn: Wie Meisterhände Sie Schufen* (Berlin: Deutsche Buch-Gemeinschaft GmbH, 1925), 219.

were dashed when, despite two rebuildings, a prototype engine failed repeatedly to move up a one-half-degree incline.[5]

A private horse-powered line of equally modest length opened in the Ruhr in 1827. Two additional horse colliery roads of eleven and six kilometers followed in 1828/29.[6] But the failure of three more ambitious projects served to dampen spirits. In 1826 mine officials in Essen interested superiors in Dortmund with a proposal for a Corps railroad that would wind forty-five kilometers up the Ruhr from Ruhrort to Witten. Such a line, they argued, would provide private owners of newer, deeper mines with access to the wider markets necessary to cover expanding overhead. Simultaneously, Friedrich Harkort and a group of investors petitioned the Interior Ministry for permission to construct eighteen kilometers of track connecting the middle Ruhr near Steele with the Ennepe road running between Hagen, Barmen, and Elberfeld. The partners hoped to reduce significantly the cost of Ruhr coal in the latter two industrial towns. Both designs called for horse-drawn cars suspended from tracks mounted a few meters above the ground on wooden pilings—a system patented in England by R. R. Palmer.[7]

In October 1826, the Ministry insisted that all carriers owning cars have access to the Ruhr-Wupper railroad for a small fee. This was the first sign that the state, probably on Beuth's advice, intended to treat railroads no differently than inexpensive toll highways. Harkort's investors, already somewhat skeptical about suspended railways, quickly abandoned the proposal, for without a monopoly of transport their sizeable investment would have amortized at a snail's pace. The Dortmund Mine Office rejected the Ruhrort-Witten line in early 1827, siding with mining subordinates in Bochum who feared that smaller-scale cooperative mines, especially those above Witten, would suffer a competitive disadvantage if the railroad were constructed.[8] Here was an early sign of a Corps beginning to divide against itself over the shape of the future. It

[5] See two entire volumes of official correspondence in OBA Bonn 3527 and 4085, HSA Düsseldorf.

[6] F.W.R. Kind, *Entwicklung und Ausdehnung der Eisenbahngesellschaften im niederrheinisch-westfälischen Kohlengebiet* (Leipzig: Druck von August Hofmann, 1908), 23; M. Reuss, "Mitteilungen aus der Geschichte des Königlichen Oberbergamtes zu Dortmund und des Niederrheinisch-Westfälischen Bergbaues," *Zeitschrift für das Berg-Hütten- und Salinenwesen* 40 (1892): 366–67.

[7] See the protocols of meetings of Ruhr mining officials, 15 October and 8 December 1825, Bergamt Essen-Werden 92, Bl.12–13, HSA Düsseldorf; Fürst, *Hundert-Jährige Eisenbahn*, 223–25; and Kind, *Entwicklung und Ausdehnung*, 20–21.

[8] Bergamt Essen to Oberberghauptmannschaft, 26 May 1828, Bergamt Essen-Werden 92, Bl.120, HSA Düsseldorf; Fleck, "Studien," *Archiv für Eisenbahnwesen* 19 (1896): 31–32; and Krampe, *Staatseinfluss*, 186–87.

was characteristic of the proud conservatism in Bölling's headquarters that Bochum's bias against modern mining prevailed.

An even more daring western undertaking soon became the third stillborn child of the railroad age in Prussia. Two separate companies had come forward in mid-1828 with the bold idea of constructing a railroad between the middle Ruhr, Elberfeld, and Düsseldorf—Beuth and Vincke inspected this latter portion of the route in 1829. The line would then continue from Düsseldorf, over Crefeld, to Venlo on the Dutch border. Plans also called for feeder lines to Viersen, Gladbach, and Rheydt—in all, over one hundred kilometers. The consulting engineer, Matthias Berger, rejected the Palmer system in favor of a "railroad dam." More familiar to the modern eye, a double line of track would lie atop a flat earthen ridge where horses and drivers or locomotives could easily operate. The Business Department reacted enthusiastically to the project, primarily because it provided a direct link between Ruhr mines and Barmen-Elberfeld.[9] The line was clearly part of Beuth's dream of creating a regional textile bastion which could block British cotton imports.

But the military had its plans too. 1828 was a year of crisis in Europe and the Prussian Army was bracing itself for combat. Holland was not a likely invasion route for a renascent France, but Napoleon had annexed this country in 1811 and Berger's route, cutting menacingly between the army's major northwestern fortresses at Wesel and Cologne-Deutz, worried the General Staff. As one contemporary put it, "anxious spirits shudder at the thought that some fine spring morning a hundred thousand Frenchmen, thirsting for war, will suddenly invade our peaceful valleys at bird-like speed, thanks to the new means of locomotion, and begin their old game over again."[10]

Always consulted before highway construction,[11] Karl von Müffling's General Staff demanded safeguards now with railroads. Perhaps to insure a potential line of defense, the dam had to be made of stone, thereby greatly increasing construction costs. Pioneer units would remove ties and tracks in the event of an invasion with no compensation for the owners. To prevent an enemy from moving cavalry and artillery after such demolition work, the top could be no wider than five feet, thus preempting plans for double tracks. Finally, the owners were denied in advance the right to build a highway along the same route if their railroad went

[9] See Berger's proposal, n.d. (winter of 1828/29), Bl.88–96, and Schuckmann (Beuth) to Ingersleben and Pestel, 23 October 1828, Bl.29, 33, 403/11721, LHA Koblenz.

[10] Cited in Edwin L. Pratt, *The Rise of Rail-Power in War and Conquest 1833–1914* (London: P. S. King & Son Ltd., 1915), 3.

[11] For the origins of this practice, see Hardenberg to Ingersleben, 4 February 1818, and Bülow to Regierung Trier, 25 July 1818, 402/853, Bl.1–2, 11, LHA Koblenz. See also Schön to Stägemann, 30 October 1826, printed in Rühl, *Briefe*, 3:278.

out of business. Complicating matters further, Interior again stipulated that others could place cars on the line for fees regulated by the state. In despair by late 1829, promotors invested their money elsewhere.[12]

By this time, however, the highly influential Minister of Finance, Friedrich von Motz, had elevated the railroad question to a higher plateau. Prussia's great highway builder diverted his attention to the newer technology in early 1828 after Harkort and provincial officials in Cologne and Minden convinced him that a railroad line connecting the Rhine with the Ems or Weser would permit Prussian and middle German exporters to circumvent stiff Dutch tolls on the lower Rhine. From spring 1828 to spring 1829, the Finance, Interior, and Foreign ministries debated three daring state projects: Wesel to Emden (on the North Sea); Lippstadt to Reyme; and Cologne to Reyme. Other railroad plans were on the drawing board too. In September 1828, the Bonn Mining Office proposed a seventy-five kilometer Corps line between Saarbrücken and Conz (on the Mosel), while South German newspapers claimed in July 1829 to know about a Prussian scheme to construct a route from the middle Elbe to Leipzig.[13]

The Interior Ministry was in the strongest position to effect the outcome of this debate, for it possessed the experts. And months before the South German reports, Prussia's experts had turned thumbs down. The Mining Corps squelched the optimism of its younger railroad enthusiasts with a devastating memorandum to Schuckmann about the Cologne-Reyme line in March. The high hills and narrow, winding valleys between the two cities would necessitate the use of scores of stationary steam engines, thus causing delays and driving up costs. In good weather, locomotives would encounter difficulties on gentler grades, but snowfall would reduce traction and render most parts of the route "entirely useless."[14] The big investment and high costs would be hard to recoup, moreover, due to the low volume of anticipated traffic. The Business Department went even farther than this, doubting the technical and economic potential of the shorter Reyme-Lippstadt and Emden-Wesel lines. A canal between the latter towns made more sense. Motz surren-

[12] For the army's stipulations, see War Ministry to Generalkommando Koblenz, 13 April 1829, 403/11721, Bl.107–08, LHA Koblenz; and Interior Ministry to Oberpräsident Koblenz, 18 July 1829, ZStA Merseburg. The collapse of the venture is mentioned in Kind, *Entwicklung und Ausdehnung*, 22–23.

[13] Motz to Vincke, 29 June 1828, and Motz to Vincke and Schuckmann, 28 February 1829, Oberpräsidium B-1136, StA Münster; Schuckmann (Beuth) to Motz, 9 April 1829 (as cited above in n. 2); Bergamt Saarbrücken to Oberbergamt Bonn, 28 September 1828, and Oberberghauptmannschaft to Oberbergamt Bonn, 17 July 1829, OBA Bonn 703, Bl.1–5, 69, HSA Düsseldorf; and Petersdorff, *Friedrich von Motz*, 2:305–15.

[14] Oberberghauptmannschaft to Schuckmann, 28 March 1829, Rep.121, D.VI.2, Nr.108, Bd.1, Bl.35–36, ZStA Merseburg.

dered in April, salvaging only one concession: that Interior reexamine the possibility of a railroad or canal between Minden/Reyme and Lippstadt after completion of a terrain study begun by a young railroad enthusiast, Oberbergrat Carl von Oeynhaussen, in 1828.[15]

There was little reason for Vincke to be optimistic, therefore, when he set out with Beuth on the morning of May 30. In fact, nothing came of the Minden-Lippstadt line. Beuth congratulated Oeynhaussen upon completion of the survey work in December 1829, but added soberly that "next year is still too early to think about implementing the plans."[16] With firm backing from the Business Department, Motz had returned to his roadbuilding schemes, proposing 1.3 million thaler for highways radiating from Magdeburg to Hamburg and Erfurt. Frederick William's grudging approval in January 1830, precluded any possibility of significant funding for a largely experimental railroad running through eastern Westphalia.[17] As the new decade opened, Prussia possessed less than thirty kilometers of track—all insignificant, horse-powered colliery roads.

Railroad technology in England, meanwhile, was evolving very quickly. Hardpressed by confident railway engineers like George Stephenson, the directors of the still uncompleted Manchester-Liverpool Railroad decided to conduct tests of locomotives along a moderate gradient which rose about one foot in elevation for every 140 feet (1:140). The so-called Rainhill Trials of October 1829 brought mixed results. Stephenson's "Rocket" hauled its twenty-ton load easily at speeds of 25–50 kilometers per hour, while two competing engines broke down during the tests. But the lesson seemed clear. Stephenson's multiple firetube-boiler generated more steam—and power—than the older, single-flue types, for it doubled the area of heating space relative to engine weight. When the prototype of all modern railroads completed its second year of business in 1832, Stephenson's locomotives possessed better boilers, a third set of wheels to distribute weight, and twice as much tractive power as the original Rocket.[18] Seven years after he first described it in 1825, Harkort's vision of "smoking colossuses" seemed more realistic.

These breakthroughs presented those wishing to invest in transportation technology with difficult decisions. Engines could now negotiate grades of 1:70 or more, enabling track routes to traverse and ascend hilly country. Initial investment could therefore be much lower, and the am-

[15] Schuckmann (Beuth) to Motz, 9 April 1829 (as cited above in n. 2).

[16] Beuth to Oeynhaussen, 15 December 1829, Rep.121, D.VI.2, Nr.108, Bd.1, Bl.51, ZStA Merseburg.

[17] Petersdorff, *Friedrich von Motz*, 2:298–99.

[18] G. F. Westcott, *The British Railway Locomotive* (London: Her Majesty's Stationary Office, 1958), 5–12; Ferguson, "Steam Transportation," in Kranzberg and Pursell, *Technology in Western Civilization*, 1:297.

ortization of capital much quicker, as deep cuttings and high embankments were avoided on many parts of the line. The trade-off, however, was higher operating costs and lower profits. Steeper gradients meant heavier, costlier engines, increased fuel consumption, and slower speeds. It was obvious that sharper curves would add to wheel and track wear, increase the likelihood of accidents, or, as a precaution against these, reduce speeds even more. But the "flat versus steep" decision was part of a larger dilemma, for railroads certainly did not exhaust the list of attractive transportation options for rational investors. Horses could pull six to ten times more load on rails than on roads, while river improvement schemes, canals, and toll roads all found convincing advocates in every industrializing country during the early 1830s.

Although well informed about the Rainhill Trials, Prussian experts were justifiably skeptical. The Bonn Mining Office forged ahead with survey work on its long, Saarbrücken-Conz colliery road in early 1830, but, pessimistic about locomotives since its own fruitless experiments of the 1820s, stayed wedded to horse power. Some mining officials in the Rhineland believed river dredging would be a more practical improvement for the economy.[19] Similarly, Business Department analysts continued to suggest a mixture of canals and straight, horse-drawn railroads. The high cost of boiler explosions, track repairs, coke consumption, and weather-related delays argued against the Manchester-Liverpool approach.[20] Not surprisingly, when the Business Department revived its pet railroad project between the Ruhr and Elberfeld, horses won over steam. Baurat Pickel of the Düsseldorf District Government began terrain studies in July 1830.[21]

The third Westphalian Diet also looked askance at locomotives. In January 1831, Harkort united this body around the old plan of connecting Minden and Lippstadt. Aware of the controversy surrounding the new English technology, however, the indefatigable promoter did not even ask that the route open with steam power. Rather, he proposed construction of a double track made from durable wrought iron rails and stone ties for the day when heavy, fast locomotives would be substituted for horses. He also favored an eventual, 375-kilometer extension of this line to the Oder in eastern Prussia. The delegates logically refused to walk

[19] Bergamt Saarbrücken to Oberbergamt Bonn, 28 September 1828, and Saynerhütte to Oberbergamt Bonn, 24 December 1832, OBA Bonn 703, Bl. 1–5, 111–12, HSA Düsseldorf.

[20] See the articles by Wedding and Hagen in: *Verhandlungen* 9 (1830): 86, and 13 (1834): 229, TUGB Berlin; and Camphausen's correspondence with Krüger of 1835/36, printed in Mathieu Schwann, *Ludolf von Camphausen* (Essen: G. D. Baedeker Verlagshandlung, 1915), 2:386, 409–10.

[21] Fleck, "Studien," *Archiv für Eisenbahnwesen* 19 (1896): 36, 41.

with Harkort into an uncertain future, agreeing only that the shorter railroad was important for Westphalia—and that the state should pay for it.[22]

While the Westphalians deliberated, other steam railroad projects were in the western air. Anxious to establish a rapid, secure route for its industrial exports to Germany, the young Belgian state approached Prussia for permission to survey routes from Antwerp to Cologne. Holland, determined to maintain at least an economic advantage over its former Belgian provinces, petitioned Prussia with similar plans for a line from Rotterdam to Cologne. By early 1832, suspicious officials in Berlin were reassessing means of locomotion, examining the terrain, querying the army about strategic considerations, and grappling with the critical question of state versus private ownership of the expensive new technology.[23]

As Pickel's work along the twenty-three-kilometer horse-car route to Elberfeld progressed that year, the financial question arose again. And the Finance Ministry and new Ministry of Commerce and Business came to essentially the same conclusion reached by the Westphalian Diet. This particular route presented the technical challenge of building through very hilly country. Because industrialists were unfamiliar with risky, high-overhead investments of this sort, government had to take the initiative, thereby providing encouragement for private construction of other railroads. On such an expensive line, moreover, only the state could ensure freight rates low enough to benefit businesses throughout the entire region. The two ministries therefore advocated public borrowing to cover the costs of a "showcase" state railroad.[24] As noted above (chap. 2), public borrowing may have been accepted as inevitable by civilian officials as state coffers emptied during the Belgian crisis.

Indeed, without borrowing there was no room in the budgets of the early 1830s for state-funded railroads. The Belgian and Polish crises strapped Prussia's treasury as fortresses were provisioned, army corps mobilized, and troops shifted anxiously between threatened provinces. Emergency road construction further depleted state funds. In addition to monies already committed to Motz's last wave of road building, Prussia's frugal king approved over a million thaler of extra expenditure for strategic highways connecting Mainz, Aachen, and Jülich; Minden and Wesel; and Königsberg and Tilsit. The army requested these roads to speed

[22] Ibid., 37–38.

[23] Delius to Schuckmann, 13 April 1832, and Schuckmann, Maassen, and Ancillon to the king, 2 October 1832, 403/3581, Bl.6, 17, LHA Koblenz; Freymann to Delius, 18 June 1832, Reg. Köln, Abt.1, 2045, HSA Düsseldorf; Schuckmann (Beuth) to Hake, 19 September 1832, and Witzleben to Schuckmann, 11 October 1832, 403/3585, Bd.3, Bl.233–38, LHA Koblenz.

[24] See Gleim, "Zum dritten November 1888," *Archiv für Eisenbahnwesen* 11 (1888): 800–801.

South German units to the northwest and Prussian divisions to the west or east.[25] There could be no talk of more extra funds for civilian purposes.

Frederick William came to decisions on the Dutch, Belgian, Elberfeld-Ruhr, and Minden-Lippstadt railroad lines during 1832 and 1833. Construction of all four was approved. Appropriate assistance from the state or Seehandlung would be forthcoming, but *private* joint-stock corporations would have to bear the major financial burden. Following earlier precedents—and foreshadowing the practice of today's trans-European connections—the king also announced that railroad companies had to give all car owners access to the tracks.[26] The decrees were characteristic of a sovereign who liked to make moderate compromises. Because he wanted material prosperity and equality of economic opportunity, railroads of this type were acceptable. But progress would not occur at the expense of his army or the monarchical principle.

Still hopeful of more substantial state assistance, Westphalian businessmen around Harkort proposed amalgamation of the "smaller" Minden-Lippstadt and Ruhr-Elberfeld projects into a more ambitious Rhine-Weser line. Between the first two towns, tracks would follow the route studied by Oeynhaussen in 1829; but southwest of Witten, promoters rejected Pickel's recommendations. The road engineer had charted a direct route from Burg Kemnade that called for gradients as steep as 1:72. Harkort's group opted for a longer course with gentler gradients, but sharper curves. Meandering and slithering up and down three river valleys, this portion of the railroad would still be cheaper to construct, yet costlier to maintain as tracks and wheels wore and cracked in the turns. Nor were horses to pull these cars: on the prospectus cover appeared a locomotive like Stephenson's Rocket.[27]

Beuth was not at all pleased. Aroused by the industrialists' challenge to his department's technical expertise and convinced of the economic unfeasibility of the monstrous, 250-kilometer Rhine-Weser project—eleven times the length of Pickel's Ruhr-Elberfeld line—he fired off a characteristically venomous missive to Vincke in February 1833. His ministry was always prepared to further undertakings which promised to enhance the general welfare, but these entrepreneurs possessed "neither sufficient intelligence nor sufficient capital"[28] and could therefore count

[25] Paul Thimme, *Strassenbau und Strassenpolitik in Deutschland zur Zeit der Gründung des Zollvereins 1825–1835* (Stuttgart: W. Kohlhammer, 1931), 67–69.

[26] Fleck, "Studien," *Archiv für Eisenbahnwesen* 19 (1896): 39–43.

[27] See ibid., 47–48; and the original report of the Rhine-Weser Railroad Committee, November, 1832, Rep.121, D.VI.2, Nr.108, Bd.1, Bl.56, ZStA Merseburg.

[28] Schuckmann (Beuth) to Vincke, 23 February 1833, Oberpräsidium B-1136, StA Münster.

on no help from the state. But, intelligently or not, local investor committees were forming from the Rhine to the Weser by early 1834.

Railroad fever spread to other parts of the Rhineland that year. Thus ironmakers in Siegen were lobbying for a line running sixty-five kilometers through the rugged foothills of the Siegerland and Sauerland to Hagen. Puddling and pig iron firms needing cheaper coke fuel were the major backers of this railroad, but charcoal iron producers, aware that less expensive coals could reduce the demand for wood, were pushing hard too.[29]

County and mining officials warmed to the project immediately. The Siegen-Hagen railroad found another influential champion in Count Beust, Captain of the Bonn Mining Office. Although concerned about high construction costs, he reported to his colleagues in Berlin that "nowhere could one execute a more useful project."[30] Then his apparent enthusiasm cooled. The Siegerland, he claimed in 1835, lacked "the dense population"[31] required to make such a technically complex rail line profitable.

Behind the mask of such pessimistic expressions, one suspects, were deeper suspicions and fears. For if the two regions were linked, private coal production in the Ruhr could take off in a wild, irresponsible spiral. Because Beust was always more committed to the interests of the Mining Corps than those of private industry, it is actually not surprising that his excitement waned. That same year, in fact, the proud captain recommended construction of railroads through the even less densely populated Eifel and Saarland as a means to promote coke iron, increase demand for Saar coal, and enhance the long-run profits of the Corp's mining section.[32]

Beust's reports, however, were rebuked at headquarters in Berlin. Swayed by the arguments of Carl Karsten, head of the metallurgical section, Corps leaders adopted the cautious position that railroad engines were powerful enough to carry passengers, but not bulk commodities like coal. Logical in 1836, this opinion would appear ridiculous in the 1840s as freight traffic increased three-hundredfold. Karsten, the reader will recall, prevailed in the mid-1830s with proposals of his own for Corps

[29] Oberbergamt Bonn to Oberberghauptmannschaft, 27 March 1833, Rep.121, A.XX.1, Nr.102, Bd.6/7, Bl.85, ZStA Merseburg; and supporting documentation from 1833/34 in Reg. Arnsberg Iv, Nr. 475, StA Münster.

[30] See his report of 23 March 1834, Rep.121., A.XX.1, Nr.102, Bd.7/8, Bl.135, ZStA Merseburg.

[31] Beust to Oberberghauptmannschaft, 4 April 1836, Rep.121, A.XX.1, Nr.102, Bd.9/10, Bl.58, ZStA Merseburg.

[32] Ibid., Bl.57.

puddling works to exploit the demand for wrought iron rails.[33] Like Prussia and Germany, the divided Mining Corps sought to deal with forces that were wrenching the country into a new era.

Skepticism in official circles notwithstanding, railroad pamphlets, petitions, and prospectuses were multiplying with a quickening tempo in the Germany of mid-decade. English promoters in Hamburg wanted to construct lines to Magdeburg and Berlin, while Prussian entrepreneurs in both cities were making inquiries about railroads to and from the capital. Dissatisfied with unreliable river transportation on the Oder, moreover, merchant-manufacturers in Breslau petitioned the state to build a railroad to Frankfurt (Oder). And in Leipzig, investors were excited about linking their city with Dresden and Magdeburg.[34] Friedrich List, the insufferable, overconfident visionary from Leipzig, raced far ahead of his committee with publications describing a railroad network for all of Germany. "As if by magic," he wrote Prussian officials, "Berlin will see itself elevated to the central point of 30 million people." The "moral, political, and military as well as industrial, commercial, and financial" effects of having Hamburg, Bremen, Munich, Strassburg, and Cologne only a day or two away would be "immeasurable."[35]

Pressed on all sides to reconsider his earlier pronouncements on state aid for railroads, Frederick William turned to Christian Rother for advice. From complicated loan transactions in London to the rehabilitation of Silesia's textile industry, the peasant's son was now his sovereign's most trusted economic aide. This confidence was manifest in the bureaucratic empire which Rother had amassed by the mid-1830s. Director of the Prussian Bank, the Public Debt Administration, and the Seehandlung, he also oversaw the Business Department and Highway Construction Division in his own new ministry.

Not surprisingly, perhaps, Rother's report of August 1835, was the most damning official word on railroads to date. Widespread publicity about the British and American experience with long-distance rail transport, he informed the king, was often exaggerated. Railroads required a huge capital outlay in locomotives, rolling stock, roadbed construction, and station houses. Only those lines running through densely populated areas and near mines would begin to amortize their initial investment, but even here the process would be slow and uncertain due to stiff com-

[33] Oberberghauptmannschaft to Oberbergamt Bonn, 25 May 1836, Rep.121, A.XX.1, Nr.102, Bd.9/10, Bl.154–55, and Oberberghauptmannschaft (Karsten) to Oberbergamt Brieg, 16 July 1835, Rep.121, F.IX.3, Nr.103, Bd.1, Bl.65, ZStA Merseburg. For statistics on passenger and freight traffic in Prussia, see Fremdling, *Eisenbahnen*, 17.

[34] Fleck, "Studien," *Archiv für Eisenbahnwesen* 18 (1895): 11 and 19 (1896): 861–62.

[35] List to Maassen, 14 October 1833, printed in Alfred v. der Leyen et al. (eds.), *Friedrich List: Schriften zum Verkehrswesen* (Berlin: Reimar Hobbing, 1931), 3 (Pt. 2): 826–27.

petition from roads and canals. The Budweis-Linz Railroad, he pointed out, carried little freight and returned less than 3 percent on the initial investment. In France, the St. Etienne line was struggling to survive against older means of transport. Moreover, while freight rates on highways and waterways averaged eight pennies per centner in eastern Prussia, the Manchester-Liverpool Railroad charged nine and a half. These facts, Rother concluded, made state support for railroads—or investment in them—extremely ill advised.[36]

Railroads were admittedly a very new technology in continental Europe when Rother drafted his report. Two state lines in Austria and one private railway in France were open for business, but there were no intercity companies operating in Germany. The six-kilometer line between Nuremberg and Fürth was not finished until December 1835, while observers had to wait until April 1837 to see the first engines chug out of Leipzig and yet another eighteen months before tracks reached the Prussian capital from nearby Potsdam. But long before these dates, railroads were perceptibly accelerating the pace of industrialization and altering the nature of this process. Productive forces were tooling up in the mid-1830s to supply millions of tons of coal and iron—and these technologies were taking new forms. Deeper, more ambitious mine shafts powered by bigger steam engines; ingenious new coking ovens; heavy-duty lathes and milling machines, and taller blast furnaces using coke iron technology better suited for rails were beginning to appear. Industrialism uncovered its modern face during this decade as the mix of producer and consumer goods began to shift from latter to former. At 7–8 percent of net industrial investment in the early 1800s, the value of metals, fuels, and industrial leather (e.g. for pulleys, machine belts, and harnesses) increased to 11 percent between 1831 and 1842. The scale of operations was also changing. Whereas 15,000–70,000 thaler was sufficient to build most factories in the early 1830s, joint-stock companies commanding millions of thaler were the rule with railroad ventures. In order to construct the lines, unprecedented work crews employing thousands of laborers over many years were required.[37]

Much of this was already apparent—and worrisome—to Prussia's technocrats. Mining and metallurgical experts in the Corps planned to benefit from the coming of railroad-induced coal markets, coke-burning locomotives, and iron rails, but were threatened by visions of overheated

[36] Rother to the king, 16 August 1835, 2.2.1., Nr. 29517, Bl.9–26, ZStA Merseburg.

[37] Tilly, "Capital Formation in Germany," Mathias and Postan, *Cambridge Economic History*, 7 (1): 419, 438; Wehler, *Gesellschaftsgeschichte*, 2:97; Fremdling, *Eisenbahnen*, 100, 165. The ratio of producer and consumer goods is taken from Tilly's "list of twelve" industries.

growth in the private sector.[38] Such uncontrollable industrial development, both sections feared, would undermine revenues and irrevocably deplete natural resources. To railroad advocates in the Corps like Beust, the only responsible solution were Corps lines to the state's own mines.

Peter Beuth's concerns were of a similar sort. Although he believed that certain railroad connections would be useful and beneficial, most appeared to him to be foolish attempts to replace roads and canals which were far more practical. The unacceptable result would be a squandering of scarce capital—monies desperately needed to establish those model joint-stock corporations which would rejuvenate Prussia's linen and woolen industries. The threat to such an obviously sound economic development seemed to him a conspiracy, as we saw in chapter 3, of allegedly sinister Jewish bankers. Following his bouts with western textile manufacturers over tariffs, railroads were worsening Beuth's relationship with the bourgeoisie.[39]

Beuth's new chief issued similarly anticapitalistic warnings against the railroad promoters. Rother claimed that unscrupulous stock jobbers eager for a quick profit had lied about the profitability of railroads. Unless the state regulated these investments, brokers would perpetrate a cruel "stock market game"[40] on an unsuspecting public. Inextricably mixed with such alarming statements, however, was a nagging anxiety that many railroads—more than Beuth believed—were indeed solid propositions. Thus to ensure safety, he reported to the king, rail lines would have to deviate from the course of roadways and canals. This would pit the livelihood of one set of towns and villages against another and create struggles for the state to mediate. If railroads began to win this competition, the state, too, would lose as income on highways built recently at great expense was drained away. Rother also knew that railroad construction slated for regions west of Berlin would draw labor and capital away from eastern agriculture and rural industry. This was a threatening development at a time when he, as chief of the Highway Construction Di-

[38] See Beust's report of 4 April 1836 (as cited above in n. 31); and Karsten's report of 16 July 1835 (as cited above in n. 33).

[39] See Beuth's letters to Rother of 20 February, 24 March, and 21 August 1835, Ca-Nr.17, Bl.12, 18–19, 32, Rep.92 Rother, ZStA Merseburg. For Beuth's general concerns about unworthy projects drawing capital away from projects closer to his heart, see his position paper opposing limited liability corporations, 17 April 1838, cited in Paul Martin, "Die Entstehung des preussischen Aktiengesetzes von 1843," *Vierteljahresschrift für Sozial- und Wirtschaftsgeschichte* 56 (1969): 537.

[40] Rother to the king, 16 August 1835 (as cited above in n. 36). For the remainder of this paragraph—Rother's road-building activities in northeastern Prussia and concern about rural labor shortages—see Thimme, *Strassenbau und Strassenpolitik*, 70; and Rother to Crown Prince Frederick William, 3 April 1837, Cb-Nr.19, Bl.11–20, Rep.92-Rother, ZStA Merseburg.

vision, was pouring 2.7 million thaler annually into new roads, most of which were located in the eastern provinces of Pomerania and Prussia.[41] Like his fellow technocrats, Rother realized that railroad technology presented officials with clear choices about the future course of industrialization.

Frederick William approved Rother's report in September 1835. The Ministerial Cabinet was instructed to draft a set of general guidelines for the regulation and supervision of railroad affairs.[42] In the meantime, no new concessions would be granted for specific lines. The cars, it seemed, were not going to roll forward as quickly as railroad enthusiasts wanted.

The ministries deliberated the general railroad legislation during the winter of 1835/36. Rother and Beuth approached this task with what one western industrialist described as outright "opposition." Because none of the other ministers was "energetically behind"[43] the railroad cause, the law which took shape gradually assumed draconian proportions. In order to curb speculation and ensure stable ventures, subscribers were liable for all amounts pledged and shares could not be distributed until flotations were 25 percent subscribed. Once operational, railway companies had to rent space on the line to other car owners at rates closely regulated by the state. The price charged by owners and users for freight and passenger travel was also state-controlled, and both rents and prices were subject to downward revision if dividends exceeded 10 percent—all this to prevent "abuses"[44] of the public. Special contracts of this sort had to be negotiated separately, moreover, between every company and the Postal Department. Owners were liable for all personal injuries, fire damage, and other destruction caused by their trains and were required to sell expropriated land back to previous owners if the latter could prove that this property was not "indispensable"[45] to the railroad company. As an added measure of control, state commissars were placed on every board of directors. There would be no state assistance of any kind, finally, for the beleaguered companies—even negotiations with foreign governments would have to be conducted independently. The king agreed to all of these proposals in decrees of February and June 1836.

Disappointment in railroad circles was very widespread. "Closer insight to the Prussian state machinery has not been very pleasant for

[41] Tilly, "Capital Formation in Germany," Mathias and Postan, *Cambridge Economic History*, 7 (1): 412.

[42] Helmut Paul, "Die preussische Eisenbahnpolitik 1835–38," *Forschungen zur Brandenburgischen und Preussischen Geschichte* 50 (1938): 260.

[43] Camphausen to Krüger, 10 July 1836, printed in Schwann, *Ludolf von Camphausen*, 2:419.

[44] For the following discussion, see Gleim, "Zum dritten November 1888," *Archiv für Eisenbahnwesen* 11(1888): 807–13.

[45] Ibid., 812.

me,"[46] observed a depressed Ludolf Camphausen in Cologne. Indeed, the government seemed to be adopting a cynical "build-them-if-you-can" attitude. As protests mounted that summer and fall, however, the king deemed it appropriate that the Ministerial Cabinet investigate all complaints. With an eye on their political future under the Crown Prince, three of the younger ministers—Albrecht von Alvensleben in Finance, Gustav von Rochow in Interior, and Heinrich Mühler in Justice—now rallied around the heir apparent's stymied crusade for railroads.[47]

It was indeed an enigma, as Treitschke observed, that one as inclined to view the world romantically as Crown Prince Frederick William of Hohenzollern had become the kingdom's most highly placed advocate of railroads by the mid-1830s.[48] As explained in chapter 3, the seemingly inevitable advance of industry, the practical advice from his father about public finance, trips to the provinces to meet business leaders, and a fashionable Anglophilia all worked toward a greater—albeit superficial—appreciation of the Prussian economy. His adjutant of ten years, Adolf von Willisen, was another potent influence. A liberal Anglophile, Willisen admired England's union of "freedom, power, riches, and fame." He was also a railway enthusiast and generally known as the person to contact when seeking to gain the Crown Prince's support for specific railroad ventures.[49] It could also be true, as so often reported, that an exhilarating ride on the Prince William Railroad in October 1833 captured Frederick William's lively imagination.[50] The appearance on the horse-drawn line named after his uncle was part of a month-long "public relations" visit to Rhineland-Westphalia which brought the prince into contact with many businessmen, town councils, and provincial officials who depicted the new technology as a stimulus to prosperity. All pleaded with their royal patron to remove bureaucratic obstructions and speed concessions in Berlin.[51]

[46] Camphausen to Krüger, 9 March 1836, printed in Schwann, *Ludolf von Camphausen*, 2:408.

[47] Treitschke, *Deutsche Geschichte*, 4:591; Paul, "Die preussische Eisenbahnpolitik," *Forschungen zur Brandenburgischen und Preussischen Geschichte* 50 (1938): 260–61; Gleim, "Zum dritten November 1888," *Archiv für Eisenbahnwesen* 11 (1888): 814.

[48] Treitschke, *Deutsche Geschichte*, 4:591.

[49] For Willisen himself, see Priesdorff, *Soldatisches Führertum*, 6:312–14. For the quote, see Varnhagen Diary I, 9 December 1826, 4:152 (see also the entry of 29 June 1826, 4:88). Later, Varnhagen noted that Willisen and his brother, Karl Wilhelm, were two key liberals in the prince's entourage (28 January 1830, 5:264–65). For Willisen as the man behind Frederick William's railroad cause, see Wittgenstein to Rother, 24 September 1837, Nr./Ca-9, Rep.92 Rother, ZStA Merseburg, and Camphausen to Krüger, 29 May 1836, printed in Schwann, *Ludolf von Camphausen*, 2:419.

[50] For the ride, see Albert von Waldthausen, *Geschichte der Steinkohlen-bergwerks Vereinigte Sälzer und Neuak* (Essen: G. D. Baedeker, 1902), 285–86.

[51] See Eduard Schmidt's circular of January 1834, Regierung Arnsberg Iv Nr. 475, StA

Presented with an opportunity to demonstrate the alleged irresponsibility of the hated ministries, Frederick William was spurred on to action when he returned to the capital that November. The Crown Prince concentrated initially on expediting the Rhine-Weser line. The father, however, did little more than humor the son. One commission gave way to another as 1834 drew on to 1835, and the final report—coming two years after Frederick William's western trip—recommended no more state aid than the king had promised to the Minden-Lippstadt promoters in 1832.[52] After the Ministerial Cabinet repeated its economic and technical doubts to the king in February 1836, the Rhine-Weser cause was buried again in paperwork.[53]

The businessmen's criticism of the ministries' railroad guidelines of 1836 revived the Crown Prince's crusade that autumn as Alvensleben, Rochow, and Mühler supported their future sovereign. Heated, at times nasty, debate in the Ministerial Cabinet finally led to an important revision of the provisional regulations: railroad firms could set their own rates and operate without other carriers on the line for three years. Beuth, who had boasted earlier to Rother about drawing up a "war list" and "eliminating opposition,"[54] saw victory slip away as the spring of 1837 approached. Upset and offended over a series of encounters and disagreements with the Crown Prince, Rother asked to be relieved of railroad affairs, while the gloating prince dispatched Captain Adolf von Willisen to a special commission which began a complete redrafting of the railroad law.[55]

Although his allies doubted Rother's wisdom, the seasoned political veteran may have known better. For Willisen, a General Staff Captain and instructor at the War School and combined Artillery and Engineering School, belonged to a growing prorailroad movement in an army which usually got its way in Prussia. Early fears that railroads might facilitate enemy invasions had gradually disappeared, and by the mid-1830s an intense debate was raging in the technical branches about the military advantages of railways as road-supplements. Skeptics like Ernst

Münster; and Landrat Opladen to Oberpräsident Koblenz, 22 January 1834, 403/3675, Bl.253, LHA Koblenz.

[52] Beuth to the other departments of the Finance Ministry, 8 December 1834, Rep.121, D.VI.2, Nr.108, Bd.1, Bl.63, ZStA Merseburg; Regierungspräsident Düsseldorf to Oberpräsident Koblenz, 10 February 1834, and Rother to Krüger, 19 June 1836, 403/3675, Bl.268, 355, LHA Koblenz.

[53] Fleck, "Studien," *Archiv für Eisenbahnwesen* 19 (1896): 235–36.

[54] Beuth to Rother, 6 June 1836, Rep.92 Rother, Ca Nr.17, Bl.75, ZStA Merseburg.

[55] Lottum to Rother, 5 June 1837, Rep.92 Rother, Ca Nr.10, Bl.104, ZStA Merseburg; Gleim, "Zum dritten November 1888," *Archiv für Eisenbahnwesen* 11 (1888): 816; Dietrich Eichholtz, *Junker und Bourgeoisie vor 1848 in der Preussischen Eisenbahngeschichte* (Berlin: Akadamie-Verlag, 1962), 97.

Ludwig Aster, Commandant of the Fortress of Ehrenbreitstein, and Friedrich Leopold Fischer, a talented General Staff Captain, abounded. There would never be enough lines, it was said, or they would be too costly to build, then too easily destroyed. Others feared that troops would become soft and lazy without the toughening experience of marching across hill and dale.[56] The new mode of transportation was therefore not free of those dilemmas which many Prussian soldiers associated with the use of modern industrial technology in warfare. But many more overcame such doubts, tantalized by the possibility that railroads would realize "the military fantasy," as Fischer put it, of "entire armies flying across vast territories in a few days."[57] Infantry units could be concentrated or evacuated with relative ease and heavy artillery and mortars moved and supplied in a fraction of the time.

Indeed the list of railroad supporters in uniform was impressive. One group was concentrated, not surprisingly, in the artillery: Eduard Peucker of the War Ministry's Artillery Division; Eduard Kunowski of the artillery's industrial complex; young visionaries like Albert du Vignau of the Experimental Department; and undoubtedly Prince August himself.[58] Willisen represented an important link between the artillerists and his own colleagues in the General Staff. One was his brother, Karl Wilhelm, who saw railroads as the answer to Prussia's western defenses. His response to the Belgian crisis was to utilize Rhenish fortresses and carefully laid-out rail lines to outflank superior French forces pushing east.[59] Another was Karl von Müffling, commanding the Seventh Corps in Münster in 1835, who believed that a Rhine-Weser connection had become "an indispensable tool of war against France."[60] More important, they had the ear of the Chief. Johann Wilhelm Krauseneck knew that roads would always be indispensable—certainly a modern point of view—but he also emerged as an early (1835/36) and consistent supporter of strategic railways. Enhancing the political chances for railways, moreover, was

[56] For Aster, see Treitschke, *Deutsche Geschichte*, 4:592; for Fischer and others, see Bernard Meinke, "Die ältesten Stimmen über die militärische Bedeutung der Eisenbahnen 1833–42," *Archiv für Eisenbahnwesen* 42 (1919): 53–55, 48, 61–63. Meinke's argument that Fischer authored two anonymous tracts on railroads in 1836 and 1841 seems confirmed by Fischer's letter of 31 May 1841 to Rochow (Rep.77, Titel 258, Nr.27, Bl.14, ZStA Merseburg) wherein the officer mentions his 1841 brochure on "the use of railroads for military purposes."

[57] Meinke, ibid., 60, cites Fischer's tract of 1841.

[58] Ibid., 47–49; and Priesdorff, *Soldatisches Führertum*, 4:422.

[59] A. D. von Cochenhausen, *Von Scharnhorst zu Schlieffen 1806–1906 Hundert Jahre preussisch-deutscher Generalstab* (Berlin: E. S. Mittler, 1933), 140–41.

[60] The remark was attributed to Müffling by Provincial Tax Director Krüger in a letter to Rother, 20 June 1835, Oberpräsidiium B 1138, StA Münster.

the fact that conservatives like Duke Carl of Mecklenburg were just as excited.[61]

The military use of railroads came before the Ministerial Cabinet in April 1836.[62] The protocols unfortunately do not reveal which ministry raised the issue, but there can be little doubt that War, still under Job von Witzleben, was responsible. As noted above, the military debate in Prussia was escalating. Throughout the previous winter, moreover, Prussia had received official and unofficial advice from South Germany to pay closer attention to the military advantages of a German railroad network. These materials were given to Rühle von Lilienstern for review, and, after receiving Rühle's cautiously positive report, Witzleben probably wanted to see a more thorough investigation.[63] Will railroads benefit German armies, and if so, what steps should be taken to prepare for their most purposeful use? Should a network of lines be prepared for consideration by the Military Commission of the German Confederation?[64] In May his civilian colleagues agreed to establish an investigating committee under the auspices of the General Staff.

The final report represented a compromise between the two military antagonists on the committee—Fischer and Peucker—but was essentially a victory for the latter.[65] Railroads under locomotive power would neither replace highways nor make them partially dispensable. This was particularly the case for cavalry or when large numbers of infantry had to be transported very short distances. The savings in time increased tremendously over longer stretches, especially for movement of horses, weapons, ammunition, and supplies. These advantages could not be realized,

[61] "Die Entwicklung des Militäreisenbahnwesens vor Moltke," n.a., *Beiheft zur Militär-Wochenblatt* 5 (1902): 238, 241, 243. Krauseneck was clearly not as enthusiastic about railroads as the author of this article; but by no means as unenthusiastic about them, nor as backward, as portrayed here. Showalter, *Railroads and Rifles*, 23–26, tends, like the earlier works, to underestimate the army support for railroads in the mid-1830s. For Duke Carl, see Fleck, "Studien," *Archiv für Eisenbahnwesen* 19 (1896): 246.

[62] A copy of the Staatsministerium protocol of 26 April 1836 is located in Rep.90, Nr.1674, GStA Berlin.

[63] Meinke, "Die ältesten Stimmen," *Archiv für Eisenbahnwesen* 41 (1918): 923, 931–32, and 42 (1919): 46–47; "Militäreisenbahnwesen vor Moltke," n.a., *Beiheft zum Militär-Wochenblatt* 5 (1902): 239–40. Both articles exaggerate Rühle's opposition to the military use of railroads. List, on the contrary, considered Rühle a far-sighted ally during a visit to Berlin in May 1835 (List to Duke Ernst I of Saxony-Coburg-Gotha, 24 December 1840, printed in Erwin V. Beckerath and Otto Stühler [eds.], *Friedrich List: Schriften zum Verkehrswesen* [Berlin: Reimar Hobbing, 1929], 3 [Pt.1]: 37). See also above, n. 61.

[64] See the protocol of 26 April 1836 cited above in n. 62.

[65] The report of 4 and 15 July 1836, including Fischer's and Peucker's signatures, is located in Rep.90, Nr.1674, Bl.13–46, GStA Berlin. See also the sources cited above in notes 56 and 58 documenting Fischer's skepticism regarding, and Peucker's advocacy of, military uses for railroads.

however, without locomotives safe enough to avoid ammunition explosions, cars and station platforms large and durable enough to carry animals and war materials, tracks of standard gauge, stations which were located in good terrain equidistant from one another, and a system of rail lines which met the army's needs. Private planning and construction of railroads should therefore receive detailed military scrutiny and approval before concession. Where private interests did not pursue lines desired by the military, the state should offer companies "appropriate advantages"[66] to ensure that there were no gaps in the system.

Unfortunately, there is very little information about the reaction to Peucker's report in the War Ministry, especially from Job von Witzleben. But it is highly unlikely that a man who was working so closely and enthusiastically with Peucker to perfect the needlegun would reject his opinion on railroads and fail to promote his recommendations. Witzleben placed Peucker on the commission, after all, because he respected the artillerist's technical expertise. One indication that Peucker's report had advanced quickly past his immediate ministerial superiors came in February 1837, when Rother informed the Rhine-Weser Railroad Company that tracks in Prussia would have to be of standard gauge.[67] Thus Witzleben probably joined Alvensleben, Rochow, Mühler, and the Crown Prince to create a highly persuasive and influential coalition as the politics of railroads heated up that winter and spring.

The Ministerial Cabinet reported to the king on the work of Peucker's and Willisen's commissions in June and July 1837. The former found unanimous backing for every one of its recommendations except one. In order to avoid creating false hopes of state aid and thereby inhibiting private investment in railroads, it was decided to postpone any promise of "appropriate advantages" for lines deemed important by the army but not seen as lucrative by private industry.[68] Willisen's group recommended—and the above majority of ministers approved—three further improvements for the railroad companies. They were freed from the regular tax on industrial firms (*Gewerbesteuer*) in return for a special levy which would both aid in the amortization process and compensate the Postal Department for lost revenues. Like the concession approved the previous winter, this provision would also take effect after three years of tax exemption. Second, once other carriers were permitted on the lines, the Postal Department would abide by the same rates and regulations as everyone else, thus removing the real possibility of obstructionist sepa-

[66] Ibid., Bl.45–46.

[67] Rother to the Rhein-Weser Railroad Company, 3 February 1837, 403/3675, Bl.411, LHA Koblenz.

[68] Staatsministerium to Frederick William, 18 June 1837, Rep.90, Nr.1674, Bl.72–77, GStA Berlin.

rate contracts and cutthroat competition for passenger travel. Last, a reasonable arrangement was proposed for the option of state purchase after thirty years.[69]

The draft legislation of 1837 was not at all what corporate railway promoters wanted. It was essentially a combination of Rother's antirailroad bill and those ameliorations which a military-civilian coalition advocating railway development was able to exact. The king was not satisfied either. Although military fears of a decade earlier still concerned him, Post Office and Seehandlung arguments that state rights and revenues were in jeopardy seemed irrefutable and worthy of further investigation. At the dinner table Frederick William also illustrated an indifference toward rapid intercity travel and discomfort with the "democratic" effects this could have on the common folk.[70]

But Frederick William III was neither a political fool nor, in fact, a diehard opponent of railroads. He accepted the hybrid railway bill grudgingly after the Crown Prince swung the Council of State behind the document in May 1838. The aging monarch also approved concessions for those rail lines—Aachen-Cologne, Cologne-Minden, Magdeburg-Leipzig, and Berlin-Potsdam—which had been in limbo since 1835. When he noticed the train from Potsdam to Berlin passing his carriage, moreover, Prussia's pragmatic monarch began to patronize the line himself. In fact, his last testament would include one million thaler for a railroad connection between Berlin, Prussian Saxony, and the middle Rhine.[71] Sensing his father's change of feelings, the Crown Prince could afford displays of public optimism. "These cars rolling across the earth," he proclaimed jubilantly in October 1838, "will be delayed no longer by the hands of man."[72]

The same could be said for industrialization. In terms of wrought iron production, for example, the four lines given final approval in 1837/38 would require nearly *four times* as much as Prussia produced in 1837. Unless purchased abroad, this iron would take almost a million tons of Prussian coal or charcoal to smelt and puddle, thus placing tremendous strains on the country's resources of energy and capital. Enhanced productive forces were also needed to make hundreds of bridges, stations,

[69] Gleim, "Zum dritten November 1888," *Archiv für Eisenbahnwesen* 11 (1888): 815–16.

[70] Ibid., 816; Frederick William to the Staatsministerium, 12 August 1837, 2.2.1., Nr.29517, Bl.178, ZStA Merseburg; Eylert, *Charakter-Züge*, 3A:205; Treitschke, *Deutsche Geschichte*, 4:591.

[71] Eylert, *Charakter-Züge*, 3A:205; Paul, "Die preussische Eisenbahnpolitik," *Forschungen zur Brandenburgischen und Preussischen Geschichte* 50 (1938): 262–63; Fleck, "Studien," *Archiv für Eisenbahnwesen* 19 (1896): 867; Treitschke, *Deutsche Geschichte*, 4:594; Eichholtz, *Junker und Bourgeoisie*, 101.

[72] Cited in Waldthausen, *Vereinigte Sälzer und Neuak*, 287.

Figure 3. Spectators view train on a viaduct near Aachen-Burtscheid, 1845.

locomotives, rolling stock, and myriad small parts. Rother figured these and other costs of construction for the four lines at 8.5 million thaler.[73] This amount was equivalent to 6.5 percent of all transportation investment in Prussia in the 1830s—and requests for the licensing of other lines were multiplying quickly. In order to appreciate more fully the concern in Berlin, we should not forget that the Prussian economy was already growing rapidly. Agricultural investment was up 156 percent, building construction up 278 percent, and road and canal expansion up 256 percent from the 1820s.[74] It was the unavoidably accelerated pace of economic change and the shifting pattern of investments that would result from railroads which convinced the Beusts, Beuths, and Rothers in a Prussian bureaucracy which had always favored growth that "restraint was urgently necessary."[75]

Now that Frederick William had given his grudging approval, restraint would have to be indirect. Eager to capture a greater share of the railroad market for domestic producers, western coal and iron magnates like Franz Haniel were lobbying for higher tariffs on British iron rails. "It is impossible to compete with this country,"[76] he pleaded in a letter to Peter Beuth. But to no avail. Haniel was exasperated to learn from Ludwig von Vincke, who had taken up the cause in Berlin, that the free traders were considering lowering, not raising tariffs. Such a measure would have cooled industrial growth in Prussia by shifting rail production to England. Apparently nothing came of either proposal, however, for the tariff on rails remained at 20 thaler per ton in the late 1830s.[77] Beuth achieved more success in the ministries in the area of laws of incorporation, turning both his superior in Finance, Albrecht von Alvensleben, and the Minister of Justice, Heinrich Mühler, against the concept of limited liability dear to modern capitalists. The state should apply the principle to industries which served "the common good" as defined by "the general interests of political economy," he wrote, but not universally lest

[73] For the combined distance of the four lines, see G. Stürmer, *Geschichte der Eisenbahnen* (Bromberg: Mittler, 1872), 32–37; for production in 1837 and the tons of wrought iron required per kilometer of track during this period, see Beck, *Geschichte des Eisens*, 4:693, 715. The actual cost approached 30 million thaler (see cost per Prussian mile figures for each line in Friedrich Wilhelm von Reden, *Die Eisenbahnen Deutschlands* (Berlin: E. S. Mittler, 1846), 386.

[74] Tilly, "Capital Formation in Germany," Mathias and Postan, *Cambridge Economic History*, 7 (1): 427. For methods used in calculating percentages of net investment, see chap. 1, n. 63.

[75] Rother to the Crown Prince, 3 April 1837, Rep.92 Rother, Cb Nr.19, ZStA Merseburg.

[76] Haniel to Beuth, n.d. (1836/37), Portfolio 210, Haniel Museum Duisburg.

[77] See Vincke to Haniel, 20 July 1837 and 31 January 1838, Ludwig von Vincke Papers, Q4, Bl.26, 31, StA Münster.

the door be opened to swindles and speculation "at the expense of solid industry."[78]

While civilian hands tried to slow down the railroad cars, the army which had played such a crucial role in getting the coaches rolling wanted to push in the other direction. To Eduard Peucker and Eduard Kunowski, chiefs of the military's industrial establishment, the construction of railroads was too important for the nation's security to entrust to private industry. For one thing, the technical decisions of private engineers seemed dubious. In an attempt to get maximum use out of scarce capital, Prussian companies were deviating from English and Belgian construction practices. Rail routes tended to wind through river valleys and up and around hills, thus driving up the cost of operations and—critical from the military purview—slowing down trains. If the profit motive of private industrialists determined which stretches were built, moreover, the construction process would proceed *far too slowly and incompletely.* Peucker and Kunowski saw the solution to such problems in state construction and ownership of the railroads.[79] Their conclusion was consistent in an organization which increasingly favored "command" over the economy.

Gustav von Rauch, the former Chief of the Corps of Engineers and Witzleben's successor as Minister of War since July 1837, knew that nationalization plans were politically impossible under the parsimonious Frederick William III. But there were other, less objectionable ways to gain a measure of control over a technology so potentially advantageous to army mobility. During the Council of State railroad debates (1837/38), his ministry offered amendments which would have forced railway lines to move troops during international crises—and at considerably reduced rates. Rauch repeated the demand with more urgency in memoranda to his civilian colleagues of November 1838 and July 1839. Because the new railroad legislation did not compensate railroads for damage during wartime, companies had either a disincentive to transport soldiers, or, if units were actually moved, no incentive to charge lower rates. In the latter case, Rauch figured that the cost of moving just one division from Berlin to Cologne would be 60,000 thaler higher than a normal forced march. Yet Prussian law demanded that private interests yield to the needs of the nation in times of emergency. Having sounded this anticapitalistic alarm, Rauch rested his case.[80]

[78] See Beuth's position paper of 17 April 1838, cited in Martin, "Die Entstehung des preussischen Aktiengesetzes," 537.

[79] See the report of Peucker's commission (July 1836), Rep.90, Nr.1674, Bl.23ff, GStA Berlin; and Priesdorff, *Soldatisches Führertum*, 4:422.

[80] Rauch's position paper of 11 July 1839 refers to his paper of 28 November 1838 (Rep.

Alvensleben, Rochow, and Mühler disagreed for reasons, once again, which seem quite opportunistic. The Crown Prince who would soon head their state was very skeptical in the late 1830s that railroads would ever connect Prussia's divided provinces. This disbelief undoubtedly weakened the army's arguments when they were presented before him. Despite Willisen's best efforts, moreover, the transportation needs of the army's technical corps were normally far from Frederick William's religious-oriented thoughts.[81] The prince fought consistently for railroads out of a growing conviction that he, not the unresponsive bureaucracy, perceived the needs of businessmen and understood the best means to improve the nation's economy and standard of living.[82] What mattered to the politically prudent ministers, therefore, was that Rauch, whether he appreciated it or not, was advocating measures which would reduce returns on massive investments, attenuate property rights, and generally make railroads less attractive to Prussian businessmen. The three ministers responded in this vein to Rauch throughout 1839.[83]

The Near Eastern crisis of 1839/40 postponed further debate until after the ascension of Frederick William IV. Not surprisingly, the Ministerial Cabinet finally decided against the War Ministry's requests in December 1841.[84] Anticipating defeat, however, the army had already found another way to achieve its goals. For War announced in December 1840 that it would move no troops on railroads if the price exceeded the cost of a normal march. Within six months the Berlin-Anhalt, Berlin-Potsdam, and Magdeburg-Leipzig companies succumbed to the pressure and offered special rates.[85]

In Camphausen's words, such practices reflected "the immense pretensions of the military authorities."[86] Indeed, private property rights

---

84a, Nr.11203, GStA Berlin). For the Staatsrat effort, see the Staatsministerium to Frederick William IV, 31 December 1841, Rep.77, Titel 258, Nr.27, Bl.17–20, ZStA Merseburg.

[81] For the Crown Prince's overriding interest in church reform and doubts about the coming of a railroad network, see, respectively, Petersdorff, *König Friedrich Wilhelm*, 6–12, 46–58, and Pfannenberg to Meding, 26 November 1837, cited in Eichholtz, *Junker und Bourgeoisie*, 134.

[82] Petersdorff, *König Friedrich Wilhelm*, 6–12, 46–58. For an example of the prince's promotional efforts, see his letter to Rother of 11 November 1835, cited in Fleck, "Studien," *Archiv für Eisenbahnwesen* 19 (1896): 54. For his critique of the unresponsive bureaucracy, see Petersdorff, *König Friedrich Wilhelm*, 58.

[83] For their position papers (August–October 1839), see Rep.84a, Nr.11203, Bl.19–21, 27–32, GStA Berlin.

[84] See the report of 31 December 1841 cited in n. 80 above.

[85] See Berlin-Frankfurt (Oder) Railroad to Oberpräsident Koblenz, 16 December 1843, 403/1996, LHA Koblenz; and Meinke, "Die ältesten Stimmen," *Archiv für Eisenbahnwesen* 42 (1919): 68–69.

[86] Camphausen to Henz, 16 September 1838, printed in Schwann, *Ludolf von Camphausen*, 2:440.

were narrowly circumscribed in the mind of Prussia's soldiers and, consistent with this, their expectations were arrogantly high. Locomotives, cars, platforms, and stations—all were constructed with Prussia's eight corps commanders looking on. When fortress commandants deemed it necessary, companies were told to spend enormous sums to fortify nearby tracks. When the Corps of Engineers required it, moreover, bridges were made of wood, not stone. The army also had the right to destroy railroad property without indemnifying the owners.[87]

Even the most influential organization in Prussia, however, would not find it easy to glare at industrialists and command that they invest millions in unpromising rail lines connecting the nation's capital with out-of-the-way garrison towns. Yet these connections were what an ever-increasing majority of Prussia's finest wanted. Former skeptics in the General Staff like Leopold Fischer, young pragmatic colleagues like Helmuth von Moltke, and open-minded cavalry veterans like Wilhelm von Reyher joined the railroad faithful in a systematic campaign to convince ministers and doubting colleagues of the need to eliminate gaps in Prussia's railway network.[88] About eight hundred kilometers of track were either open for business or assured of construction by 1842, but none of these lines tied Berlin to forward depots, fortresses, and corps headquarters. Thus military pressure for state subsidies or actual nationalization mounted.[89]

[87] In general, see Eichholtz, *Junker und Bourgeoisie*, 104. Clause 43 of the Railroad Law of 1838 forbade indemnification after wartime destruction (see the text of the law in H. Höper, *Die Preussische Eisenbahn-Finanz-Gesetzgebung* [Berlin: Carl Heymann's, 1879], 11). Also, the General Staff's report of July 1836 (see above, n. 65) recommended close scrutiny by corps commanders over railroad construction. The report was approved and the overseeing occurred, particularly on lines which approached fortresses (see LHA Koblenz, 403/3585, Bd.4, Bl.281ff., for negotiations over the Fortress of Cologne).

[88] For Fischer's conversion, see the sources on him cited above in n. 56. For his lobbying, see Fischer to Rochow, 31 May 1841 (n. 56); Fischer to Müffling, 2 February 1844, Rep.92 Müffling, A.19, ZStA Merseburg; and Treitschke, *Deutsche Geschichte*, 5:494 (n. 1). Fischer and Moltke were friends since their service in Turkey during the late 1830s (see Moltke's "Selbstbiographie," printed in Helmuth von Moltke, *Gesammelte Schriften und Denkwürdigkeiten des Generalfeldmarschalls Grafen Helmuth von Moltke* [Berlin: E. S. Mittler, 1892], 1:23), thus it is likely that Moltke's advocacy of state lines (see Moltke to his brother Adolf, 13 May 1844, cited in Moltke, *Gesammelte Schriften*, 2:234; and Treitschke, *Deutsche Geschichte*, 5:494) stemmed from Fischer. Reyher, who had taken over command of the Guard Corps from a sick Prince William (the future Kaiser) in September 1839, was responsible for moving the entire unit by train from Potsdam to Berlin (see General von Ollech, *Carl Friedrich Wilhelm von Reyher* [Berlin: Ernst Siegfried Mittler und Sohn, 1879], 4:86; and Meinke, "Die ältesten Stimmen," *Archiv für Eisenbahnwesen* 42 [1919]: 65). Reyher's work with railroads as Chief of the General Staff after 1848 probably originated, therefore, from a conviction about their usefulness held for nearly a decade.

[89] For the data on serviceable railroad lines in 1842, see Stürmer, *Geschichte der Eisenbahnen*, 32–33; and the Staatsministerium's Report to the United Committees of the Diet, October 1842, printed in Höper, *Eisenbahn-Finanz-Gesetzgebung*, 22. The Report (p. 22)

Frederick William IV was not unmoved by the entreaties of his generals. With the mantle of supreme warlord had come a somewhat greater awareness and appreciation of what was required for military preparedness. His martial brother, Prince William, strengthened this tendency to suppress a natural unsoldierliness and give responsible consideration to the army's technological needs. But there were other, stronger forces at work on the enigmatic sovereign. The first twenty-four months of his reign brought a succession of trips to the outlying provinces, and here, along with the hurrahs of a people who anticipated political reform, he must have heard pleas for governmental aid for the railroads. The emphasis placed on this theme in numerous private audiences was reinforced by explicit public resolutions from most of the provincial diets which met throughout 1841.[90]

Political, economic, and military considerations were therefore wrapped together in the king's final decision. Twelve representatives from each diet would assemble in Berlin during the fall of 1842. The "United Committees" of the Diets would be asked to lend sanction to a program creating a special state railway fund and guaranteeing private companies 3.5 percent return on monies invested in five troubled lines—(1) Minden-Cologne, (2) Halle-Mainz, (3) Frankfurt (Oder)-Austrian border, (4) Berlin-Königsberg-Russian border, and (5) a north-south spur over Posen linking lines three and four. If completed, 1,650 kilometers of *new* track would tie together the Prussian West, East, and South. Although generally opposed to statism, most of the delegates still believed an exceptional case should be made for railroad lines built and operated by the state. In an overwhelming majority, however, the Committees settled for the program which Frederick William offered them.[91]

The Railroad Law of 1842 accelerated a moderate railway expansion already underway. The enticements of the new legislation were certainly not great enough to attract adequate funds to the Posen and Königsberg spurs, yet within three years the guaranteed western and southern lines were licensed and either completed or under construction. Their 750 kilometers would double the length of operational track in Prussia. Anticipating a greater volume of traffic throughout the kingdom, moreover, investors poured money into nonguaranteed lines as well. Thus the

and Treitschke, *Deutsche Geschichte*, 5:494, refer to the military rationale, and military pressure, respectively, for railroads. See also "Die Entwicklung des Militäreisenbahnwesens," *Beiheft zum Militär-Wochenblatt* 5 (1902): 242.

[90] Petersdorff, *König Friedrich Wilhelm*, 8; Treitschke, *Deutsche Geschichte*, 5:168–84; Staatsministerium Report of October 1842, printed in Höper, *Eisenbahn-Finanz-Gesetzgebung*, 21.

[91] See the Staatsministerium Report of October 1842 and other documentation in Höper, ibid., 22–26.

number of new railroad issues listed on the Berlin Stock Exchange jumped from six to twenty in two years and the price of most listings shot upward. It now appeared, as the Crown Prince had predicted in 1838, that no man's hand could halt the cars. The state contributed to the boom by investing 3.8 million thaler from its new fund.[92]

This impression of inevitable railroad expansion helps to explain the dark mood of those who had tried to halt the cars—and apparently failed. A young Rudolph von Delbrück observed Peter Beuth in the fall of 1843 making "occasional sarcastic comments" about the new railroad technology. Beuth was unenthusiastic about it "because he had returned from his English experience [in 1826] with the conviction that Prussian commerce was far from being developed enough to require railroads, and because he was not capable of divorcing himself from this perception."[93] With Prussia's leading technologist, however, the motivation was more complex. It was general knowledge by 1843 that German railroads were developing quickly and cheaply in "American" fashion, avoiding the straight tracks and expensive viaducts and tunnels which only English trade and industry could amortise rapidly. Like Rother, Beuth dreaded this rapid development because, even though cheaper than the English lines and economically feasible, it would still draw capital away from projects closer to his heart. "The energy of this nation is not to be controlled: it is at present exclusively applied to the acquisition of wealth and to improvements of stupendous magnitude. . . . I would have proposed a gradual, slow, and more secure progress." These lines were penned in 1836 by an early American proponent of manufacturing, Albert Gallatin, who had also grown dissatisfied with the shape of present and future. But the Prussian technocrat would not have changed a word. It was a healthier alternative, less riddled, Beuth believed, with Semitic influences, which was displayed in Berlin for the great Industrial Exhibition of 1844.[94]

[92] Eichholtz, *Junker und Bourgeoisie*, 114; Stürmer, *Geschichte der Eisenbahnen*, 32–34; Hanns Leiskow, *Spekulation und öffentliche Meinung in der ersten Hälfte des 19. Jahrhunderts* (Jena: Gustav Fischer, 1930), 8–11; Wolfgang Klee, *Preussische Eisenbahngeschichte* (Stuttgart: W. Kohlhammer, 1982), 108; Fremdling, *Eisenbahnen*, 126.

[93] Delbrück, *Lebenserinnerungen*, 1:136, 135.

[94] For Beuth's opposition to "insecure" progress, see his position paper of 17 April 1838—and other sources—cited above in n. 39; for the Gallatin quote, see John R. Nelson, *Liberty and Property: Political Economy and Policymaking in the New Nation, 1789–1812* (Baltimore: The Johns Hopkins University Press, 1987), 170; for the artistic motive behind the exhibit, see Delbrück, *Lebenserinnerungen*, 1:135. See also the letter of Viebahn (of the exhibit's organizing committee) to the Regierung Köln, 5 July 1844 (Abt.1, Nr.28, Fasz.1, R-W WA Köln) which denied the charge that the fair was only for products of "Fabrikindustrie." Also welcome were agricultural, mining, and metallurgical products as well as those produced with "special precision and artistry" (*besonderen Sorgfalt und Kunstfertigkeit*).

The Mining Corps was in a better position to delay the encroachment of an ugly new world—or barring this, survive in it. Under Ernst von Beust's conservative stewardship, the mining section amended Karsten's liberal mining law reform in the 1840s, restraining somewhat the momentum of productive mining forces which were finally freed in 1865. An anxious Corps could also protect itself from the onslaught of private industry by preserving and advancing a technological expertise it once monopolized. Modern puddling furnaces in Silesia and coking ovens in the Saar were examples of a strategy which provoked both anger and admiration among industrialists. So, too, were Corps railroads. Central headquarters did not finally warm to this technology until 1845, preferring instead to foil the plans of private railroad companies—somewhat like the army—by refusing to transport coal or sell coke without special privileges. But by early 1848 crews from the Bonn Mining Office were breaking ground on the first "state" line in Prussia. Running thirty-two kilometers between Saarbrücken and Bexbach, the railroad was, in effect, Corps property.[95]

Rother's Seehandlung was unable to make such an easy transition to the new age. The old king's trusted adviser had warned the new king's ministers throughout 1842 against proceeding too quickly with railroad building, only to see the law accepted over his objections.[96] But worse news came that December. In what amounted to a cruel revenge, Frederick William IV directed Rother to place Seehandlung resources behind construction "of those railways recognized as a requirement for the state."[97] The corporation was in no position to invest in five railroads without drastically curtailing its high-priority textile projects in Silesia. The sixty-six-year-old financial wizard was pushed much farther this time than he had been when resigning from his ministry in 1837. For now his citadel, the autonomous Seehandlung, was under attack. "Ranting old Rother,"[98] as the king called him, would have to decide whether to exit the walls and break the siege.

Paradoxically, the very rapidity of railroad development in the 1840s played into Rother's hands. The 800 kilometers of track open for business or nearing construction in 1842 grew to 2,850 operational kilometers before the lower classes struck in March 1848.[99] Railroad companies poured

[95] Oberbergamt Bonn (Dechen) to Oberberghauptmannschaft, 20 December 1841 and 12 September 1845, OBA Bonn 705a and 705b, and Ministry of Finance to Oberbergamt Bonn, 10 March 1848, OBA Bonn 705b, HSA Düsseldorf.

[96] Radtke, *Seehandlung*, 291.

[97] Frederick William to Rother, 31 December 1842, Rep.109, Nr.3845, GStA Berlin. See also ibid., 282.

[98] Cited in Treitschke, *Deutsche Geschichte*, 5:503.

[99] Schreiber, *Die preussischen Eisenbahnen und Ihr Verhältnis zum Staat 1834–1874* (Berlin: Verlag von Ernst & Korn, 1874), 17.

about 134 million thaler of net investment into their lines between 1841 and 1849—roughly 5.5 percent of all net investment in the kingdom. As the cars accelerated, the nature of modern industrialism became even clearer. Within five years, for instance, the share of producer goods in the industrial sector jumped from 11 percent to 16 percent—double the level of 1830. Once representing about 71 percent of industrial output, Prussia's linen and woolen business had fallen to 36 percent by 1842. Within a decade, moreover, agriculture's share of total investment declined from 57.8 percent to 28.6 percent as construction, transport, and industry increased proportionately.[100] As the expansion continued, Cassandras of the industrial age like Rother reiterated their warnings—now with striking effectiveness. The Seehandlung Chief pointed with anxious alarm to the scores of petitions for new licenses which lay before the Ministry of Finance and the frenzy of speculation in railway shares on the Berlin Stock Exchange.[101] The new minister, Ernst von Bodelschwingh, needed little convincing. Perhaps because Frederick William IV had not genuinely and thoroughly divorced himself from anti-industrial prejudices of the 1820s and was therefore frightened by the mighty industrial boom which his railroads were unleashing—he agreed.

The antirailroad ordinances of 1844 were the first result. An April decree condemned the disadvantageous, "harmful" flight of capital away from agriculture and solid industry and forbade the licensing of all railways which could not demonstrate that they served an "overwhelming common interest."[102] May ordinances banned civil servants from investing in railroads, forbade the flotation of new shares without ministerial permission, and outlawed trading in railroad futures. "It is a mistake," snapped Johannes Kupfer, owner of a sugar refinery in Berlin, "when a government believes it has a clearer and deeper insight than its subjects into things of this sort."[103] There can be no doubt that these bombshells drastically curbed railroad growth. Of the fifty-six concession requests lying before the ministry in April 1844, only seventeen were licensed, and most of these were insignificant feeder spurs or lines like the Minden-Cologne, Thuringian, and Stargard-Posen which were already sanctioned by the United Committees of 1842. The value of most railroad stocks plummeted so quickly that Bodelschwingh and Rother had to intervene in the market to prevent a general calamity. Still, railroad stocks

[100] Tilly, "Capital Formation in Germany," Mathias and Postan, *Cambridge Economic History* 7 (1): 416, 419, 427.

[101] Radtke, *Seehandlung*, 291–92.

[102] Cited in Leiskow, *Spekulation*, 12.

[103] Cited in Kaelble, *Berliner Unternehmer*, 239. For the remainder of this paragraph, see Leiskow, *Spekulation*, 12–15; Radtke, *Seehandlung*, 291–93; Schreiber, *Preussische Eisenbahnen*, 17.

averaged 15 percent lower in August 1844. Railway shares generated 9.16 million thaler less in 1845 than the previous year and the market never fully recovered. Frederick William had shown how quickly, in fact, man's hand could slow the cars.

The forces of reactionary monetarism soon moved forward to other anticapitalistic victories. In March 1845, the Ministry of Finance issued instructions that were designed to prevent a new Securities Law of November 1843 from accelerating economic growth. Officials threatened to refuse rights of incorporation under the law unless undertakings were "advantageous and worthy of promotion from the standpoint of the common good."[104] Investors in Cologne were among the first to suffer from the amended policy when the ministry refused to grant concessions for a corporate mining firm and a joint-stock bank.[105]

In April 1846, the cause of joint-stock banking received another major setback. Christian Rother's administration of the Royal Bank after the death of Karl Ferdinand Friese in 1837 has not figured prominently in this study.[106] It is sufficient to observe that his strict, proprietary control of the country's money supply was another aspect of the economic conservatism which characterized Prussia in the last fifteen years of the Pre-March. Not surprisingly, these policies received increasing criticism in business circles, especially during the industrial expansion of the 1840s. "Men who are educated for state service and are familiar with it," quipped Berlin's leading banker, Joseph Mendelssohn, "do not know the fabric of industry and are not accustomed to it in as exacting a manner as is absolutely necessary."[107] But Rother's most persistent critic was Ernst Gottfried von Bülow-Cummerow, a Pomeranian nobleman. In fact, Cummerow's plan for a private central bank intrigued the king and won the support of Hermann von Boyen, the Minister of War, as well as Boyen's nemesis in the army, Prince William (the king's brother). But an odd coalition of opponents kept Frederick William's ear. Rother repeated his fears about a runaway economy and was supported by Friedrich von Rönne, the conservative Head of the Board of Trade (see chap. 3), and General Ludwig Gustav von Thile, undoubtedly the most influential "zealot" (*Frömmler*) at court. It was Thile, the reader may recall, who

[104] Cited in Karl Rauch, "Die Aktienvereine in der geschichtlichen Entwicklung des Aktienrechts," *Zeitschrift der Savigny-Stiftung für Rechtsgeschichte*, Germanistische Abteilung 69 (1952): 284.

[105] For the Cologne setbacks, see Ralph Sonnemann and Hans-Reinhard Meissner, "Einige Rechtshistorische Aspekte der Industriellen Revolution im Mitteldeutschen Raum," *Jahrbuch für Wirtschaftsgeschichte* (1977/4): 89.

[106] For an evaluation of Prussian banking policy, see Tilly, *Financial Institutions*, passim.

[107] Cited in Kaelble, *Berliner Unternehmer*, 240.

regarded industry as a "cancerous growth on the land."[108] Joint-stock banking in Prussia would wait until Prince William and Bismarck took power.

Thile also did his best to warn Frederick William against the public financial dangers posed by the new state aid for railroads.[109] But here Frederick William drew the line, especially when it came to one particular railroad, the great "Ostbahn" stretching six hundred kilometers from Berlin to the Russian border. Farmers in the eastern provinces needed it; the army wanted it; and so it would be built.[110] Thus when private investment in conjunction with the state's railway sink fund proved inadequate, Frederick William decided in November 1846, to proceed with state construction. As is well known, the need to borrow 32 million thaler for this massive project triggered a constitutional crisis in the spring of 1847 when the first United Diet refused to sanction the necessary loans.[111] The state could not have its technology and its absolutism too.

Thus railroads were generating a tremendous amount of tension between Prussian officialdom and private industrialists. The Railroad Law of 1842 had preserved the buoyant, harmonious optimism generated by the ascension of Frederick William IV in 1840, but a falling out came abruptly and severely after 1844 as worried ministers implemented draconian monetary measures, forbade new railroads in most areas of the kingdom, yet expected private credit for the Ostbahn at no political expense.

It is important to realize, however, that the railroad crises of the 1840s were merely part of a wider struggle over the ownership and control of industrial technology. Never unequivocally behind private industry after Waterloo, the Prussian Army shifted toward autarky after 1830, constructing more of its own powder mills and rifle factories, otherwise bullying private owners for what it could not have. The Mining Corps also maintained an industrial empire which it knew how to preserve through successful bureaucratic infighting and timely technological modernization. Private efforts like deep-shaft mining were simultaneously frustrated by Corps harassment and delay. The Seehandlung's industrial ventures provoked similar controversies, especially after the late 1830s as Rother abandoned the Silesian bourgeoisie in a drive to revitalize that

[108] Cited in Alphons Thuns, *Die Industrie am Niederrhein und ihre Arbeiter* (Leipzig: Duncker & Humblot, 1879), 190. For the internal political struggle over the Royal Bank, see Varnhagen Diary II, 17–18 February 1846, 3:304–8; and Radtke, *Seehandlung*, 123–28.

[109] Treitschke, *Deutsche Geschichte*, 5:495.

[110] See the sources cited above in notes 88 and 89.

[111] See Richard Tilly, "The Political Economy of Public Finance and the Industrialization of Prussia, 1815–1866," *Journal of Economic History* 26 (1966), passim.

province's linen and woolen industries. And while Beuth may have criticized his friend's "Egyptian" approach to the promotion of technology, the Business Department's programs and politics also caused considerable friction. Many factory owners resented Beuth's arrogant insistence that they integrate his machines into their operations, while simultaneously preventing them through low tariffs from constructing the modern plants which he professed to desire for Prussia. Like railroads and parliaments, these were all highly charged, highly politicized issues.

Some accommodation between feuding elites would have to be reached if the kingdom was to survive the Great Power struggle in Europe.

Conclusion

A CRISIS WAS DEVELOPING among Prussian elites well before the firestorm of lower class revolution struck in 1848. Central to this worsening political relationship between bureaucrats, soldiers, and businessmen was a struggle over ownership, control, and promotion of the forces of production. With the empirical evidence of previous chapters behind us, it is now time to engage in a more extended and explicit discussion of the themes introduced at the outset. (1) What was the role of the Prussian state in early industrialization? (2) How did Prussia's technocrats view past, present, and future? (3) What can be said about the nature of state and society in the Prussian Pre-March? (4) Finally, how can these observations enrich our understanding of Prussia during the historically critical years between the Revolution of 1848 and the establishment of the German Empire in 1871?

. . . . .

Historians are sharply divided over the economic contribution of the state to early Prussian industrialization. One grouping of nationalist scholars, modernization theorists, and neo-Marxists argues for an indispensable role in the initiation of industrial growth, while another school of economic historians rejects these arguments. In beginning our quest for a more balanced assessment, it would be useful to begin with the quantitative record, then proceed in the next segment to qualitative considerations.

Wolfram Fischer stated his belief rather early in this debate that the state's share of total investment in many of the German states during the Pre-March exceeded the impressive level of 25 percent.[1] His assessment reflected the widespread belief among historians in the 1960s that the state had played an indispensable role in initiating industrialization. If we combine our figurings on Prussian state investment with the more recent statistical estimates of total net investment by Richard Tilly, Fischer's percentage seems far too high. Concentrated mainly in transportation, construction, and agriculture in the 1820s, Berlin's outlays amounted to about 38 million thaler or 13 percent of the total. With around 47 million thaler of net investment in the expanding economy of the 1830s, the state's share slipped to 6.2 percent. Shifting somewhat to transportation,

[1] Fischer, "The Strategy," 436.

construction, and industry in the 1840s, Borussia's contribution rose to about 57 million thaler, but declined in percentage terms to 2.3 percent in the burgeoning railroad age.[2] Although the industrial sector was increasing only slightly in importance—from 2.2 percent of total net investment in the 1810s to 3.3 percent in the 1840s—the state's share in this sector was rising very rapidly from 4–5 percent in the early 1820s to 11–12 percent during the 1830s and 1840s, the era of Seehandlung, Army, and Mining Corps expansion. If we isolate the years just before 1850, net state investment in the industrial sector was about 17 percent of the total. While this amount was significant and growing, it was not as massive, even in this smaller sector, as Fischer posited. It was great enough, on the other hand, to generate private sector criticism of state intervention in the industrial economy.

The direct contribution of the state to investment seems even less if we employ "counterfactual" analysis—that is, if we take a look at "hypothetical alternatives." Obviously there is no way to determine how much private investment was *thwarted* by state policies, but collectively they probably took a huge toll. Investors waited for over two years (1835–

[2] Richard H. Tilly, "The 'Take-Off' in Germany," in Erich Angermann and Marie-Luise Frings, *Oceans Apart: Comparing Germany and the United States* (Stuttgart: Klett-Cotta, 1980), 52 n.15, gives total investment figures in current marks which combine easily with our figurings when divided by three for thalers. I have used the older and less tentative figures he cites here; use of the newer, radically higher figures presses the state percentage even lower.

For state investment in the 1820s in current prices, see (A) Tilly, "Capital Formation in Germany," Mathias and Postan, *Cambridge Economic History* 7 (1): 412–13—for 23.27 million thaler in roads and waterways; (B) Pruns, *Staat und Agrarwirtschaft*, 1:291–96—for 3 million thaler of Seehandlung investments in agriculture; and (C) Treitschke, *Deutsche Geschichte*, 3:464—for 11.72 million thaler of all investments outside of transportation—most of which must have gone into construction of government buildings (see Tilly, "Capital Formation," 435, for the fact that 15,844 government buildings were constructed between 1819 and 1834).

Treitschke cites a privy cabinet document of 1835 which gives an actual total of 27.9 million thaler for the years 1820–1834. I have used Treitschke's annual average of 1.86 million thaler, multiplied by 0.63 to arrive at annual net investment (1.172 million thaler) for 1820–34. I have used the annual average of 2.17 million, reflecting the higher budgets of the 1830s and 1840s (Wehler, *Gesellschaftsgeschichte*, 2:376), multiplied by 0.65 to arrive at net investment (1.41 million thaler) for the years 1835–1849.

For state investment in the 1830s in current prices, see (A) Tilly, "Capital Formation"—for 31.91 million in roads and waterways; (B) chap. 6 above—for 1.76 million thaler of Seehandlung industrial investments (2.8 x 0.628); and (C) 13.15 million thaler for state investments outside of transportation.

For state investment in the 1840s in current prices, see (A) Tilly, "Capital Formation"—for 33.85 million thaler in roads and waterways; (B) Klee, *Preussische Eisenbahngeschichte*, 108—for 3.83 million thaler in railroads; (C) 14.10 million for state investments outside of transportation; and (D) chap. 6 above—for 4.99 million thaler of Seehandlung industrial investments (7.7 × 0.648).

1837/38), for instance, while the Prussian bureaucracy fought over railroad policy. The postponed construction of 470 kilometers of line cost the nation at least 10 million thaler for every year of the delay, annually exceeding the likely amount of state investment by 7–8 million.[3] It is also reasonable to blame the antirailroad laws of 1844 for a large share of that year's 9.16 million thaler shortfall in purchases of railroad stocks over the previous year. We have argued repeatedly, moreover, that Prussia's unwillingness to protect "infant industries" forestalled investment in iron and textiles. And there are other significant examples. Army secrecy; the competition of state enterprises; and throttling mining and monetary policies accounted for additional millions which were lost. Perhaps on balance the state's investment contribution was still a positive one, but we would do well not to exaggerate the net gain.

If we consider what private railroad companies were accomplishing, moreover, we can place this "gain" in perspective. Indeed, the railroads assumed a tremendous economic significance during the 1840s. With no state subsidization at first, Prussian railroad companies sank 134 million thaler of net investment into their lines in the 1840s—or about 5.5 percent of the Prussian investment total. The iron horses were also major employers. Using a "multiplier-effect" model, Rainer Fremdling estimates that the 17–22 million thaler paid out annually in wages to railway crews generated over 35 percent of incremental income in Germany from 1845 to 1849. By this time, admittedly, the Prussian government was purchasing between 6–8 percent of all railway shares, thereby slightly enhancing its own contribution to national investment.[4] As we have seen, moreover, the state performed economic and political spadework by creating the Zollverein and passing the first railroad laws (1838 and 1842). While the state certainly lent a hand, however, it was primarily the private railroad companies which pulled Prussia into the modern industrial age. There is little irony here, for the various agencies and factions which made up the Prussian state never intended to create a "modern industrial age" in the first place.

. . . . .

It is mainly in this area of alternative visions that our study has attempted to reshape thinking and enhance understanding of the Pre-March. Indeed, it is misleading—and somehow unfair—to praise or criticize the

[3] See chap. 7, n. 73.

[4] Tilly, "Capital Formation in Germany," Mathias and Postan, *Cambridge Economic History* 7 (1): 416, 427; and Tilly, "The 'Take-Off' in Germany," 52. I have subtracted 3.83 million thaler of state railway investment from the Prussian total given in Tilly, "Capital Formation."

programs of Prussian bureaucrats without carefully considering the original goals and mentalities which stood behind these policies. Instead of judging Prussia's soldiers and civilians with posterity's criteria, in other words, we should include in our assessment the types of questions which contemporaries asked of themselves. To what extent did the competing governmental factions which have been the central focus of this study achieve their own goals? Second, did these policies trigger, intentionally or unintentionally, what we understand today by industrialization?

The economic liberals in Hardenberg's chancellery and the Business Department envisioned a free, productive, rural-oriented society. Modern, mechanized factories and workshops would emerge gradually but steadily in every corner of a kingdom shorn of its medieval social and political trappings. Parliamentary institutions were to advance with country industry, while many in the Business Department wanted technocrats to wield greater influence in the new state. Hardenberg and his men were able, in fact, to legislate a great many of their ideas before 1848. Agrarian reform, abolition of the guilds, and other economic measures during the chancellor's first flurry of legislation (1810–1811) led after Vienna to new patent (1815), tariff (1818), and customs union (1828–1834) policies. To this we must add perennial road-building activity as well as the technological and educational empire which Peter Beuth created in the 1820s. In Beuth's case, we have a unique variation on the theme of rural industrialization—a highbrow, aesthetically pleasing industrialization.

But frustrations mounted along the way. The partial rescinding of agrarian reform in 1816, the compromises which were made over tariffs in 1818, the abolition of the Ministry of Business and Commerce in 1825, and the repeated failure of technocratic and parliamentary schemes were major political setbacks. The restoration of the Mining Corps and Seehandlung to their former influence created headaches of a similar sort as neo-mercantilists struggled successfully for intrastate influence and angered businessmen who felt that they were ready to fend for themselves. After Maassen's death in 1834, moreover, a more conservative Ministerial Cabinet stiffened guild reform provisions and rejected a more liberal set of mining laws.

Beuth's oft-praised transfer of British and American textile machinery and machine tools was undoubtedly a positive contribution to the Prussian economy. In many years, machinery worth 30,000–40,000 thaler was donated or lent to the business sector. But the swimming instructor, as Ulrich Peter Ritter described him, was struggling against a strong tide. His gifts of machines were not always appreciated, overshadowed in many cases by anger and disappointment over tariffs. A new, modern Prussia was rising, yet it bore increasingly little resemblance to the country envisioned by Beuth and an earlier generation of technocrats. Linen and woolen goods were yielding to cotton; heavy mining, metallurgy, and producer goods were beginning to overshadow light manufacturing and railroads were rapidly absorbing capital from agriculture and other preferred forms of investment.

Statistics allow us to measure what the future held in store. Heavy industry's share of industrial output doubled from 8 percent to 16 percent from the 1830s to the 1840s, then increased to 26 percent by the 1870s and 36 percent by 1913. The industrial sector was absorbing a greater portion of the nation's net investment, meanwhile, increasing from the 2–3 percent range of the Pre-March to 13–22 percent in the 1850s and around 43 percent after 1900. Net investment, moreover, represented a greater share of national income, doubling from 8 percent at midcentury to 16 percent on the eve of World War I.[5]

We have seen that Prussia's leaders sensed the general trend and did not like it. The greatest frustration of all, perhaps, was the fact that their own policies had helped to open Pandora's box. The agrarian reforms were not completed after 1816, but they created a broad strata of prosperous eastern farmers whose purchases fueled rapid industrial growth in Berlin, a city which Hardenberg's team had wanted to downsize.[6] Moderate textile tariffs undercut rural industrialization by facilitating entry of foreign yarns woven in Prussian finishing factories whose owners had no incentive to locate in the countryside near flax, wool, and other domestic raw materials. Laws designed to further rural industry by taxing it at lower rates could not counterbalance the effect of tariffs, but provided an ideal fiscal environment for modern heavy industry by capping urban taxes at a moderate absolute amount which decreased in percentage terms as profits increased.[7] The abolition of the guilds and successful parrying of guild restoration efforts did not produce the reformers' ideal society, but quickened modern industrial trends by accelerating the

[5] Friedrich-Wilhelm Henning, *Die Industrialisierung in Deutschland 1800–1914* (Paderborn: Verlag Ferdinand Schöningh, 1973), 137; Tilly, "Capital Formation," 419; Walther Hoffmann, *Das Wachstum der Deutschen Wirtschaft seit der Mitte des 19. Jahrhundert* (Berlin: Springer Verlag, 1965), 104, 143.

[6] Hartmut Harnisch, "Die Bedeutung der kapitalistischen Agrarreform für die Herausbildung des inneren Marktes und die industrielle Revolution in den östlichen Provinzen Preussens in der ersten Hälfte des 19. Jahrhundert," *Jahrbuch für Wirtschaftsgeschichte* (1977): 63–82. See also Harnisch's more recent formulation of the argument, "Zum Stand der Diskussion um die Probleme des 'preussischen Weges' kapitalistischer Agrarentwicklung in der deutschen Geschichte," in Gustav Seeber and Karl-Heinz Noack, *Preussen in der deutschen Geschichte nach 1789* (Berlin: Akademie-Verlag, 1983), 135–38.

[7] W. R. Lee, "Tax Structure and Economic Growth in Germany 1750–1850," *Journal of European Economic History* 4 (1975): 164; Vogel, *Allgemeine Gewerbefreiheit*, 183, 230.

death of an antitechnological institution.[8] There can be no doubt, moreover, that the elimination of internal toll barriers, completion of a road network, and founding of the Zollverein created the market prerequisites for something akin to "industrial take-off." In fact, these market improvements may have stimulated the first railroad promotion schemes.[9] By refusing to promote railways or revamp outmoded monetary institutions and policies during the 1830s and 1840s, the "liberals" were attempting to slam Pandora's box shut. These later efforts certainly retarded industrial growth and, in so doing, contributed to a worsening political tension.

The top priority of the Mining Corps during the first years after Jena was organizational survival. Ludewig Gerhard and his fellow captains accomplished this goal quickly, warding off many threats from economic liberals in the 1810s and early 1820s. One legacy of this effort, however, was the constant need to highlight the Corps' public service role. This presented no problem to the mining section, for its technological and managerial services were largely appreciated by small shareholders whose main source of livelihood lay outside mining. Indeed, the early 1800s were good years for the Corps and its sector of the Prussian economy. One mining center after another was restored to its former prosperity as tunnel-mining techniques perfected centuries earlier in central Europe were introduced by the proud black coats. Like medieval monks copying priceless manuscripts, the "mining clergy" preserved, perpetuated, and occasionally improved upon a centuries-old technology. The metallurgical record was also impressive. Early Modern *charcoal* ironmaking techniques were vastly improved in the Corps' model works, then transferred to private producers who had shared none of the risks and costs of these innovations. It is true that daring—albeit self-serving—attempts to spread coke smelting technology were rebuffed by ironmasters who knew that iron quality suffered; but the organization's widely emulated coking ovens were for a time the best in Europe.

This bright portrait began to darken during the 1830s. The captains found themselves scrambling to catch up with Rhenish and Silesian iron firms which had perfected coke puddling techniques. By succeeding in doing so, Karl Karsten generated considerable friction in turn with private industrialists like Count Hugo Henckel von Donnersmark. By this time the mining section was reacting with equal anxiety to Franz Haniel and other coal merchants who wanted to sink deep, vertical shafts into the earth's finite resources of ore and coal. The Corps' campaign of harassment and discrimination against mining tycoons—one which owed

[8] Vogel, *Allgemeine Gewerbefreiheit*, 161.

[9] Wehler, *Gesellschaftsgeschichte*, 2:132–39, especially p. 134.

part of its determination to anger over an unforgivable liberal challenge from within—was deeply resented by entrepreneurs whose resources were at great risk. When the 1840s dawned, the Corps was no longer a boon to economic and technological progress in Prussia. Reacting negatively to deep shafts, railroads, and deregulatory mining laws, the captains also had a trembling hand on the lid of Pandora's box.

The Seehandlung was another postwar success story. Under Rother's talented leadership, a public financial empire came to life in less than a decade. By the early 1830s, his organization was beginning to make its presence known on the economic scene with oceangoing vessels, river steamers, flour and paper mills, chemical works, machine-tool plants, road construction, and a wool purchasing and resale business. Rother's humanitarian complex of enterprises was never entirely free of controversy, for many of these early projects stood in direct competition with private businessmen. His road-building contracts of the 1820s, for example, were won by outbidding private investors.

After the peasant's son joined Beuth in opposing railroads, however, his economic role turned more negative. And as it did, the grumbling of industrialists who saw the state increasingly as a barrier to their economic and political fortunes increased. When Rother placed the massive resources of the Seehandlung behind the heartfelt cause of Silesia's dying linen industry, entrepreneurial passions spilled over from the private to the public arena as numerous scathing indictments of the Seehandlung were published. Motivated by the progressive, charitable impulses emanating from Freemasonry, Rother's huge textile investments accomplished one goal by employing thousands of poor Silesians. But his controversial factories were overcapitalized, inefficient, and ultimately unsuccessful. Most of Rother's works passed into private hands after 1848.

For many years after 1815 the Prussian Army was ambivalent about industrialization as well as the military use of modern technology. Beginning with the series of crises which spread across Europe in the late 1820s, however, options became more limited. The military grew more appreciative of the need to modernize and less willing to rely on outsiders for the supply of new ideas and weapons. Feelings remained mixed as witnessed by the tendency to revel in the glory of ancient warriors. Nevertheless, army investment in powder mills and gun factories increased. The benefits of this expenditure to the private sector remained slight—especially in the area of machine tools—due to the secretiveness of army operations.

By warming to railroads after 1836, however, the military played a major part in furthering industrialization throughout the land. Indeed, although it was arrogant, demanding, and heavy-handed about it—a haughty demeanor which alienated private railroad promoters—the Prussian Army was still the most significant force within government promoting the railroad cause during the late 1830s. Like the defeat of guild restoration efforts in the 1820s, moreover, the historical importance of Frederick William III's decision to back his soldiers and his son and move cautiously forward with railroad construction in 1838 cannot be overestimated. Frederick William IV's enthusiasm for the iron horse wavered in the 1840s as the hectic pace of development lent credibility to Rother's arguments. But government purchase of shares in lines of strategic importance never stopped after 1843.[10]

It is not a simple matter, therefore, to draw conclusions about the role of the state during early industrialization in Prussia. While the direct contribution of the state in the form of investments was quantitatively slight, the qualitative record was better—but clearly mixed. Hardenberg's staff, the Business Department, the Mining Corps, and the Seehandlung made highly important contributions early in the process, then ceased to play positive roles on balance in the 1830s and 1840s as economic reality departed radically from their visions. Ironically, industrialization received renewed positive impulses in the last decade of the Pre-March from an army which was highly suspicious of private industry. Prussia's warriors nevertheless facilitated railroad legislation in 1838 and 1842 because they wanted to exploit the technology which was accelerating industry's growth.

As the veil lifted from industry revealing its modern face, attitudes toward contemporary and ancient times changed. During the exhilarating decades when dreams were still a potential reality, Prussia remained in antiquity's shadow in the minds of her most progressive officials. Respect for the world of the heroes was both an indication of humility for the present age and a sign of ambition for change in the future. "Our new world is really nothing at all," Wilhelm von Humboldt had written. "It consists of nothing more than a yearning for the past and an always uncertain groping for that which is yet to be created." By immersing themselves and their students in the languages and literature of antiquity, advocates of classical education like Humboldt, Wolf, Schulze, Rühle, and Peucker wanted to build a liberal Prussia worthy of the ancient freedoms and constitutions. Similarly, Hermann von Boyen urged cadets to study Greek and Roman history as a means to "carry great ideals" and comport themselves bravely in future wars for the Fatherland. Peter Beuth wanted to recapture the greatness of antiquity through a combination of industrial art forms and classical education, using both to elevate the middle class aesthetically and politically. The captains of the Mining

[10] Fremdling, *Eisenbahnen*, 126.

Corps strove to rebuild the country's oldest industry and measured their achievement against the architectural wonders of the ancient world. The Freemasons sought to find and preserve the scientific mysteries of Classical Greece and other ancient civilizations in order to advance society in the modern era.

By the 1830s, however, the outlines of a modern world more impressive than the old were evident. As Prussia emerged from the shadow of antiquity, some, like Heinrich von Brandt's students in the War School, appeared "completely indifferent" to the ancients. For others, modern developments spawned hyperbole for past accomplishments. "The precision and solidity of Roman road-building," observed a military engineer in 1842, "still arouse the admiration of all the experts. Even the most ingenious works of our modern railroads," he continued, "hardly withstand comparison with that magnificent road construction."[11] We have seen how pronounced this exaggerated, escapist pattern became among the generation of officers who had fought Napoleon. Similarly, captains of the Mining Corps like Ernst von Beust reacted to modernity by retreating into "the old ways" protected by a *Direktionsprincip* of an earlier age. Civilian officials like Peter Beuth also drifted backward, in his case into a world where ancient standards of artistic excellence were a barricade against unpleasant modern realities. Theodor von Schön, the first Head of the Business Department, commented in 1835 on this tendency among his colleagues. While Schön felt the widespread praise for antiquity was generally exaggerated, he admitted that a period of decline had set in after the death of Christ—with no times as "vulgar"[12] as the present.

Ever sensitive to his epoch, Varnhagen von Ense made a similar observation in 1838. Having visited the construction site of the Berlin-Potsdam railway line, he confided to his diary that "the world will change if these continue to spread! One can see it!"[13] In fact, the railroads were only one aspect of a modern world that seemed increasingly strange and foreign to aging progressives like Varnhagen. As street violence, popular demagoguery, atheism, and anticlassicism grew more intense, he withdrew into an intellectual haven which was marked off by the classics he loved to read. Homer, Xenophon, Caesar, Tacitus, the Bible, and Goethe, who had devoted much of his career to praise of antiquity, brought him great peace of mind. But realities always interfered. "The world is laboring greatly," he wrote in 1838, "to remove two of the main pillars of

[11] Karl Eduard Pönitz, *Die Eisenbahnen als militärische Operationslinien betrachtet und durch Beispiele erläutert* (Adolf: Verlags-Bureau, 1842), 17.

[12] Schön to Stägemann, 27 June 1835, Rühl, *Briefe*, 3:560.

[13] For the quotes in this paragraph, see, respectively, Varnhagen Diary II, 23 June 1838, 1:102 and 19 November 1838, 1:114.

life as we have known it—the Bible and [the works of] classical antiquity." When these props fell, as he believed they must, "then we will see one of the greatest revolutions, the first stages of which have already begun." Varnhagen was thinking mainly of a revolution in mentality—from the humble, scholarly regard for the past which characterized his generation, to the arrogant intoxication with the present which he felt all around him. Waxing more optimistic and contradictory at the end of the same entry, the diarist noted that "the Bible and the classics will be pushed aside, but never pass away—on the contrary, they will convert new brothers and sisters." But he must have known that the new, modern weltanschauung, like the railroads which were so inextricably bound up with it, was more potent.

One could see it. Already growing much faster than classical gymnasiums in Varnhagen's time, modern, vocational secondary schools spread to every corner of Prussia after 1871 as parents and administrators opted for practical curricula more attuned to the needs of the marketplace. Numbering only 51 in 1853, there were 523 of these schools in 1911. As we have seen, moreover, the Prussian army turned from humanistic to technical coursework after Peucker's rearguard defense of the classics as Inspector of Military Schools (1854–1867). It was perhaps some vindication for Varnhagen's predictions that Prussia's classical gymnasiums also expanded from 149 in 1853 to an impressive 367 in 1911. It warrants repeating, however, that antiquity was no longer a source of progressive inspiration. For men who became content with the social and political gains of the cultivated bourgeoisie in the second half of the century (see below), but recoiled, like Beuth and Varnhagen, from modern industrialism, study of the ancients was transformed from a politically emancipatory to a politically sterile exercise which served a reactionary, anti-industrial function. Small wonder that student responses evolved from the indifference which Brandt observed to the widespread alienation evident by 1900.[14]

. . . . .

Although closely related, the debate among historians concerning the nature of the state in the Prussian Pre-March is in many ways more complicated than the discussion over the state's economic and technological

[14] For the statistics, see Petersdorff, *Friedrich von Motz*, 2:353; and James C. Albisetti, *Secondary School Reform in Imperial Germany* (Princeton, New Jersey: Princeton University Press. 1983), 288. In general, see Albisetti, *Secondary School Reform*, 43–118, 306–13. For the argument that the transition from liberal to conservative classicism was facilitated by the state—as opposed to the inner transition emphasized by the present study—see Peter Jelavich, *Munich and Theatrical Modernism: Politics, Playwriting, and Performance 1890–1914* (Cambridge, Massachusetts: Harvard University Press, 1985), 15–21.

role. For this debate has been tinged by much more overtly ideological and political beliefs about the course of Prussian and German history in the nineteenth and twentieth centuries. According to many historians of different political persuasions, Prussia before 1848 was a state controlled by a dominant aristocracy which acted in its own interests—a *Junkerstaat*. Others have challenged this model, preferring that of an autonomous state which cut across outside class interests. These historians project the image of a bureaucratic estate—or *Beamtenstand*—occupying its governing citadel. It is now the appropriate time in our study to draw together the evidence of earlier chapters, and in so doing, offer a more extended critique of the reigning models. Finally, we shall build on previous chapters and proceed to an alternative view of the state.

The argument for an aristocratic state weakens when the declining social and economic position of the nobility is considered. As we observed in chapter 1, the Junker was losing his land in the eighteenth and nineteenth centuries at an alarming rate. As he did, bourgeois owners swept into the vacuum, purchasing 42 percent of noble estates by 1856. The wealthiest aristocratic families and clans also benefitted from the rapid turnover of property, but they no longer dominated the state and it did not act exclusively in their interests. While it is true that many of Hardenberg's frustrations resulted from a political tack which ran against the wind of "the great families,"[15] as one contemporary put it, aristocratic victories like the partial rescinding of peasant emancipation, the preservation of aristocratic tax exemption, and the maintenance of local police rights must be viewed in a larger context. We should not forget that Hohenzollern monarchs had succeeded in enervating former Junker strongholds like the old diets by the late seventeenth century. The attempts of the influential clans to revive these estatist institutions after Jena—and again in the early 1820s—were largely unsuccessful.[16] Simultaneous efforts to implement entailment laws, revive guild privileges, and undercut Hardenberg's industrial legislation of 1810/11 were complete failures. This pattern of rearguard victories amidst general political retreat is consistent with the loss of economic predominance presented above.

A major problem facing the aristocrats was the lack of sympathy shown

[15] Benzenberg to Gneisenau, 3 January 1823, printed in Heyderhoff, *Benzenberg*, 153.

[16] Wehler, *Gesellschaftsgeschichte*, 2:156–58, stands practically alone in emphasizing the aristocratic gains which came with the restored diets. For more pessimistic accounts, see Helmuth Croon, "Die Provinziallandtage im Vormärz unter besonderer Berücksichtigung der Rheinprovinz und der Provinz Westfalen," in Peter Baumgart (ed.), *Ständetum und Staatsbildung in Brandenburg-Preussen: Ergebnisse einer Internationalen Fachtagung* (Berlin: Walter de Gruyter, 1983), 456–59, 475–77; Manfred Botzenhart, "Verfassungsproblematik und Ständepolitik in der preussischen Reformzeit," in Baumgart, *Ständetum*, 432–37, 450; and Koselleck, *Preussen*, 337–46.

for their cause by Frederick William III. This deeply troubled monarch never completely abandoned his nobles, leaving many of them in high office, appropriating millions of thaler to bolster the sagging economic position of others. As we saw in chapter 1, however, he also played an important role in allowing the new, unpredictable industrial age to advance unchecked by advocates of the guilds and defenders of a medieval restoration. Indeed, no Hohenzollern resented aristocratic pretense and resisted estatist intrigues as consistently as this unpretentious occupant of the throne. Nor did the proudest Junker families forgive him the disrespectful caprice with which he flaunted class distinctions in public, opened court, high society, and the state to the middle classes, and ennobled scores of commoners. While Frederick the Great raised only a few persons from the lower estates to the nobility, for instance, his irreverent great nephew elevated 140 between 1797 and 1806 alone.[17] Together with chancellors and privy councillors who sometimes promoted separate agendas, Frederick William III attempted to mold and maintain control of the "ruling class" (Mosca) for purposes he deemed worthy, not exclusively for the interests of Junkers.

There were other factors working against the success of reactionary politics. Many of the nobility's best talents, for example, were lost to competing causes. Stein and Hardenberg, while not Junkers, were German noblemen who turned their backs on class considerations to erase feudalistic and mercantilistic institutions restricting agriculture, commerce, and industry. Friedrich von Motz and Ludwig von Vincke moved in the same direction, promoting the Zollverein, railroads, metallurgy, and innovative machinery. Varnhagen von Ense described recently elevated noblemen like Schön and Stägemann as well as scions of older aristocratic families like Motz, Vincke, and Rühle von Lilienstern as noblemen who were "decidedly bourgeois-minded" (*entschieden bürgerlich* gesinnt).[18] And much the same could be said for Mining Corps blue bloods like Friedrich Wilhelm von Reden, Ernst von Beust, Toussaint von Charpentier, and the young Alexander von Humboldt. We are reminded here of Antonio Gramsci's notion of an "aristocratic vanguard" which abandons its own class and forms a leadership cadre for less experienced members of the industrial bourgeoisie.[19]

[17] Rochow, *Vom Leben*, 87–88, 131–32, 135; Koselleck, *Preussen*, 678.

[18] Varnhagen Diary I, 5 February 1825, 3:230.

[19] Antonio Gramsci, *Selections from the Prison Notebooks*, edited and translated by Quintin Hoare and Geoffrey N. Smith (New York: International Publishers, 1971), 19, 106–20, 269–70. In general, see Anne Showstack Sassoon (ed.), *Approaches to Gramsci* (London: Writers and Readers, 1982); and Walter L. Adamson, *Hegemony and Revolution: A Study of Antonio Gramsci's Political and Cultural Theory* (Berkeley, California: University of California Press, 1980).

One weakness of Gramsci's approach, however, is that it overlooks the extent of bourgeois penetration of the state. Thus Stein and Hardenberg relied very heavily on technocrats like Theodor von Schön, Christian Scharnweber, Albrecht Thaer, Johann Hoffmann, Johann Sack, and Friedrich August [von] Stägemann. Motz usually turned for advice to his successor as Minister of Finance, Georg [von] Maassen; to Peter Beuth and Christian Kunth of the Business Department; or to Ludewig Gerhard and Carl Karsten of the Mining Corps. Christian [von] Rother was another valued commoner within the vanguard. It was "the higher purpose" of the Seehandlung, wrote Vincke to him admiringly, to lift Germans "out of their misery."[20] The reader will recall (chap. 1) that bourgeois membership in the ministries, institutes, agencies, corps, and services which made up the state stood, variously, between 46–100 percent. When we consider all of these arguments, the concept of Prussia as a *Junkerstaat* in the Pre-March loses much of its meaning.

In turning to the thesis of a monolithic bureaucratic estate, it is important to begin with the intrastate fragmentation which has highlighted our analysis. Permitted by an often passive monarch who was wedded to no single school of political economy, internal wrangling characterized relations within government in the Pre-March. The Business Department conducted a running battle with the Mining Corps from the 1810s into the 1830s, eventually intriguing within the Corps to liberalize mining law. On occasion Beuth also butted heads with the Seehandlung as witnessed by his disagreement with Rother over promotion of the Silesian linen industry after 1839. Army secretiveness worked at cross-purposes, moreover, with the Business Department's campaign of diffusing technology widely throughout the kingdom, while the Mining Corps was also treated harshly by soldiers who wanted to control the forces of production. It was the army, finally, which pushed for railroads over the objections of the other technical agencies. While this study has not engaged in a comprehensive analysis of the scores of departments and divisions which comprised the state, our examples of internal discord are no doubt representative. Thus Varnhagen, a man who had observed the internal workings of state for decades, summarized his observations with a metaphor. The bureaucracy reminded him of "the many petty states, imperial knights, and corporate entities" of the Holy Roman Empire. "It is a war of all against all—a feud between the parts. . . . Most of the best and noblest forces in the state are consumed by this internal struggle."[21] As we shall see below, this fragmentation was not limited to the departmentalism described here.

[20] Vincke to Rother, 28 June 1824, Rep.92 Rother, Nr.18, ZStA Merseburg.

[21] Varnhagen Diary I, 1 March 1825, 3:244–45.

It is also important to appreciate class divisions within the state. Of the numerous allegiances (see below) which competed with loyalty to the bureaucracy as a monolithic whole, consciousness of class or estate was one of the most potent.[22] "It is the rule," Wilhelm Riehl wrote in 1851, "that people move from every real [social] estate into the so-called *Beamtenstand*." Underlying class differences always persisted, however, because they were "so deeply embedded in a person's inner nature." A bourgeois civil servant could let himself be ennobled, "but in a social sense he will very seldom become a nobleman." It was even more difficult, he concluded, "for a nobleman to become bourgeois."[23]

Riehl's statements have considerable merit. Thus while Beuth could bridge the class divide with aristocratic friends like Vincke and Humboldt, this does not appear to have been the case with Beuth and a close associate like Motz. Similarly, Stägemann had close ties with Boyen from their university days, but was not sure he could trust the politics of moderates like Vincke and Ferdinand von Schönberg. "After all," he wrote a bourgeois friend, "they are also noblemen." Even Stein, his former political mentor, had "always been an aristocrat, although in a better sense [of the word]."[24] Johann [von] Krauseneck provides another good example. The longtime Chief of the General Staff harbored grudges for decades against perceived class slurs from noble officers closely aligned to him politically and only reluctantly allowed himself to be ennobled. Unlike many members of the Mining Corps, moreover, Carl Karsten never felt comfortable with aristocrats and was deeply resentful when Count Einsiedel was chosen to head the Silesian Mine Office.[25] Instances like this lend weight to Joseph Schumpeter's observations concerning class interaction:

> Social intercourse within class barriers is promoted by the similarity of manners and habits of life, of things that are evaluated in a positive or negative sense, that arouse interest. In intercourse across class borders, differences

[22] For a contrasting view, see Gillis, *Prussian Bureaucracy*, 29–30. Although an adherent of the Beamtenstand model, Gillis's work can be seen as an early western departure from this school, for he highlights the conflict between youth and age in the bureaucracy and rejects "the oversimplified, organic image of the Prussian civil service presented in much of the historical literature" (p. 212).

[23] Wilhelm Heinrich Riehl, *Die Bürgerliche Gesellschaft* (Stuttgart: J. G. Cotta, 1851), 233–34.

[24] For Beuth's relationship with Motz, see Petersdorff, *Friedrich von Motz*, 2:206, 244, 312, 384. For Stägemann's remarks, see, respectively, Stägemann to Cramer, 13 February 1822, printed in Assing, *Briefe an Chamisso*, 2:77; and Stägemann to Benzenberg, 23 June 1827, printed in Rühl, *Briefe*, 3:360.

[25] For Krauseneck, see Brandt, *Aus dem Leben*, 2:31–32; and E. Folgermann, *Der General der Infanterie von Krauseneck: Ein Lebensabriss* (Berlin: E. S. Mittler, 1852), 74. For Karsten, see Karsten, *Umrisse*, 88–90.

> on all these points repel and inhibit sympathy. There are always a number of delicate matters that must be avoided, things that seem strange and even absurd to the other classes. The participants in social intercourse between different classes are always on their best behavior, so to speak, making their conduct forced and unnatural. The difference between intercourse within the class and outside the class is the same as the difference between swimming with and against the tide.[26]

We should conclude, therefore, that unless successfully suppressed by friendship, fraternity, or another more potent allegiance, class differences could surface, cleaving the bureaucracy in a meaningful way. Within the vanguard, nobleman and commoner sometimes worked toward their joint goals perceiving themselves as allied elements of different classes.

Another drawback of the bureaucratic estate model concerns the assumed relationship between state and society. While adherents of the *Beamtenstand* emphasize its tax and pension privileges and its aloofness from the people, this study has drawn attention to the ties between bureaucrats and Prussia's struggling entrepreneurial elite. Private correspondence, bureaucratic tours of the provinces, frequent visits between businessmen and industrial councillors, common discussion groups, overlapping friendship circles, masonic connections, occasional intermarriage, joint lobbying efforts, and a common belief in constitutionalism—all this brought state and civil society closer together. "The real democrats," recalled Heinrich von Brandt, were found in the bureaucracy, for the bureaucrats "were in constant touch with the people."[27]

Class factions also extended hands within the army's industrial establishment, in joint ventures of the Seehandlung, during the showcase operations and regulatory activities of the Mining Corps, and in the model factory efforts of the Business Department. Beuth's Association for the Promotion of Technical Activity in Prussia was even more important in this regard. Joining over nine hundred bureaucrats and businessmen of varied sorts, it gave solid institutional form to the state-society nexus. While often feuding bitterly among themselves over the means to modernize industry during the 1810s and 1820s, the bureaucrats nevertheless preserved their senior station in this partnership, respected as they were by extragovernmental subordinates with less education, expertise, and social status. Beginning in the 1830s, however, ugly disputes over tariffs, deep-shaft mining, state-owned technology, army secretiveness, and rail-

[26] Joseph Schumpeter, "The Problem of Classes," printed in Reinhard Bendix and Seymour Martin Lipset (eds.), *Class, Status, and Power: Social Stratification in Comparative Perspective* (New York: The Free Press, 1966), 43.

[27] Brandt, *Aus dem Leben*, 2:54.

roads weakened bonds. By the 1840s, state aid and economic intervention were perceived widely in business circles as unnecessary, inappropriate, and unacceptable. The Prussian bourgeoisie felt ready to overtake its vanguard.

Thus, instead of an aristocratic vanguard, it is more meaningful to posit a state vanguard which was comprised of separate bourgeois and aristocratic elements temporarily allied, in turn, to junior or subordinate bourgeois partners outside of government. The king, joined sometimes by the chancellor or chief privy councillor, presided above the fray, attempting to preserve royal prerogatives and mediate between conflicting forces inside and outside of government. It has not been the purpose of this study, however, to offer an amalgam of sorts between the two reigning models—bourgeois mixing with nobleman in a class alliance under an autonomous "bourgeois" monarch. Rather, it is presented as a basic framework for understanding a state whose structure was really much more complex.

We have seen that the cultivated middle class (*Bildungsbürgertum*) was divided along numerous professional and vocational fault lines. Each of these "moral forces" (Mosca) formed its own distinct enclave, variously coexisting and competing with other occupational class factions within the middle class and within the state. Thus Beuth's cameralists were unable to restructure bureaucracy to serve businessmen. His proposals for technocratic reform crashed repeatedly against the imposing "formal culture" (Collins) of a legal profession which successfully asserted its claim to intellectual supremacy. Altenstein's attempt to place liberal arts credentials on a par with legal degrees as training for state service also failed completely. After the late 1820s, moreover, the philologists around Johannes Schulze encountered stiff opposition from the natural scientists. Justus Liebig, Prussia's famous chemist, equated efforts to promote classical education with "the fight of soap-makers against gas lights, the protestations of innkeepers against express mail coaches or the teamsters against canals and railroads."[28] Long before he made this comparison (1840), allies of science and industry like Alexander von Humboldt had found the ear of Frederick William III and were founding two to three *Realschulen* for every new gymnasium. In assessing this intracultural friction, there is every reason to conclude that class struggle was becoming "irreparably multi-sided" (Collins) as factional energies were invested in establishing and guarding enclaves or dissipated in feuding with rivals in and out of government.

As noted above, Varnhagen described the competition between ministries, institutes, and departments as a "feud between the parts." This

[28] Cited in Schnabel, *Deutsche Geschichte*, 3:320–21.

"war of all against all" where "each ministry is a castle" was also conducted along the cultural plane identified above. Given this fragmented condition characterized by intrastate divisions along *many* planes, it follows that there were many separate access routes to the state. Aspiring members of each profession studied the same subjects together, entered their respective niches and substructures in the bureaucracy, then sought entry to—or were recruited into—the luncheon, discussion, or social circuits which promised to place junior civil servants into the company of senior bureaucrats in the field.[29] One observes a similar process operating in the Prussian Army and the Mining Corps as young persons advanced from specialized academies to the first official assignments and social associations. "It seems as if certain bureaucrats understand one another with half-words even though they are not personally well-acquainted," observed Hans von Bülow, Minister of Commerce and Business until 1825. This was so because "they agree over accepted systems and [on the other hand] are ill-disposed toward principles and persons who, in their view, do not come from the right school."[30] The seasoned Bülow was describing a "gatekeeping" phenomenon where certain civil servants and soldiers were "potential friends because of a common culture," while others were quickly shunned if they did not honor "shared symbols" (Collins).

But Bülow's gaze took in more than lawyers, educators, mining captains, and members of highly structured *formal* cultures at the doors of their respective "consciousness communities" in Berlin. For *indigenous* cultures with "a highly formalized group identity" (Collins) operated in a similar fashion. Secret orders like the Freemasons, politicized discussion clubs like the Spanish Society, Prince August's luncheons, or the Christian-German Table Society and salons like those of the Rochows and Varnhagens were extremely sensitive about the cultural calling cards of persons entering the inner sanctum. Thus Rahel loved to cite Goethe's words: "Come in friends and kindred spirits." Outside the salon, "half-words" sufficed with those who were welcome on the inside. The small, completely indigenous, political friendship circles which permeated the state were also shut off to strangers. When loved and respected comrades like Gneisenau and Valentini "recommended" young friends, however, they were usually welcome. And when Vincke brought friends like Josua

[29] For a good description of this phenomenon, see Delbrück, *Lebenserinnerungen*, 1:127–28, 130–31.

[30] Bülow to Frederick William, 5 March 1819, cited in Barbara Vogel, "Beamtenkonservatismus: Sozial- und verfassungsgeschichtliche Voraussetzungen der Parteien in Preussen im frühen 19. Jahrhundert," in Dirk Stegmann et al. (eds.), *Deutscher Konservatismus im 19. und 20. Jahrhundert: Festschrift für Fritz Fischer zum 75. Geburtstag und zum 50. Doktorjubiläum* (Bonn: Verlag Neue Gesellschaft, 1982), 17.

Hasenclever to Berlin, liberal circles welcomed the entrepreneur from Remscheid like one of their own.[31]

It was at this very private level that conversational politics flourished. It was well known to liberal insiders that Hegel's friend, Eduard Gans, "held forth"[32] at the Varnhagens, while on many occasions women like Bettina von Arnim and Rahel herself were unmatched. Beuth sometimes found it difficult to animate his Sunday gatherings, but the conversation "moved along excellently" when the art world came under discussion. A regular at Ludwig Kühne's Sunday parties recalled how the host mixed humor and gravity to impart knowledge and experience to younger bureaucrats. "A more engaging personality was scarcely imaginable."[33] Alexander von Humboldt was another giant of the salon and table circuit, mesmerizing listeners at the king's palace, Prince August's luncheons, the Varnhagen's salon, the Bernsdorff's home, and elsewhere. The Foreign Minister and his wife were also regularly impressed with the intellectual depth and analytic prowess of Clausewitz, their good friend.[34] And Heinrich von Brandt remembered that "there was always a thorough and exhaustive discussion of politics" in General Staff and War School circles where "General Müffling, Rühle von Lilienstern and Radowitz were the masters."[35] Clearly, these elocutionary "masters" shaped images of reality for junior friends during normal times. At critical moments like the aftermath of Hardenberg's death in 1823 or the Belgian crisis of 1830/

[31] See Trende, *Im Schatten*, 18, "The Duties of a Freemason" (1738), printed in Ulrich von Merhart, *Weltfreimaurerei* (Hamburg: Bauhütten Verlag, 1969), 43–51, and Ernst Mannheim, *Aufklärung und öffentliche Meinung* (Stuttgart, 1979), 90–97, for the exclusivity and mutual favors of the masons; Johannes Bachmann, *Ernst Wilhelm Hengstenberg: Sein Leben nach gedruckten und ungedruckten Quellen* (Gütersloh: C. Bertelsmann, 1876–92), 1:175–78, for the closed nature of the Spanish Society and other Berlin clubs; Achim von Arnim's rigid "anti-Philistine" criteria for members of the Christian-German Table Society (1811), printed in Baxa, *Adam Müllers Lebenserzeugnisse*, 1:603; "Der Salon Frau von Varnhagen," in Varnhagen von Ense, *Vermischte Schriften*, 19:184–85, for one man's "recommended" path to Rahel's salon, and Otto Berdrow, *Rahel Varnhagen: Ein Lebens- und Zeitbild* (Stuttgart: Greiner & Pfeiffer, 1902), 266, for the "come in" quote; Varnhagen Diary I, 14 January and 9 February 1827, 4:172, 186, for the conservative prerequisites at the Rochow's mansion; Gneisenau to Boyen, 20 December 1815 (recommending O'Etzel); and Gneisenau to Clausewitz, 23 September 1818 (recommending Eichler), printed in Delbrück, *Gneisenaus Leben*, 5:91, 340; Brandt, *Aus dem Leben*, 2:1–51, for numerous examples of Brandt's "entree" through Valentini; and Josua Hasenclever's "Memoirs," written in 1841, printed in Hasenclever, *Josua Hasenclever*, 64–65, for a visit to Berlin with Vincke in 1822.

[32] Brandt, *Aus dem Leben*, 2:51.

[33] The visitor at Beuth's house and Kühne's apartment was Rudolf von Delbrück (for both quotes, see his *Lebenserinnerungen*, 1:136, 130).

[34] "Der Salon der Frau von Varnhagen," printed in Varnhagen von Ense, *Vermischte Schriften*, 19:198–99, 203, 207–8; Bernsdorff, *Ein Bild*, 2:101–3.

[35] Brandt, *Aus dem Leben*, 2:26.

31, caucuses, conferences, and meetings of friendly equals to build consensus on common political action were more typical. At these conspiratorial and revolutionary junctures, indigenous culture became the motive force of politics.

Our study has shown that politics in the Pre-March was conducted in a *party-political* setting. Accustomed as we are to the functioning of political parties in modern parliamentary systems, we should not blind ourselves to the fact that every "house of power" (Weber) spawns parties whose structures vary according to the political surroundings. Prussia, like authoritarian systems in our century, drove party politics to the private level. Here, largely tucked away from public view, contending factions decided the great issues of the day. Here, within the cliques and circles that made up the state, one found "the playing field of the political parties,"[36] as Otto Hintze put it. Because these groupings interlocked politically as well as socially—working, that is, for many of the same general political goals—it is meaningful to refer to *party blocs*.

We can identify two party blocs in the Pre-March—one furthering reform, the other reaction. Each cut across class planes, penetrated occupational enclaves, dissected the Ministerial Cabinet and Council of State, bridged the various divisions of the civilian bureaucracy and army, and extended from state into society. Each bloc also possessed obvious leaders and, in some instances, received cohesion from the formal cultures ensconced within it. Thus philologists were known for their liberalism.[37] But the real structure of parties on both sides was formed by bases of indigenous culture like overlapping friendship circles, clubs, salons, religious gatherings, and Freemasonic lodges. On the conservative side, we should add families, for the Gerlachs and Rochows provided a definite structure to reactionary politics.

The reactionary parties have been identified in earlier chapters. The compatriot leaders were Baron Kottwitz, Gustav and Caroline von Rochow, Otto von Voss-Buch, Duke Carl of Mecklenburg, and Prince Wittgenstein. The latter three used their social standing at the apex of the nobility to gain frequent personal access to the chambers of Frederick William where they presented their estate's case against egalitarianism, democracy, and industrial freedom. The bloc's goals and modus operandi were served well by a loose, decentralized structure of interlocking aristocratic coteries and a highly indigenous culture produced at prayer meetings, dinner parties, card-playing evenings, court functions, or during the autumn maneuvers of Duke Carl's elite Guard Corps. For more

[36] Otto Hintze, *Der Beamtenstand* (1913), reprinted in Kersten Krüger (ed.), *Beamtentum und Bürokratie* (Göttingen: Vandenhoeck & Ruprecht, 1981), 42.

[37] See Lenore O'Boyle, "Klassische Bildung und soziale Struktur in Deutschland zwischen 1800 und 1848," *Historische Zeitschrift* 207 (December 1968): 590–608.

structured indigenous and formal cultural counteroffensives against liberalism, the reactionary cause relied on influential publicists like Adam Müller, prominent political philosophers like Karl Ludwig von Haller, and fundamentalist theologians like Wilhelm Hengstenberg. It is possible that the lodges of the "Zealous Friends of the Cross," the masonic heirs of the antirevolutionary Rosicrucian order of the eighteenth century, also joined the reactionary parties. For the most part, however, it seems that conservatives like Barons Kottwitz, Stolberg, and Haugwitz exited the Freemasons after 1815, leaving the orders even more one-sidedly liberal than before.[38]

Ludwig von Gerlach, one of Kottwitz's devotees and a prominent conservative in his own right, grew increasingly dissatisfied with the informal structure of the reactionary parties during the 1830s and 1840s. Thus in June 1840 he complained to his brother Leopold that they were "not at all prepared," lacking "trust and unity among ourselves as a party." In January 1845, he bemoaned the final passing of the Industrial Code by the Council of State: "As a conservative party we could bring nothing forward because of a lack of agreement and organization." Finally in May 1847, Ludwig "poured out [his] political heart" to Leopold. "Since 1840, perhaps since the July Days of 1830, we should have held regular party assemblies, whenever possible opened with a prayer."[39] His brother's response is a strong indication that Ludwig's plans for party reorganization were inappropriate for the times. Leopold agreed that he and Karl von Voss had failed to impose their own personalities on the monarch, but rejected the idea of structural reforms. "We would not have conquered the world the way you say," he wrote, citing "the personality of the king"[40] as one reason. Ludwig saw only a "lovable childishness"[41] in Leopold's reply, but was himself overestimating the influence of modern party structures in the waning era of absolutism. It is true that informally structured parties were being subjected to more stress and strain in an age of great transformation, particularly after the shocking events of 1830. But the monarch's personality was still more important than party assemblies in determining a political outcome in the Pre-March.

The liberal reformers were guided by Gerhard [von] Scharnhorst in

[38] Ludwig Keller, *Die Freimaurerei: Eine Einführung in ihre Anschauungswelt und ihre Geschichte* (Leipzig: B. G. Teubner, 1918), 81–83; and Kottwitz to Frederick William, 14 June 1834 and 24 June 1834, printed in Eduard Emil Eckert, *Geschichte meiner persönlichen Anklage des Freimaurer-Ordens als einer Verschwörungs-Gesellschaft bei dem Ministerium zu Berlin und meiner Behandlung als Verbrecher darauf* (Schaffhausen: Freidrich Hurter, 1858), 109–16.

[39] Diary entries of 22 June 1840, 3 January 1844, and 11 May 1846, cited in Gerlach, *Aufzeichnungen*, 270–71, 343, 447.

[40] Leopold to Ludwig von Gerlach, 18 May 1846, cited in ibid., 447.

[41] Ibid., 447.

the army and Karl August von Hardenberg in the civilian bureaucracy. Hermann von Boyen, Neidhardt von Gneisenau, and Wilhelm von Humboldt attempted to fill the huge leadership voids left by the respective deaths of their predecessors in 1813 and 1822. Like the reactionary factions, political and cultural structure was provided by encounters in numerous friendship circles, politicized salons, luncheon societies, and the more "humanitarian" Freemasonic lodges of the Royal York, the Three Globes, and the Grand National.

But there were significant differences. The indigenous culture of party gatherings was colored to a much greater extent by the higher premium placed upon learned discourse and formal education by liberal reformers. The liberals were also more bourgeois, or in the case of noblemen, more "bourgeois-minded." Moreover, while they matched the aristocrats with firm ties to parties outside of the state, the reform factions outnumbered reactionaries inside government. Thus Varnhagen was not greatly exaggerating when he wrote that "the legislation of the administration, all of the lower-level bureaucrats, some of the higher [civil servants], much in the legal sytem and all things from the world of money and industry [are] thoroughly democratic."[42] The absolute numbers of those in and out of government favoring limited parliamentarism and economic progress offered comparatively little advantage, however, without privileged entry to the king's privy cabinets—and after Hardenberg's passing, no liberal party leader possessed such access. Accordingly, his successors were forced to work through cautious friends like Witzleben, who rarely approached the king as a party man, and sympathetic members of the royal family like Prince William (Frederick William III's brother) and Prince August. But, with public financial worries greatly reduced by 1832, the king had no compelling reason to listen.

The liberal parties also suffered from deeper internal divisions over the new industrial age. The general eighteenth-century notion that economic advance was good for society deteriorated into endless intrabloc controversy as Stein and Hardenberg moved forward with specific industrial reforms. During the 1810s and 1820s enervating disputes occurred over guilds, tariffs, patents, vocational versus classical education, statism versus nonstatism—and eventually, even the desirability of industry itself. Liberals in the medical profession, for instance, were comfortable with the concept of scientific and political progress, but many regarded lower class poverty and sources of industrial pollution like acid factories and metallurgical plants as threats to public health which government had to eradicate. Prominent theologians like Georg Nicolovius agreed with the most reactionary noblemen that the old economic order was better

[42] Varnhagen Diary I, 12 January 1825, 3:212–13.

than the new.[43] Nor was anti-industrialism absent from the legal profession. Thus district judge Karl Immerman wrote in his novel, *Die Epigonen*, that "factories should perish and the land be given over again to cultivation. . . . [Mechanization's] passage we cannot hinder, yet we are not to be blamed if we fence off a little green tract for us and ours."[44] His middle-class colleagues in the district government of Trier fought such a rearguard action against the gas-spewing towers of coke ovens desperately requested by bourgeois technocrats in the Saar Mining Office.[45]

Opposition to the unfolding technological modernization of Prussia also characterized the campaign of Wilhelm Süvern and Johannes Schulze, privy councillors in the Ministry of Education and Ecclesiastical Affairs, to further the study of classical languages at the expense of technical subjects. Like many of their early liberal counterparts outside of government, Süvern and Schulze wanted government opened to intellect and talent, but resisted the seemingly chaotic and unpredictable departure from traditional societal structures which beckoned with modern industry.[46] With help from classical enthusiasts—and unlikely allies like Peter Beuth—the two succeeded in blocking modern vocational schools until almost 1830.

As the work of Lothar Gall and others has shown, moreover, the kingdom's "material forces" (Mosca) were split between a pioneering group of daring entrepreneurs like Cockerill, Harkort, Haniel, and Borsig, and a dominant faction of smaller capitalists who harbored socially conservative, anti-industrial and petty bourgeois views.[47] Johannes Schuchard of

[43] For doctors, see Wehler, *Gesellschaftsgeschichte*, 2:234–35; the report of the Royal Medical Council, n.d. (fall, 1817), 441/17834; and Dr. Röchling, a county physician, to Regierung Trier, 11 October 1839, 442/4427, LHA Koblenz. For the theologians, see Baron Kottwitz to Frederick William, 27 February 1819, Rep.74, J.XVI.14, Bd.2, ZStA Merseburg; Ludwig von Gerlach's "Familiengeschichte," in Schoeps, *Aus den Jahren*, 300, 301–2; Görlitz, *Die Junker*, 184, 209–10; and Robert M. Bigler, "The Social Status and Political Role of the Protestant Clergy in Pre-March Prussia," in Wehler, *Sozialgeschichte*, 176, 184.

[44] Quoted in Ernest K. Bramsted, *Aristocracy and Middle-Classes in Germany* (Chicago: University of Chicago Press, 1964), 65–66.

[45] See Oberbergamt Bonn to Regierung Trier, 26 February 1839, 442/4427, LHA Koblenz.

[46] Varrentrapp, *Johannes Schulze*, 238, 350, 357, 363–69, 410–15.

[47] For the opposition, or in some cases, ambivalence to industry among early German liberals, see Lothar Gall, "Liberalismus und 'Bürgerliche Gesellschaft': Zu Charakter und Entwicklung der Liberalen Bewegung in Deutschland," *Historische Zeitschrift* 220 (April 1975): 324–56; James J. Sheehan, *German Liberalism in the Nineteenth Century* (Chicago: University of Chicago Press, 1978), 19–34; and Hans-Ulrich Thamer, "Zur Ideen- und Sozialgeschichte von Liberalismus und Handwerk in der ersten Hälfte des 19. Jahrhunderts," in Wolfgang Schieder, *Liberalismus in der Gesellschaft des deutschen Vormärz* (Göttingen: Vandenhoeck & Ruprecht, 1983), 55–73.

Barmen was representative of this latter group. A textile manufacturer himself, Schuchard nevertheless bemoaned the rapid proliferation of western factories and opposed the higher tariffs which promised to accelerate this development. Favoring a more natural balance between agriculture, trade, and industry, he sent a petition in 1840 to Christian Rother, a man who could be expected to sympathize with Schuchard's railings against "the commercial-industrial overexcitement which exists in England, promises palaces to us . . . and [figures to provoke] a struggle of poverty against wealth."[48] Like the liberal bureaucrats and professionals, many businessmen reeled backward away from the modern industrial order as it began to emerge.

This internal bickering among liberals played squarely into the hands of a king who did not like to deal with political parties and staunchly resisted the western idea of parliaments. When Frederick William III and Witzleben decided in the mid-1820s that domestic peace could not survive, nor public finance withstand, a rescinding of Hardenberg's economic legislation,[49] the consequences were: continued industrialization, which produced further turmoil in the liberal camps and tremendous strain on party structures; and budget surpluses, which undermined liberal parliamentary schemes. This decision to uphold Hardenberg's reforms also engendered bitterness and frustration among reactionaries who wanted full, not partial victory over party foes. Indeed, neither the fall of Witzleben, nor the deaths of Humboldt, Motz, Maassen, and Kunth, nor the rising fortunes of a "younger, stronger party"[50] around Wittgenstein and Rochow in the 1830s provided reactionaries with an opportunity to undermine industrialism.

As we have seen, the more confident, well-advised Frederick William of the postwar period sought to balance contending party forces in the state. By the 1820s and 1830s, the liberal ascendency of the early Hardenberg years had yielded to an equilibrium of sorts as individuals of various party stripes were admitted to court, privy cabinets, ministries, Council of State, and key army posts. Judging from the frustration in both blocs by the 1830s, the balancing act between reactionary and progressive appointments and policies had been performed with great skill.[51]

[48] Schuchard to Rother, December, 1840, cited in Hans Höring, "Johannes Schuchard," in *Rheinisch-Westfälische Wirtschaftsbiographien* (Münster: Verlag der Aschendorffschen Verlagsbuchhandlung, 1932), 1:12.

[49] See Frederick William's letter to the Crown Prince, 30 November 1824, Müffling Papers 92, A-9, Bl.2, ZStA Merseburg.

[50] Rochow, *Vom Leben*, 263.

[51] For liberal disappointment, see Varnhagen Diary II, 23 December 1836, 1:28. But for reactionary descriptions of the 1830s as "liberal," see Rochow, *Vom Leben*, 411; Leopold von Gerlach, *Denkwürdigkeiten aus den Leben Leopold von Gerlachs* (Berlin: Wilhelm Hertz), 1:106; and Carl Ludwig von Haller's commentary of 1840, cited in Hans Riegel-

The usually perceptive Varnhagen, however, doubted the wisdom of promoting contradictory tendencies by continually pitting one side against the other. "The difficulties presented by this predicament are increasing," he noted in 1825, "and the dissolution of the whole accelerating."[52] While he probably underestimated the old king's ability to sustain this delicate effort, Varnhagen was ultimately correct. For here was a hegemonic game of chance which always led to disaster when royal fathers played far better than their sons.

In an obvious effort to play one class off against the other, Frederick William IV returned to the proaristocratic politics of Frederick II. But this policy was not merely a cold calculation of raison d'etat, for, as we have seen in this study, such a policy was quite consistent with a man whose basic sympathies were aristocratic. Accordingly, an abrupt deceleration occurred in the number of commoners ennobled, dropping to half the annual pace of the 1820s and 1830s. Moreover, unlike Frederick William III, who drew over 90 percent of his new titleholders from the ranks of unpropertied army officers and bureaucrats, Frederick William IV elevated mostly those bourgeois who already owned noble estates. The new monarch's advisers also drafted a royal patent in 1846 which envisioned a revitalized landowning nobility. They argued that bourgeois estate owners possessed a commercial, antiaristocratic spirit and constituted a threatening presence on the land and in the provincial diets. Therefore an extremely careful process of cooptation was necessary in order to prevent the impending ruin of the old nobility. "In this way," as one of the king's commissioners put it, the landowning estate "will become the flower and the spokesman of the educated nation, instead of a noble caste that is neither loved nor desired and that, in the course of the century, is threatened with poverty and extinction." The draft law strove to "introduce a healthy accommodation [and] a moderation of the apparent separation between noble and non-noble estate owners."[53]

In the meantime, a less "accommodating" face was turned toward the rest of the middle class. By 1848, the percentage of leading bourgeois bureaucrats had fallen precipitously from 60 percent to 36 percent.[54] The

---

mann, *Die europaeischen Dynastien in Ihrem Verhaeltnis zur Freimaurerei* (Berlin: Nordland Verlag, 1943), 182. "Throughout his lifetime," wrote Haller, "Frederick William III was ruled by the humanitarian concepts of the Enlightenment. These principles combined in an unhealthy fashion with a hesitant character which did not dare to stand up against the progressive democratization of the Prussian state."

[52] Varnhagen Diary I, 12 January 1825, 3:212–13.

[53] Berdahl, *Politics of the Prussian Nobility*, 326–33. The quotes are from Christian Karl Bunsen's memorandum of April 1844, and the "Motive" of the draft patent, n.d. (late 1846 or early 1847), cited on pp. 330–31, 332.

[54] Nikolaus von Preradovich, *Die Führungsschichten in Preussen und Österreich (1804–1918)* (Wiesbaden: Franz Steiner, 1955), 105.

monarch also dashed the initial enthusiasm of industrial and financial circles by turning a deaf ear to their increasingly assertive legislative pleas. There would be no independent Ministry of Trade, no adequate central bank, no easing of monetary policy, and no limited liability for corporations. Simultaneously, the once-cordial intrabloc relations between technocrats and industrialists in the liberal parties were deteriorating as a result of disputes over tariffs, Seehandlung investments, mining and manufacturing technology, army secrecy and heavy-handedness, and railroads. An already tense situation was exacerbated, in other words, and the consequence was a political crisis of the first order of magnitude.

The final image of the Pre-March Prussian state resembles neither a *Junkerstaat* nor a *Beamtenstand*. Frederick William III and his chief advisers schemed fairly effectively against two loose party blocs to preserve a precarious personal power during decades when Prussia underwent the beginnings of a turbulent and unpredictable socioeconomic transformation. Along the way, concessions were parceled out to noble families longing to turn back the clock, and to bourgeois jurists, theologians, physicians, and philologists seeking to feather professional nests. But the most important deals—those which reinforced the economically modernizing tendencies of the times—were struck with the vanguard of bureaucrats who were allied for a long time with Prussia's rising industrialists. The political equilibrium began to fall apart in the 1830s when the vanguard ceased to lead. It disintegrated in the 1840s when Frederick William IV leaned too far in the direction of the reactionary parties.

. . . . .

Already separating at numerous seams during the last years of the father and the first years of the son, Prussia's "whole" finally split apart amidst peasant unrest and mob violence in March 1848.[55] Now Frederick William IV, William I, and Otto von Bismarck were compelled to forge a different—and far greater—Prussian state out of the old "parts." It was something of a *Flucht nach vorn*, or a retreat which takes the form of a desperate offensive. Bismarck, the maverick Junker, worked hardest and

[55] The following works have informed these conclusions: Hamerow, *The Social Foundations of German Unification*; Michael Gugel, *Industrieller Aufstieg und bürgerliche Herrschaft: Sozioökonomische Interessen und politische Ziele des liberalen Bürgertums in Preussen zur Zeit des Verfassungskonflikts* (Cologne: Pahl-Rugenstein, 1975); Bleiber (ed.), *Bourgeoisie und bürgerliche Umwälzung*; Diefendorf, *Businessmen and Politics in the Rhineland*; David Blackbourn and Geoff Eley, *The Peculiarities of German History: Bourgeois Society and Politics in Nineteenth-Century Germany* (New York: Oxford University Press, 1984); and Michael John, *Politics and the Law in Late Nineteenth-Century Germany: The Origins of the Civil Code* (Oxford: Oxford University Press, 1989).

most effectively to cajole and coerce upper- and middle-class factions and strike the necessary deals and compromises among them.

Drawn into the vortex of more intense class and estate conflict after the July Days of 1830, the party blocs of the Pre-March were thrust into a new political era after 1848. Emerging from the intimate world of friendship circles, private gatherings, and salons, the old groupings rearranged themselves publicly within the political clubs and parliamentary cliques of the late 1840s and 1850s. The influence of older political customs persisted, however, into the new age. Theodore Hamerow writes, for example, of "the cozily haphazard conglomeration of factions" which distinguished midcentury politics. "While ideology provided a criterion for the separation of liberals from conservatives," he continues, "there was a constant process of alignment and realignment [within each camp]. . . . Old combinations disintegrated and new ones arose through the decision of a few influential politicians." While more modern party structures began to appear in the 1860s, the restrictions of Prussia's local and state suffrages allowed the old Pre-March political mold to survive largely unbroken. In Marburg, for instance, bourgeois sociability and town politics were so closely linked that one dinner club, the Käsebrot, functioned like a "shadow" city council. An observer of diet affairs in Berlin during the 1860s noted similarly that "politics was of the greatest significance . . . for social intercourse."[56] Nor did the importance of personal-political networks and factional alliances disappear quickly with the coming of universal manhood suffrage in 1867. Thus, a decade later, Bismarck described the typical German party as a "theoretical caucus-grouping" or a "kind of parliamentary joint-stock company," while Heinrich von Treitschke was repelled by a Reichstag "honey-combed with intriguing cliques." And as late as 1892, Conservative parliamentarians bemoaned the fragmenting influence of "the clubs" within their party.[57] For approximately a half century after the Revolution of 1848, nevertheless, Germany's party-political "caucus groupings" were well suited for meeting challenges and accomplishing goals in this semiparliamentary age.

Indeed, the new state was more "liberal" (Mosca) than the old. Constitutionalism and a more representative government of the plutocratic variety favored by men of property came to Prussia soon after the Revo-

[56] For the quotes, see Hamerow, *The Social Foundations of German Unification*, 1:309, 308; Rudy Koshar, *Social Life, Local Politics, and Nazism: Marburg, 1880–1935* (Chapel Hill: The University of North Carolina Press, 1986), 187; and Kaelble, *Berliner Unternehmer*, 189–90.

[57] Both the Bismarck and Conservative Party quotes here are cited in James Retallack, *Notables of the Right: The Conservative Party and Political Mobilization in Germany, 1876–1918* (Boston: Unwin Hyman, 1988), 20, 96. For Treitschke, see Andreas Dorpalen, *Heinrich von Treitschke* (New Haven: Yale University Press, 1957), 183.

lution of 1848. Bismarckian defiance in the early 1860s set a definite limit to the diet's powers, but crown and parliament had negotiated a political peace by the autumn of 1866. Reflecting the new political equilibrium in state and society, many of the kingdom's largest banks formed a syndicate with the Seehandlung in 1868 to underwrite government loans.[58] Thus Prussia joined the ranks of modern states like England and France which, through political compromise, had earned the ability to borrow on a regular, long-term basis from the wealth-holding public. The same arrangement was carried over into the Empire on an even grander scale after 1871.

The new state was also more "democratic" (Mosca) than the old. Thus non-nobles in the ministries rose from 36 percent to 72 percent between 1848 and 1911—and the noble category included many new noblemen who were never assimilated into the ranks of the old aristocracy.[59] Indeed, as David Blackbourn argues, there is certainly sufficient cause for a healthy skepticism concerning the alleged "feudalization" of the middle classes in Imperial Germany.[60] Frederick William IV's royal patent on the new nobility was never issued and the "separation between noble and non-noble estate owners," so evident to observers in the Pre-March, continued. Given the uninterrupted influx of bourgeois parvenus onto the land, it could hardly have been otherwise. The figure for middle-class ownership of noble estates marched inexorably upward from 42 percent, where it had stood in 1856, to 68 percent at the end of a prolonged depression in 1889. Nor were the largest landowners immune from this process, for 31.9 percent (2,061 of 6,454) of latifundia over one thousand hectares had fallen into common hands in the seven provinces of eastern Prussia by 1885, while hundreds more were owned by the recently ennobled.[61]

Simultaneously, those disenfranchised elements of the *industrial* middle class which had grown increasingly dissatisfied with the performance of class compatriots and aristocrats in government were incorporating themselves within the state. The first significant "democratic" rearrangement after the March bloodshed came with the establishment of a Min-

[58] Jacob Riesser, *The Great German Banks and their Concentration in Connection with the Economic Development of Germany* (Washington: GPO, 1911), 62.

[59] Berdahl, *Politics of the Prussian Nobility*, 278; Gillis, *Prussian Bureaucracy*, 212.

[60] See Blackbourn's convincing critique of feudalization, in Blackbourn and Eley, *The Peculiarities of German History*, 228–37.

[61] In general, see Hamerow, *Restoration, Revolution, Reaction*, 225–26. For the 68 percent figure, see John R. Gillis, "Aristocracy and Bureaucracy in Nineteenth-Century Prussia," *Past and Present* 41 (December 1968), 113; for the 32 percent figure, see Hans Rosenberg, "Die Pseudodemocratisierung der Rittergutsbesitzerklasse," in Hans Rosenberg, *Machteliten und Wirtschaftskonjunkturen: Studien zur neueren deutschen Sozial- und Wirtschaftsgeschichte* (Göttingen: Vandenhoeck & Ruprecht, 1978), 86, 88.

istry of Trade. Headed for fourteen years (1848–1862) by August von der Heydt, a banker from Elberfeld, the office eliminated a major source of dissatisfaction in the business world. Similarly popular were the long administrations of David Hansemann in the Royal Bank (1848–1864) and Otto Camphausen in the Seehandlung (1855–1869)—institutions once controlled by Christian Rother.

The years after 1848 also witnessed an end to the struggle between private industry and the Prussian state for control over the forces of production. While Frederick William IV looked the other way, the diet approved the sale of 63 percent of the Seehandlung's industrial assets between 1851 and 1854. Eleven enterprises valued at 6.6 million thaler changed to private hands—a truly massive transfer of technology. A similar fate awaited the Mining Corps. The diet passed a series of laws between 1851 and 1865 which terminated the *Direktionsprincip*, abolished local mine offices, and ended the Corps' metallurgical responsibilities. It was a sign of the laissez-faire times that the proud Saynerhütte was sold to Krupp in 1865, while the mighty Königshütte went to Henckel von Donnersmark in 1871. The Prussian Army amended its ways, moreover, after the expiration of Dreyse's patent in 1865. Within a decade, private armaments works like Krupp and Loewe were enjoying multimillion mark contracts with military armories aware of their own productive limitations. The government demonstrated a newfound flexibility, finally, toward railroads. In the 1860s the state yielded to the prevailing preference for private construction, while after the business crash of 1873 steps were taken toward nationalization policies which had regained popularity. By 1879 Prussia boasted about twenty thousand kilometers of track.[62] The railroad network—like the political accommodations and rearrangements of the previous thirty years—was complete.

So was the process of German unification. The powerful Second Empire stood tall in central Europe, a warning to other nations that "the Germanies" were no longer pawns on the continental board. The formation of Modern Germany, however, brought more than psychic rewards for the middle class. A national currency, a central bank, a standard set of weights and measures, a unified postal system, a new all-German industrial code, and elimination of the last internal toll barriers helped to complete the institutional framework for a capitalistic economy initiated

[62] For this paragraph, see Radtke, *Seehandlung*, 350–53; Schulz-Briesen, *Preussische Bergbau*, 84, 93–96, 100; Bertold Buxbaum, "Der deutsche Werkzeugmaschinen- und Werkzeugbau im 19. Jahrhundert," *Beiträge zur Geschichte der Technik und Industrie* 9 (1919): 119–20; Hans-Dieter Götz, *Militärgewehre und Pistolen der deutschen Staaten 1800–1870* (Stuttgart: Motorbuch Verlag, 1978), 300; William Manchester, *The Arms of Krupp* (Boston: Bantam, 1970), 109, 116, 157; and Klee, *Preussische Eisenbahngeschichte*, 132–147.

by Stein, Hardenberg, and their South German counterparts. The modern business order received further underpinning from enactments legalizing bills of exchange, freeing corporations from the restrictive practice of bureaucratic approval, and sanctifying rights of private property. The antiguild and strict trade union provisions of the industrial code made the German settlement of 1867/71 even more palatable to wealthy men fearful of a new struggling class of workers. It was more than a consolation, finally, that the last strongholds of reactionary Junkerdom fell as the modern society of city councils, civil codes, parliaments, and joint-stock banks arose—what remained of serfdom disappeared in 1849, aristocratic tax exemption in 1861, local police powers and restrictive mortgage laws in 1872. Having opened its first assembly with a prayer in 1861, moreover, the Old Conservative movement around Ludwig von Gerlach was in complete disarray a decade later, losing nearly all of its seats during the Prussian and Imperial elections of 1873 and 1874.[63] This bourgeois-oriented, authoritarian and Prussianized Germany enjoyed a heyday in the 1870s and 1880s. It descended from this zenith in the 1890s when the unfamiliar challenges of mass politics began to create trouble and generate anxiety. The new German order would survive, however, until 1918.

[63] Hamerow, *The Social Foundations of German Unification*, 1:326; Retallack, *Notables of the Right*, 14, 21.

Bibliography

Primary Sources from Archives and Special Repositories

The following sources are alphabetized by city.

Archiv und Bibliothek des Deutschen Freimaurer-Museums, Bayreuth (A-BDDFM)
1. Membership Lists of the Grand National Mother Lodge of the Three Globes
2. Membership Lists of the Grand Lodge of the Royal York Friendship
3. Rare Book Holdings on Freemasonry

Universitätsbibliothek der Technischen Universität Berlin, Abteilung für Gartenbau, Berlin (GB/TU Berlin)
1. *Verhandlungen des Vereins für die Förderung des Gewerbefleisses in Preussen*, Bd. 1–27 (1822–1848)

Geheimes Staatsarchiv Preussischer Kulturbesitz, Berlin-Dahlem (GStA Berlin)
1. Records of the Prussian Ministry of Justice (Rep. 84)
2. Records of the Prussian Ministerial Cabinet (Rep. 90)
3. Personal Papers (Rep. 92)
   A. Hermann von Boyen
4. Records of the Seehandlung (Rep. 109)
5. Brandenburg-Preussisches Hausarchiv
   A. Papers of Frederick William III (Rep. 49)
   B. Papers of Friedrich Karl zu Sayn-Wittgenstein-Hohenstein (Rep. 192)

Landeshauptarchiv Rheinland-Pfalz, Coblenz (LHA Koblenz)
1. Records of the Provincial Governor of the Rhineland (Bestand 402 & Bestand 403)
2. Records of the District Government of Coblenz (Bestand 441)
3. Records of the District Government of Trier (Bestand 442)

Rheinisch-Westfälisches Wirtschaftsarchiv zu Köln E.V., Cologne (R-W WA Köln)
1. Records of the Cologne Chamber of Commerce (Abt. 1)
2. Records of the Duisburg Chamber of Commerce (Abt. 20)
3. Records of the Elberfeld-Barmen Chamber of Commerce (Abt. 22)

Hessisches Staatsarchiv Darmstadt, Darmstadt (HeStA Darmstadt)
1. Personal Papers
   A. Prince William of Prussia (brother of Frederick William III)

Stiftung Westfälisches Wirtschaftsarchiv, Dortmund (SWW Dortmund)
1. Peter Beuth-Theodor Baumann Correspondence (N 7/1)

Nordrhein-Westfälisches Hauptstaatsarchiv, Düsseldorf, Schloss Kalkum (HSA Düsseldorf)
1. Records of the District Government of Aachen (Regierung Aachen)
2. Records of the District Government of Düsseldorf (Regierung Düsseldorf)

3. Records of the District Government of Cologne (Regierung Köln)
4. Records of the Prussian Mining Corps, Bonn Mine Office (Oberbergamt Bonn)

Haniel Museum, Duisburg
1. Franz Haniel-Peter Beuth Correspondence (Nr. 210)

Historisches Archiv Friedrich Krupp GmbH, Essen, Villa Hügel (HA Friedrich Krupp GmbH)
1. Records of the Krupp Works (W.A., F.A.H., Kartei)
2. Historical Studies Commissioned by the Krupp Works ("Grüne" Geschichtliche Studien)

Bundesarchiv (Aussenstelle Frankfurt/Main) (BA Aus Frankfurt/Main)
1. Letters of Eduard [von] Peucker to his Children (FSg. 1/151)

Militärgeschichtliches Forschungsamt, Freiburg im Breisgau (MgFa Freiburg)
1. *Zeitschrift für Kunst, Wissenschaft und Geschichte des Krieges*, Bd. 1–98 (1824–1856)
2. *Militär-Wochenblatt*, Bd. 2–44 (1817–1859)

Zentrales Staatsarchiv Merseburg (ZStA Merseburg)
1. Records of the Prussian Chancellery (Rep. 74)
2. Records of the Prussian Ministry of the Interior (Rep. 77)
3. Records of the Prussian Ministerial Cabinet (Rep. 90)
4. Personal Papers (Rep. 92)
   A. Karl von Altenstein
   B. Karl August von Hardenberg
   C. Karl von Müffling
   D. Christian [von] Rother
   E. Job von Witzleben
5. Records of the Prussian Ministry of Trade and Industry (Rep. 120)
6. Records of the Prussian Mining Corps, Berlin Headquarters (Rep. 121)
7. Records of the Prussian Ministry of Finance (Rep. 151)
8. Records of the Prussian Privy Cabinet for Civilian Affairs (Rep. 2.2.1.)

Nordrhein-Westfälisches Staatsarchiv, Münster (StA Münster)
1. Records of the Provincial Governor of Westphalia (Oberpräsidium)
2. Records of the District Government of Arnsberg (Regierung Arnsberg)
3. Records of the District Government of Münster (Regierung Münster)
3. Records of the Prussian Mining Corps, Dortmund Mine Office (Oberbergamt Dortmund)
4. Personal Papers
   A. Ludwig von Vincke

Historisches Archiv der Gutehoffnungshütte Aktienverein, Oberhausen (HADGHH)
1. Records of the Gutehoffnungshütte Works

Masonic Library and Museum of Pennsylvania, Philadelphia (MLM-Penn)
1. Membership Lists of the Grand National Lodge of the Freemasons of Germany
2. Rare Book Holdings on Freemasonry

Published Primary Sources

Banfield, Thomas. *Industry of the Rhine*. London, 1846–48.

Baxa, Jakob, ed. *Adam Müller's Lebenserzeugnisse*. 2 vols. Munich: Verlag Ferdinand Schöningh, 1966.

Beck, Hanno. *Gespräche Alexander von Humboldts*. Berlin: Akademie-Verlag, 1959.

Beckerath, Erwin V., and Stühler, Otto, eds. *Friedrich List: Schriften zum Verkehrswesen*. 3 vols. Berlin: Reimar Hobbing, 1929.

Bernsdorff, Elise von. *Aus ihren Aufzeichnungen: Ein Bild aus der Zeit von 1789 bis 1835*. 2 vols. Berlin: E. S. Mittler, 1896.

Brandt, Heinrich von, ed. *Aus dem Leben des Generals der Infanterie z.D. Dr. Heinrich von Brandt*. 2 vols. Berlin: E. S. Mittler, 1869.

Bülow, Eduard von, ed. *Aus dem Nachlasse von Georg Heinrich von Berenhorst*. 2 vols. Dessau: Verlag von Karl Aue, 1847.

Clausewitz, Carl von. *Politische Schriften und Briefe*. Munich: Drei Masken Verlag, 1922.

———. *Vom Kriege: Hinterlassenes Werk*. Frankfurt: Verlag Ullstein GmbH, 1980.

Delbrück, Hans, ed. *Das Leben des Feldmarschalls Grafen Neithardt von Gneisenau*. 5 vols. Berlin: G. Reimer, 1880.

Delbrück, Rudolf von. *Lebenserinnerungen*. 2 vols. Leipzig: Verlag von Duncker & Humblot, 1905.

Dilthey, Wilhelm, ed. *Aus Schleiermacher's Leben*. 4 vols. Berlin: Georg Reimer, 1863.

Eylert, Rulemann Friedrich. *Charakter-Züge und historische Fragmente aus dem Leben des Königs von Preussen Friedrich Wilhelm III*. 4 vols. Magdeburg: Heinrichshofen'sche Buchhandlung, 1842–46.

Gerhard, Dietrich, and Norwin, William, eds. *Die Briefe Bartold Georg Niebuhrs*. 2 vols. Berlin: Walter de Gruyter & Co., 1926–29.

Gerlach, Jakob von, ed. *Ernst Ludwig von Gerlach: Aufzeichnungen aus seinem Leben und Wirken 1795–1877*. Schwerin: Fr. Bahn, 1903.

Gerlach, Leopold von. *Denkwürdigkeiten aus dem Leben Leopold von Gerlachs*. 2 vols. Berlin: Wilhelm Hertz, n.d.

Hahlweg, Werner, ed. *Carl von Clausewitz: Schriften, Aufsätze, Studien, Briefe*. Göttingen: Vandenhoeck & Ruprecht, 1966.

Haniel, Franz. *Autobiographie*, n.d. (1858–1862), in: Bodo Herzog and Klaus J. Mattheier, *Franz Haniel 1779–1868*. Bonn: Ludwig Röhrscheid, 1979.

Hasenclever, Adolf. *Josua Hasenclever aus Remscheid-Ehringhausen: Erinnerungen und Briefe*. Halle: Karras, Kröber & Nietschmann, 1922.

Heyderhoff, Julius, ed. *Benzenberg: Der Rheinländer und Preusse 1815–1823*. Bonn: Fritz Klopp Verlag, 1928.

Höper, H. *Die Preussische Eisenbahn-Finanz-Gesetzgebung*. Berlin: Carl Heymann's, 1879.

Hohenlohe-Ingelfingen, Prinz Kraft zu. *Aus meinem Leben*. 2 vols. Berlin: Ernst Siegfried Mittler & Sohn, 1897.

Lange, Fritz, ed. *Neithardt von Gneisenau: Schriften von und über Gneisenau*. Berlin: Rütten & Loening, 1954.

Leyen, Alfred v. der, et al., eds. *Friedrich List: Schriften zum Verkehrswesen*. Berlin: Reimar Hobbing, 1931.

Linnebach, Karl. *Karl und Marie von Clausewitz: Ein Lebensbild in Briefen und Tagebuchblättern*. Berlin: Martin Warneck, 1925.

Moltke, Helmuth von. *Gesammelte Schriften und Denkwürdigkeiten des Generalfeldmarschalls Grafen Helmuth von Moltke*. Berlin: E. S. Mittler, 1892.

Natzmer, Oldwig von. *Unter den Hohenzollern: Denkwürdigkeiten des Generals Oldwig von Natzmer*. 2 vols. Gotha: Friedrich Andreas Perthes, 1887.

Oncken, H., and Saemisch, F.E.M. *Vorgeschichte und Begründung des Deutschen Zollvereins 1815–1834: Akten der Staaten des Deutschen Bundes und der Europäischen Mächte*. 3 vols. Berlin: Reimar Hobbing, 1934.

Pick, Albert. "Briefe des Feldmarschalls Grafen Neithardt v. Gneisenau an seinen Schwiegersohn Wilhelm von Scharnhorst," *Historische Zeitschrift* 77 (1896): 67–85, 234–56, 448–60.

Politische Correspondenz Kaiser Wilhelms I. Berlin: Hugo Steinitz, 1890.

Pückler-Muskau, Hermann von. *Briefwechsel und Tagebücher des Fürsten Hermann von Pückler-Muskau*. Edited by Ludmilla Assing. 9 vols. Berlin, 1873–76.

Raumer, Friedrich von. *Lebenserinnerungen und Briefwechsel*. Leipzig: F. A. Brockhaus, 1861.

Raumer, Kurt von. *Die Autobiographie des Freiherr vom Stein*. Münster: Verlag Aschendorff, 1955.

Richter, Wilhelm, ed. *Wilhelm von Humboldt's Politische Briefe*. 2 vols. Berlin: B. Behrs Verlag, 1936.

Rochow, Caroline von. *Vom Leben am preussischen Hofe 1815–1852*. Edited by Luise von der Marwitz. Berlin: E. S. Mittler, 1908.

Rohrscheid, Kurt von. *Vom Zunftzwang zur Gewerbefreiheit*. Berlin: Carl Heymanns Verlag, 1898.

Roon, Albrecht von. *Denkwürdigkeiten aus dem Leben des General-Feldmarschalls Kriegsministers Grafen von Roon*. 2 vols. Breslau: Verlag von Eduard Trewendt, 1892.

Rühl, Franz, ed. *Aus der Franzosenzeit*. Leipzig: Verlag von Duncker & Humblot, 1904.

———. *Briefe und Aktenstücke zur Geschichte Preussens unter Friedrich Wilhelm III*. 3 vols. Leipzig: Duncker & Humblot, 1902.

Schoeps, Hans Joachim, ed. *Aus den Jahren preussischer Not und Erneuerung*. Berlin: Haude & Spenersche Verlagsbuchhandlung, 1963.

———. *Neue Quellen zur Geschichte Preussens im 19. Jahrhundert*. Berlin:

Haude & Spenersche Verlagsbuchhandlung, 1968.
Schön, Theodor von. *Aus den Papieren des Ministers und Burggrafen von Marienburg Theodor von Schön*. 5 vols. Berlin: Verlag von Franz Duncker, 1875–76.
Schulz, Hans, ed. *J. G. Fichte: Briefwechsel*. 2 vols. Leipzig: H. Haessel, 1925.
Stadelmann, Rudolph. *Preussens Könige in ihrer Thätigkeit für die Landeskultur*. Leipzig: Verlag von S. Hirzel, 1887.
Steffens, Heinrich. *Was Ich Erlebte*. 10 vols. Breslau: Josef Max, 1840–44.
Stein, Karl Freiherr vom. *Freiherr vom Stein: Briefe und Amtliche Schriften*. Edited by Erich Botzenhart. 10 vols. Stuttgart: W. Kohlhammer Verlag, 1957.
Sybel, Heinrich von, "Gneisenau und sein Schwiegersohn, Graf Friedrich Wilhelm von Brühl," *Historische Zeitschrift* 69 (1892): 245–85.
Trende, Adolf. *Im Schatten des Freimaurer- und Judenthums: Ausgewählte Stücke aus dem Briefwechsel des Ministers und Chefs der preussischen Bankinstitute Christian von Rother*. Berlin: Verlag der Deutschen Arbeitsfront, 1938.
Varnhagen von Ense, Karl August. *Ausgewählte Schriften von K. A. Varnhagen von Ense*. Leipzig: F. A. Brockhaus, 1875.
———. *Blätter aus der preussischen Geschichte*. Edited by Ludmilla Assing. 5 vols. Leipzig: F. A. Brockhaus, 1868–69 (cited in text as Varnhagen Diary I).
———. *Briefe von Chamisso, Gneisenau, Haugwitz, W. von Humboldt, Prinz Louis Ferdinand, Rahel, Rückert, L. Tieck u.a.* Edited by Ludmilla Assing. 2 vols. Leipzig: F. A. Brockhaus, 1867.
———. *Briefwechsel zwischen Varnhagen von Ense und Oelsner nebst Briefen von Rahel*. 3 vols. Stuttgart: Verlag von A. Kröner, 1865.
———. *Denkwürdigkeiten des eigenen Lebens*. 3 vols. Leipzig: F. A. Brockhaus, 1843.
———. *Kommentare zum Zeitgeschehen: Publizistik, Briefe, Dokumente, 1813–1858*. Edited by Werner Greiling. Leipzig: Reclam, 1984.
———. *Tagebücher von K. A. Varnhagen von Ense*. Edited by Ludmilla Assing. 15 vols. Bern: Herbert Lang, 1972 (cited in text as Varnhagen Diary II).
———. *Vermischte Schriften*. 6 vols. Leipzig: F. A. Brockhaus, 1875.
Winter, Georg, ed. *Die Reorganisation des Preussischen Staates unter Stein und Hardenberg*. Leipzig: S. Hirzel, 1931.
Wolzogen, Alfred von, ed. *Aus Schinkels Nachlass*. 2 vols. Berlin: Verlag R. Decker, 1863.

Select Secondary Sources

(full citations for other secondary works in each chapter)

Anderson, Eugene Newton. *Nationalism and the Cultural Crisis in Prussia, 1806–1815*. New York: Farrar & Rinehart, Inc., 1939.
Baack, Lawrence J. *Christian Bernstorff and Prussia: Diplomacy and Reform Conservatism 1818–1832*. New Brunswick, New Jersey: Rutgers University Press, 1980.
Beck, Ludwig. *Die Geschichte des Eisens in technischer und kulturgeschicht-*

licher Beziehung. 5 vols. Braunschweig: Friedrich Vieweg und Sohn, 1884–1903.

Berdahl, Robert M. *The Politics of the Prussian Nobility: The Development of a Conservative Ideology 1770–1848*. Princeton, New Jersey: Princeton University Press, 1988.

Berdrow, Otto. *Rahel Varnhagen: Ein Lebens- und Zeitbild*. Stuttgart: Greiner & Pfeiffer, 1902.

Bergengrün, Alexander. *David Hansemann*. Berlin: J. Guttentag Verlagsbuchhandlung, 1901.

Bernal, Martin. *Black Athena: The Afroasiatic Roots of Classical Civilization*. New Brunswick, New Jersey: Rutgers University Press, 1987.

Bleiber, Helmut, ed. *Bourgeoisie und bürgerliche Umwälzung in Deutschland 1789–1871*. Berlin: Akademie-Verlag, 1977.

Blumberg, Horst. *Die Deutsche Textilindustrie in der industriellen Revolution*. Berlin: Akademie-Verlag, 1965.

Bonin, Udo von. *Geschichte des Ingenieurkorps und der Pioniere in Preussen*. 2 vols. Wiebaden: LTR Verlag, 1981.

Branig, Hans. *Fürst Wittgenstein: Ein preussischer Staatsmann der Restaurationszeit*. Cologne: Böhlau Verlag, 1981.

Brose, Eric Dorn. "Competitiveness and Obsolescence in the German Charcoal Iron Industry," *Technology and Culture* 26:3 (July 1985): 532–59.

Craig, Gordon. *The Politics of the Prussian Army, 1640–1945*. Oxford: Oxford University Press, 1955.

Dehio, Ludwig. "Wittgenstein und das letzte Jahrzehnt Friedrich Wilhelms III," *Forschungen zur brandenburgischen und preussischen Geschichte* 35 (1923): 213–40.

Demeter, Karl. *The German Officer Corps in Society and State 1650–1945*. Translated by Angus Malcolm. London: Weidenfeld and Nicolson, 1965.

Denecke, Hugo. *Geschichte der Königlichen Preussischen Artillerie-Prüfungskommission*. Berlin: Artillerie-Prüfungskommission, 1909.

Diefendorf, Jeffry M. *Businessmen and Politics in the Rhineland 1789–1834*. Princeton, New Jersey: Princeton University Press, 1980.

Dieterici, C.F.W. *Statistische Übersicht der wichtigsten Gegenstände des Verkehrs und Verbrauchs im Preussischen Staate und im deutschen Zollverbande in dem Zeitraum von 1831 bis 1836*. Berlin: E. S. Mittler, 1838.

———. *Statistische Übersicht der wichtigsten Gegenstände des Verkehrs und Verbrauchs im Preussischen Staate und im deutschen Zollverbande in dem Zeitraum von 1837 bis 1839*. Berlin: E. S. Mittler, 1842.

Dorow, Wilhelm. *Job von Witzleben*. Leipzig: Verlag von Bernh. Tauchnitz, 1842.

Eichholtz, Dietrich. *Junker und Bourgeoisie vor 1848 in der Preussischen Eisenbahngeschichte*. Berlin: Akadamie-Verlag, 1962.

Epstein, Klaus. *The Genesis of German Conservatism*. Princeton, New Jersey: Princeton University Press, 1966.

Facius, Friedrich. *Wirtschaft und Staat: Die Entwicklung des staatlichen Wirt-*

schaftsverwaltung in Deutschland vom 17. Jahrhundert bis 1945. Boppard, 1959.

Fischer, Wolfram. "Government Activity and Industrialization in Germany (1815–1870)," in W. W. Rostow, *The Take-Off into Self-Sustained Growth.* New York: St. Martin's Press, 1963.

Fleck, G. "Studien zur Geschichte des preussischen Eisenbahnwesens," *Archiv für Eisenbahnwesen* 19 (1896): 27–55, 234–52, 858–68; 20 (1897): 1073–98; 21 (1898): 653–80; 22 (1899): 234–62.

Fremdling, Rainer. *Eisenbahnen und deutsches Wirtschaftswachstum 1840–1879: Ein Beitrag zur Entwicklungstheorie und zur Theorie der Infrastruktur.* Dortmund: Ardey-Verlag, 1975.

Gall, Lothar. "Liberalismus und 'Bürgerliche Gesellschaft': Zu Charakter und Entwicklung der Liberalen Bewegung in Deutschland," *Historische Zeitschrift* 220 (April 1975): 324–56.

Genth, August. *Die preussischen Heereswerkstätten: Ihre Entwicklung, allgemeine volkswirtschaftliche Bedeutung und ihr Übergang in privatwirtschaftliche Betriebe.* Berlin: Fr. W. Universität Berlin, 1926.

Gillis, John R. *The Prussian Bureaucracy in Crisis 1840–1860: Origins of an Administrative Ethos.* Stanford, California: Stanford University Press, 1971.

Goldschmidt, Friedrich, and Goldschmidt, Paul. *Das Leben des Staatsrat Kunth.* Berlin: Julius Springer, 1881.

Gothsche, Hugo. *Die Königlichen Gewehrfabriken.* Berlin: Militärverlag der Libelschen Buchhandlung, 1904.

Gray, Marion W. *Prussia in Transition: Society and Politics under the Stein Reform Ministry of 1808.* Philadelphia: American Philosophical Society, 1986.

Gugel, Michael. *Industrieller Aufstieg und bürgerliche Herrschaft: Sozioökonomische Interessen und politische Ziele des liberalen Bürgertums in Preussen zur Zeit des Verfassungskonflikts.* Cologne: Pahl-Rugenstein, 1975.

Hamerow, Theodore S. *Restoration, Revolution, Reaction: Economics and Politics in Germany, 1815–1871.* Princeton, New Jersey: Princeton University Press, 1958.

———. *The Social Foundations of German Unification, 1858–1871: Ideas and Institutions.* 2 vols. Princeton, New Jersey: Princeton University Press, 1969.

Harnisch, Hartmut. "Die Bedeutung der kapitalistischen Agrarreform für die Herausbildung des inneren Marktes und die industrielle Revolution in den östlichen Provinzen Preussens in der ersten Hälfte des 19. Jahrhundert," *Jahrbuch für Wirtschaftsgeschichte* (1977): 63–82.

Heggen, Alfred. *Erfindungsschutz und Industrialisierung in Preussen 1793–1877.* Göttingen: Vandenhoeck & Ruprecht, 1975.

Henderson, W. O. *The State and the Industrial Revolution in Prussia.* Liverpool: Liverpool University Press, 1958.

Höhn, Reinhard. *Scharnhorsts Vermächtnis.* Frankfurt: Bernard & Graefe Verlag, 1972.

Jacob, Margaret C. *The Radical Enlightenment: Pantheists, Freemasons and Republicans.* London: George Allen & Unwin, 1981.

Jacobs, Alfred, and Richter, Hans. *Die Grosshandelspreise in Deutschland von 1792 bis 1934*. Berlin: Hanseatische Verlagsanstalt Hamburg, 1935.

Jeismann, Karl-Ernst. *Das preussiche Gymnasium in Staat und Gesellschaft*. Stuttgart: Ernst Klett, 1974.

Kaelble, Hartmut. *Berliner Unternehmer während der frühen Industrialisierung*. Berlin: Walter de Gruyter, 1972.

Karsten, Gustav. *Umrisse zu Carl Johann Bernhard Karstens Leben und Wirken*. Berlin: Druck von Georg Reimer, 1854.

Keller, Rolf. *Christian von Rother als Organisator der Finanzen, des Geldwesens und der Wirtschaft in Preussen nach dem Befreiungskriege*. Rostock, 1930.

Klee, Wolfgang. *Preussische Eisenbahngeschichte*. Stuttgart: W. Kohlhammer, 1982.

Koselleck, Reinhard. *Preussen zwischen Reform und Revolution: Allgemeines Landrecht, Verwaltung und soziale Bewegung von 1794 bis 1848*. Stuttgart: Ernst Klett, 1967.

Krampe, Hans Dieter. *Der Staatseinfluss auf den Ruhrkohlenbergbau in der Zeit von 1800 bis 1865*. Cologne: Johann Heider, 1961.

Kriedte, Peter, et al. *Industrialization before Industrialization*. Cambridge: Cambridge University Press, 1981.

Krüger, Gerhard. *Gründeten auch unsere Freiheit: Spätaufklärung, Freimaurerei, preussisch-deutsche Reform, der Kampf Theodor von Schöns gegen die Reaktion*. Hamburg: Bauhüttenverlag, 1978.

Landes, David S. "Technological Change and Development in Western Europe, 1750–1914," *The Cambridge Economic History of Europe* (Cambridge: Cambridge University Press, 1965), 6:274–601.

Lewalter, Ernst. *Friedrich Wilhelm IV: Das Schicksal eines Geistes*. Berlin, 1938.

Lotz, Albert. *Geschichte des deutschen Beamtentums*. Berlin: R. v. Decker's Verlag, 1906.

Lundgreen, Peter. *Techniker in Preussen während der frühen Industrialisierung*. Berlin: Colloquium Verlag, 1975.

McNeill, William H. *The Pursuit of Power*. Chicago: University of Chicago Press, 1982.

Matschoss, Conrad. "Geschichte der Königlichen Preussischen Technischen Deputation für Gewerbe." *Beiträge zur Geschichte der Technik und Industrie* 3 (1911): 239–75.

———. *Preussens Gewerbeförderung und ihre Grossen Männer*. Berlin: Verlag des VDI, 1921.

Meinecke, Friedrich. *Das Leben des Generalfeldmarschalls Hermann von Boyen*. 2 vols. Stuttgart: J. G. Gotta'sche Buchhandlung Nachfolger GmbH, 1899.

Mieck, Ilja. *Preussische Gewerbepolitik in Berlin 1806–1844*. Berlin: Walter de Gruyter, 1965.

Nohn, Ernst August. *Wehrwissenschaften im frühen 19. Jahrhundert unter besonderer Berücksichtigung der Wehrtechnik*. Münster: Westfälische Wilhelms-Universität, 1977.

Obenaus, Herbert. *Anfänge des Parlamentarismus in Preussen bis 1848*. Düsseldorf: Droste Verlag, 1984.

Ohnishi, Takeo. *Zolltarifpolitik Preussens bis zur Gründung des deutschen Zollvereins*. Göttingen: Otto Schwarz, 1973.

Papke, Gerhard, ed. *Handbuch der deutschen Militärgeschichte 1648–1939*. 4 vols. Munich: Bernard & Graefe Verlag für Wehrwesen, 1975.

Paret, Peter. *Clausewitz and the State*. New York: Oxford University Press, 1976.

Petersdorff, Hermann von. *Friedrich von Motz*. 2 vols. Berlin: Verlag von Reimar Hobbing, 1913.

———. *König Friedrich Wilhelm der Vierte*. Stuttgart, 1900.

Pounds, Norman J. G. *The Ruhr: A Study in Historical and Economic Geography*. Bloomington, Indiana: Indiana University Press, 1952.

Priesdorff, Kurt von, ed. *Soldatisches Führertum*. 7 vols. (Hamburg: Hanseatische Verlagsanstalt, n.d.).

Pruns, Herbert. *Staat und Agrarwirtschaft 1800–1865*. 2 vols. Hamburg: Verlag Paul Perey, 1979.

Radtke, Wolfgang. *Die Preussische Seehandlung zwischen Staat und Wirtschaft in der Frühphase der Industrialisierung*. Berlin: Colloquium Verlag, 1981.

Reden, Friedrich Wilhelm von. *Die Eisenbahnen Deutschlands*. Berlin: E. S. Mittler, 1846, 386.

Retallack, James. *Notables of the Right: The Conservative Party and Political Mobilization in Germany, 1876–1918*. Boston: Unwin Hyman, 1988.

Ritter, Gerhard. *Staatskunst und Kriegshandwerk*. 4 vols. Munich: R. Oldenbourg, 1954.

Ritter, Ulrich Peter. *Die Rolle des Staates in den Frühstadien der Industrialisierung: Die preussische Industrie-Förderung in der ersten Hälfte des 19. Jahrhunderts*. Berlin, 1961.

Rosenberg, Hans. *Bureaucracy, Aristocracy and Autocracy: The Prussian Experience 1660–1815*. Cambridge, Massachusetts: Harvard University Press, 1958.

Schiersmann, Christiane. *Zur Sozialgeschichte der preussischen Provinzial-Gewerbeschulen im 19. Jahrhundert*. Weinheim: Beltz Verlag, 1979.

Schissler, Hanna. *Preussische Agrargesellschaft im Wandel*. Göttingen: Vandenhoeck & Ruprecht, 1978.

Schnabel, Franz. *Deutsche Geschichte im Neunzehten Jahrhundert*. 3 vols. Freiburg: Herder & Co., 1933–37.

Schneider, Hans. *Der preussischer Staatsrath 1817–1918*. Munich: C. H. Beck'sche Verlagsbuchhandlung, 1952.

Schrader, Paul. *Die Geschichte der Königlichen Seehandlung (Preussische Staatsbank) mit besonderer Berücksichtigung der neuen Zeit*. Berlin, 1911.

Schulz-Briesen, Max. *Der Preussische Staatsbergbau von seinen Anfängen bis zum Ende des 19. Jahrhunderts*. Berlin: Reimar Hobbing, 1933.

Schwann, Mathieu. *Ludolf von Camphausen*. Essen: G. D. Baedeker Verlagshandlung, 1915.

Sering, Max. *Geschichte der preussisch-deutschen Eisenzölle von 1818 bis zur Gegenwart*. Leipzig, 1882.

Sheehan, James J. *German Liberalism in the Nineteenth Century*. Chicago: University of Chicago Press, 1978.

Showalter, Dennis E. *Railroads and Rifles: Soldiers, Technology, and the Unification of Germany*. Hamden, Connecticut: Archon Books, 1975.

Simon, Walter M. *The Failure of the Prussian Reform Movement, 1807–1819*. New York: Howard Fertig, 1971.

Sperber, Jonathan. *Popular Catholicism in Nineteenth-Century Germany*. Princeton, New Jersey: Princeton University Press, 1984.

Spethmann, Hans. *Franz Haniel: Sein Leben und Sein Werk*. Duisburg-Ruhrort: Lübecker Nachrichten GmbH, 1956.

Steffens, Manfred. *Freimaurer in Deutschland: Bilanz eines Vierteljahrtausends*. Flensburg: Christian Wolff, 1964.

Straube, H. J. *Chr. P. Wilhelm Beuth*. Berlin: Verlag des VDI, 1930.

———. *Die Gewerbeförderung Preussens in der ersten Hälfte des 19. Jahrhunderts*. Berlin: Verlag des VDI, 1933.

Sweet, Paul R. *Wilhelm von Humboldt: A Biography*. 2 vols. Columbus: Ohio State University Press, 1980.

Tilly, Richard. "Capital Formation in Germany in the Nineteenth Century," in Peter Mathias and M. M. Postan, *The Cambridge Economic History of Europe*. Cambridge: Cambridge University Press, 1978, 7 (1): 382–441.

———. *Financial Institutions and Industrialization in the Rhineland 1815–1870*. Madison: The University of Wisconsin Press, 1966.

———. "The Political Economy of Public Finance and the Industrialization of Prussia,1815–1866," *Journal of Economic History* 26 (1966).

Treitschke, Heinrich von. *Deutsche Geschichte im 19. Jahrhundert*. 5 vols. Leipzig: Verlag von S. Hirzel, 1879–89.

Treue, Wilhelm. *Wirtschafts- und Technikgeschichte Preussens*. Berlin: de Gruyter, 1984.

———. *Wirtschaftszustände und Wirtschaftspolitik in Preussen 1815–1825*. Stuttgart: Verlag von W. Kohlhammer, 1937.

Varrentrapp, Conrad. *Johannes Schulze und das höhere preussische Unterrichtswesen in seiner Zeit*. Leipzig: B. G. Teubner, 1889.

Vogel, Barbara. *Allgemeine Gewerbefreiheit: Die Reformpolitik des preussischen Staatskanzlers Hardenberg 1810–1820*. Göttingen: Vandenhoeck & Ruprecht, 1983.

Wehler, Hans Ulrich. *Deutsche Gesellschaftsgeschichte*. 3 vols. Munich: C. H. Beck, 1987.

———. *Sozialgeschichte Heute: Festschrift für Hans Rosenberg zum 70. Geburtstag*. Göttingen: Vandenhoeck & Ruprecht, 1974.

Wutke, Konrad. *Aus der Vergangenheit des Schlesischen Berg- und Hüttenlebens*. Breslau, 1913.

Index